Sustainability and Innovation

Technological, institutional and social innovation promotes economic development and international competitiveness, and can do much to reduce environmental burdens. Thus, innovation is an essential factor in the realisation of the principles of sustainable development.The series 'Sustainability and Innovation' aims to positively contribute to a better understanding of the connection between innovation and sustainability. It is interdisciplinary in coverage, concentrating primarily on the links between environmental and innovation research. Individual titles in the series focus on topical issues and cover both theoretical and empirical aspects: concepts of major practical relevance are also discussed. The series is addressed to researchers and policy makers, as well as to specialists and leaders in business and industry.

Jakob Edler • Rainer Walz
Editors

Systems and Innovation Research in Transition

Research Questions and Trends in Historical Perspective

Springer

Editors
Jakob Edler
Fraunhofer Institute for Systems and Innovation Research ISI
Karlsruhe, Germany

Rainer Walz
Fraunhofer Institute for Systems and Innovation Research ISI
Karlsruhe, Germany

ISSN 1860-1030 ISSN 2197-926X (electronic)
Sustainability and Innovation
ISBN 978-3-031-66099-3 ISBN 978-3-031-66100-6 (eBook)
https://doi.org/10.1007/978-3-031-66100-6

This Springer imprint is published by the registered company Springer Nature Switzerland AG
The registered company address is: Gewerbestrasse 11, 6330 Cham, Switzerland

Preface

In this volume we reflect on 50 years of what we term Systems and Innovation Research (SIR). To write and compile this anthology has been a risk, and it has been a necessity. It has been a risk because it is ambitious and enters uncharted territory. In order to do so, we needed to define and delineate the area of SIR. We therefore defined SIR as encompassing the scientific study of, first, the conditions, dynamics and impacts associated with the generation and uptake of innovations in innovation systems and, second and related to this, the development of functional systems satisfying essential needs such as the provision of energy or water. We then had to select a limited number of specific research fields within SIR in order to provide in-depth analysis, to understand the interplay of specific fields and to identify overall patterns and dynamics in SIR. All these fields are application and problem oriented. As they are interconnected in many ways, we consider it inevitable and indispensable to analyse them and their governance together.

This volume has been necessary for three reasons. First, in defining SIR, we noted that most of the fields—along with a number of important research institutes across Europe—had their origin more or less than 50 years ago. This is a milestone that deserves some recognition, and to the best of our knowledge, there is no history of the fields as such. Second, based on our own experience as researchers in the field, we developed a starting assumption that there are a number of specific patterns through which the fields have developed and interacted with each other and their stakeholders. In so doing we sought to understand the nature of change in application-oriented SIR fields and, in particular, the changing role of SIR research over time. Third, we are convinced that the SIR fields will become even more important in the future, given the urgency of the challenges and the growing complexity and uncertainty we observe. However, in order to understand how SIR research can have an impact and meet the challenges ahead, we believe it is necessary to understand the historical development first.

We hope that our risky and necessary undertaking is met with widespread interest, both in the SIR community and, equally importantly, in the stakeholder communities that we, as SIR researchers, have served for 50 years. Inspired by the 50-year anniversary of Fraunhofer ISI in 2022, our undertaking did start with an internal research project. The aim, however, was not to take the perspective of the history of a single institute, but to reflect on the history of selected SIR research fields. This book presents the result of this undertaking. The team of authors are based in Germany, and we are aware that we have a strong bias towards Europe and a certain bias towards Germany. However, we do not claim to present the ultimate global history of SIR, but to contribute to a better understanding of an area of research that has been, and we hope will continue to be, very influential.

This anthology was only possible due to the admirable endeavours of an incredible team of excellent and dedicated colleagues. We have all realised that it is not a trivial matter to delve deeply into the historical development of fields and to step out of the comfort zone of our daily practice, to zoom out and to thoroughly reflect on what it really is we have been doing and how we can draw general lessons from this development. We are convinced that all chapters have fully succeeded in their task, and we are deeply grateful to each and every contributor.

Furthermore, it would not have been possible to produce such a volume without excellent support. We would therefore like to thank our colleagues Marianne Werder and Wiebke Baumann for all their coordinating, technical and editorial support. We are also deeply grateful to Gillian Bowman-Köhler, Louise Antill-Blum and Barbara Sinnemann, who have worked excellently and tirelessly to improve our language page by page. Above all, however, we would like to thank Johanna Schuler who was the excellent and highly efficient mastermind behind all the coordination and editorial support, and who never stopped pushing all of us over the line. Without her outstanding and insightful help, this volume would not exist.

Karlsruhe, Germany Jakob Edler
Karlsruhe, Germany Rainer Walz
March 2024

Contents

Editors and Contributors

About the Editors

Jakob Edler is Managing Director of the Fraunhofer Institute for Systems and Innovation Research ISI since October 2018. The Anglo-German is also Professor of Innovation Policy and Strategy at the Manchester Institute of Innovation Research (MIoIR), which he led from 2011 to September 2018. His research focuses on analysing and conceptualising governance, policy and management of innovation. J. Edler is Speaker of the Fraunhofer Group Innovation Research, member of the German Science Council, the German Academy of Science and Engineering (Acatech) and the Royal Society for the Encouragement of Arts, Manufactures and Commerce (UK).

Rainer Walz is Deputy Director of the Institute and Head of the Competence Centre Sustainability and Infrastructure Systems at the Fraunhofer Institute for Systems and Innovation Research ISI in Karlsruhe. He obtained his PhD and "habilitation" in economics. In 1991, Prof. Dr. Walz joined the Fraunhofer ISI. Prior employment includes the University of Wisconsin and the Enquête Commission "Protecting the Earth's Atmosphere" of the German Bundestag. He is also an adjunct professor of economics at the Karlsruhe Institute of Technology (KIT) and member of the EU expert group "Economic and Social Impact of Research".

Contributors

Harald Bradke Fraunhofer Institute for Systems and Innovation Research ISI, Karlsruhe, Germany

Barbara Breitschopf Fraunhofer Institute for Systems and Innovation Research ISI, Karlsruhe, Germany

Susanne Bührer Fraunhofer Institute for Systems and Innovation Research ISI, Karlsruhe, Germany

Kerstin Cuhls Fraunhofer Institute for Systems and Innovation Research ISI, Karlsruhe, Germany

Stephanie Daimer Fraunhofer Institute for Systems and Innovation Research ISI, Karlsruhe, Germany

Ewa Dönitz Fraunhofer Institute for Systems and Innovation Research ISI, Karlsruhe, Germany

Elisabeth Dütschke Fraunhofer Institute for Systems and Innovation Research ISI, Karlsruhe, Germany

Jakob Edler Fraunhofer Institute for Systems and Innovation Research ISI, Karlsruhe, Germany
Manchester Institute of Innovation Research (MIoIR), University of Manchester, Manchester, United Kingdom

Lorenz Erdmann Fraunhofer Institute for Systems and Innovation Research ISI, Karlsruhe, Germany

Tobias Fleiter Fraunhofer Institute for Systems and Innovation Research ISI, Karlsruhe, Germany

Michael Friedewald Fraunhofer Institute for Systems and Innovation Research ISI, Karlsruhe, Germany

Rainer Frietsch Fraunhofer Institute for Systems and Innovation Research ISI, Karlsruhe, Germany

Bruno Gransche Karlsruhe Institute of Technology KIT, Karlsruhe, Germany

Anne Held Fraunhofer Institute for Systems and Innovation Research ISI, Karlsruhe, Germany

Nils B. Heyen Fraunhofer Institute for Systems and Innovation Research ISI, Karlsruhe, Germany

Harald Hiessl Formerly Fraunhofer Institute for Systems and Innovation Research ISI, Karlsruhe, Germany

Thomas Hillenbrand Fraunhofer Institute for Systems and Innovation Research ISI, Karlsruhe, Germany

Angela Jäger Fraunhofer Institute for Systems and Innovation Research ISI, Karlsruhe, Germany

Eberhard Jochem Fraunhofer Institute for Systems and Innovation Research ISI, Karlsruhe, Germany

Simone Kimpeler Fraunhofer Institute for Systems and Innovation Research ISI, Karlsruhe, Germany

Marian Klobasa Fraunhofer Institute for Systems and Innovation Research ISI, Karlsruhe, Germany

Stefan Kuhlmann University of Twente, Enschede, The Netherlands

Christian Lerch Fraunhofer Institute for Systems and Innovation Research ISI, Karlsruhe, Germany

Ralf Lindner Fraunhofer Institute for Systems and Innovation Research ISI, Karlsruhe, Germany

Frank Marscheider-Weidemann Fraunhofer Institute for Systems and Innovation Research ISI, Karlsruhe, Germany

Jutta Niederste-Hollenberg Fraunhofer Institute for Systems and Innovation Research ISI, Karlsruhe, Germany

Julia Panny Fraunhofer Institute for Systems and Innovation Research ISI, Karlsruhe, Germany

Patrick Plötz Fraunhofer Institute for Systems and Innovation Research ISI, Karlsruhe, Germany

Thomas Reiß Fraunhofer Institute for Systems and Innovation Research ISI, Karlsruhe, Germany

Christian Sartorius Formerly Fraunhofer Institute for Systems and Innovation Research ISI, Karlsruhe, Germany

Elna Schirrmeister Fraunhofer Institute for Systems and Innovation Research ISI, Karlsruhe, Germany

Ulrich Schmoch Fraunhofer Institute for Systems and Innovation Research ISI, Karlsruhe, Germany

Sarah Seus Fraunhofer Institute for Systems and Innovation Research ISI, Karlsruhe, Germany

Axel Thielmann Fraunhofer Institute for Systems and Innovation Research ISI, Karlsruhe, Germany

Rainer Walz Fraunhofer Institute for Systems and Innovation Research ISI, Karlsruhe, Germany

Philine Warnke Fraunhofer Institute for Systems and Innovation Research ISI, Karlsruhe, Germany

Martin Wietschel Fraunhofer Institute for Systems and Innovation Research ISI, Karlsruhe, Germany

Jenny Winkler Fraunhofer Institute for Systems and Innovation Research ISI, Karlsruhe, Germany

Introduction

Jakob Edler and Rainer Walz

Abstract It has become a commonplace that in order to tackle the enormous societal challenges we face throughout the globe we have to support transformation; and in order to do so, we have to mobilise innovation. As innovation and transformation are strongly intertwined, it is thus imperative essential to analyse and support their interplay.

This volume reflects on nine selected research fields in the area of systems and innovation research that have developed in the last five decades to do exactly that. It defines and delineates systems and innovation research as encompassing the scientific study of the conditions, dynamics and impacts associated with the generation and uptake of innovations and the development and transformation of functional systems satisfying essential needs such as the provision of energy or water. This allows to draw general lessons as to what drives research fields throughout their development and how their role—in particular vis-à-vis policy—changes over time. It also allows to speculate about future challenges and trends in the area of systems and innovation research. This is important because, if anything, the need to govern transformation through innovation will further grow in the future and with it the need to understand the underlying dynamics.

It is now largely undisputed that socio-technical transformations are urgently required and that innovations are key requirements for them. Policy is once again becoming much more ambitious, seeking to accelerate transformations, bringing innovation and knowledge generation in line with the directions of transformations

J. Edler (✉)
Fraunhofer Institute for Systems and Innovation Research ISI, Karlsruhe, Germany

Manchester Institute of Innovation Research (MIoIR), University of Manchester, Manchester, United Kingdom
e-mail: jakob.edler@isi.fraunhofer.de

R. Walz
Fraunhofer Institute for Systems and Innovation Research ISI, Karlsruhe, Germany
e-mail: Rainer.Walz@isi.fraunhofer.de

J. Edler, R. Walz (eds.), *Systems and Innovation Research in Transition*, Sustainability and Innovation, https://doi.org/10.1007/978-3-031-66100-6_1

and ensuring that the appropriate technological and economic foundations are in place for transformations. In addition, all of those ambitions are to be pursued in conjunction with social cohesion and innovation-driven competitiveness. This renewed ambition of the state to shape change is characterised not only by urgency, but also by growing complexity. Strategic intelligence is therefore increasingly important, and the calls for excellent, evidence-based policy advice are becoming louder and more impatient.

Against this background of rising demands, urgency and uncertainty in relation to evidence-based policy advice, we embark on a review of the history of systems and innovation research (SIR). Our intention is to reflect on the development of key research areas within SIR and its interplay with policy-making and societal developments. Based on this, we seek to draw overarching lessons concerning the development and function of that advice in the past and in the future.

We define systems and innovation research (SIR) as the area of scientific enquiry into two overlapping topic domains. First, SIR is interested in better understanding the conditions and processes through which novelties are produced and put into use and their subsequent effects as innovations on the economy, society and the environment. This approach requires thinking in innovation systems to understand the underlying conditions and processes. Second, SIR analyses the ways in which functional (sectoral) systems, such as those providing and using energy, mobility, food or water, perform and evolve over time. These simplified definitions already indicate why it is not only justified, but indispensable to consider innovation and systems together with their governance. It is impossible to fully understand the performance and transformations of functional (sectoral) systems without understanding how technological and social innovations are mobilised and diffuse in the respective systems. It is equally impossible to understand the demand for as well as the production and effects of innovations without an understanding of the role they play in the various (sectoral) systems in which they are embedded.

In our understanding, SIR is further characterised by the strong interplay between researchers and stakeholders from politics, the economy and the wider society. In fact, the impetus, the very raison d'être of SIR from the beginning, has been to contribute to a better understanding of societal problems and developing solutions to them. SIR is both application-oriented and problem-driven. This volume will show that the empirical questions, analytical approaches, scientific methods and data sources for SIR and its various fields have co-evolved with the policy demands in different policy fields and with the changing perception of problems.

Systems research in its own right had its beginnings after the Second World War in the USA, as earmarked, for example, through the RAND Corporation in the USA, or - some years later - the International Institute for Applied Systems Analysis (IIASA) in Austria. The history of Systems and *Innovation* Research starts roughly 50 years ago. At the end of the 1960s and in the early 1970s, a number of new research centres were established in several European countries both inside and outside universities. With regard to innovation studies, these included the Science

Policy Research Unit at the University of Sussex,[1] Policy Research in Engineering, Science and Technology at the University of Manchester, the Bureau d'Economie Théorique et Appliquée (BETA) at the University of Strasbourg, and the Fraunhofer Institute for Systems and Innovation Research (Fraunhofer ISI), whereby the latter was explicitly designed to span both systems and innovation research. While these institutes had very different institutional and disciplinary backgrounds, they all shared a common mission. They all sought to analyse the dynamics of the development and application of technologies and innovations more broadly and not as a purely academic exercise to advance knowledge. Instead, the protagonists of these fields were driven by a normative objective to support policy-making and decision-making in businesses by providing strategic intelligence of various kinds in order to improve economies and the welfare of societies. With the advent of environmental consciousness catalysed by the first Earth Day in 1970, the report of the Club of Rome in 1972 and the first international UN conference on the Human Environment in Stockholm 1972, this normative impetus included concerns regarding the long-term detrimental effects of technology. Researchers increasingly understood the significance of the systemic character of those processes that satisfy human needs but at the same time threaten the very foundation of human life.

We strongly believe that reflecting on the dynamics of SIR can help us to better understand how those fields can—and should—support decision-making in the turbulent times ahead. This is the reason why we are interested in the development of the scientific endeavours that started over 50 years ago.

Against this background, this volume includes historical intellectual journeys in nine selected fields within the broad area of SIR over the last five decades. We can neither cover all possible SIR fields nor give a full picture of the various international and national developments, especially for those research fields involving sectoral functional systems, which are largely shaped nationally. Therefore, some of the nine fields are analysed from a more international perspective, others somewhat focus on a German perspective.

The SIR fields in this book represent a subjective selection, but not an arbitrary one. We applied a set of conditions to justify their selection: First, they need to have been relevant in all or most of the past five decades, as our main interest is in their development over time. Second, they need to be directly relevant for policy development, engaging with policy-makers at different levels. While certain individuals and institutes within those fields have contributed to excellent academic publications and driven the development across various disciplinary fields, the bulk of the work was directed at supporting a better understanding of problems and opportunities and providing the relevant decision-makers with sound insights and evidence as a basis for policy-making. Third, while all nine fields have a variety of links to established

[1] For an impressive case study of SPRU as an example of transformational changes in the study of science, technology and innovation and related policies, see Soete, Luc. "Science, technology and innovation studies at a crossroad: SPRU as case study". Research Policy 48.4 (2019): 849–857

academic fields and sub-fields, they have always been multi-disciplinary, where different disciplines were mobilised in parallel, and in parts inter-disciplinary, where epistemic approaches of different disciplines were combined and partly integrated. All fields always have been and are strongly application-oriented at the same time, with all nine fields developing—to various degrees—their very own disciplinary norms and practices.

We deliberately do not deal with a more recent field of research, i.e. transition studies. This field has evolved in the last 20 years or so and has become an increasingly important field to understand socio-technical system transitions. A strong epistemic community of scholars has developed its own conceptual basis building on a range of conceptual sources in (transition) management, evolutionary economics and the science and technology studies. While various chapters in this volume refer to their specific relationship with this field, a history of transition studies deserves its own compilation.

We have selected research fields that cover a broad range of functions for stakeholders. Five of the nine fields are horizontal in nature, developing analytical and methodological approaches to understand, support and manage systems development and applying them across a range of (sectoral) functional systems:

The first field is Conceptualising and Analysing STI Policy. The authors base their analysis on the conceptualisation of policy paradigm change. In addition to exogenous changes in the economic, societal and political context conditions, the authors focus on primarily endogenous phenomena in three areas: changes in the main theoretical-conceptual basic understanding of the origins of knowledge and innovation, changes in STI policy objectives and changes in problem constructions and preferred policy solutions.

Our second field is concerned with the development of innovation monitoring and innovation indicators, which aims at analysing the dynamics and performance of innovation systems. The authors analyse the development of the generation and use of indicators in the context of changes in the conceptual understanding of the innovation process. In addition, they look at changes in political demands to understand innovation dynamics and the consequences for developing innovation indicators over time.

Our third field, Foresight, deals with strategic intelligence through forward-looking analytical and participatory approaches on a systems level. The authors analyse how foresight has changed drastically over time from more deterministic views to open and diverse future perspectives, and towards participative approaches and the integration of AI in a flood of "big data".

The fourth field focuses more narrowly on the evaluation of public research and innovation policies. The authors discuss the interplay between Research and Innovation (R&I) policy and evaluations that aim to capture the function and influence of such policies as well as considering the goals, instruments and approaches to evaluating research institutions. The chapter focuses on the German and the European level with regard to evaluation practice, but also includes Anglo-American literature with regard to evaluation theory.

The fifth field, Technology Assessment, emerged out of the awareness of the ambivalence of technologies increased in politics and public life at the end of the 1960s. The authors describe the development of the research field by focusing on the paradigmatic changes in Technology Assessment from an expert-based approach to a participatory one and on towards the current pragmatic approach.

The remaining four research fields are vertical in nature and focus on selected (sectoral) functional systems:

The sixth field is concerned with the research on the development of production technologies and their systemic embedding. Although the process of deindustrialisation started in the 1970s, manufacturing is still considered a driver of the high innovativeness and competitiveness in particular of the German economy. The authors analyse the development in this research field by looking at changes in the central production paradigms over the last 50 years and the systemic embedding of changing production technologies and highlight the major issues steering German manufacturing industries.

The seventh field focuses on research on renewable energy. Since the Limits to Growth report in 1972, renewable energies research has made key contributions to the debate surrounding the scarcity of resources and energy transition, with the topic continuing to become ever more important, not least due to the rise of climate policy. The authors identify the main research topics and assess project types, key methods and research approaches and how these have changed over time.

The eighth research field, energy demand and modelling of energy systems, has been a cornerstone of systems analysis ever since the publication of the Limits to Growth report and the first oil embargo by the OPEC countries in late 1973. The authors have organised this chapter by decades and show how the research field has developed, driven by the interplay of economic and political context factors, increasing energy efficiency options and policies, and the progress achieved in modelling approaches.

The last field is concerned with water systems research and focuses on the sustainable use of water. The authors have structured this chapter by distinguishing three phases over the last 50 years. They analyse the interplay of pressure to act, changes in research activities and important measures and regulations. The underlying analysis focuses on the national development in Germany, but includes links to developments at EU level.

Each of the nine chapters on the research fields explores their origin, how the associated methods, data sources and concepts have developed and, most importantly, how the major research questions and policy decisions have interacted with each other over time. Based on this, each chapter speculates about the future demands and opportunities to inform policy-makers and decision-makers more broadly in the future.

In a final chapter, we develop an analytical model to draw overall conclusions of the development of the SIR fields. In particular, we look at what we regard as most critical, i.e. the dynamic interaction between policy demands and research practice, concepts and methods over time. In doing so, we argue that it is absolutely vital for the future that each of these fields opens up and integrates perspectives and insights from other fields to a much greater extent. Given the complexity of system development, and the need for productive contributions of science, technology and technological as well as social innovation to tackle our societal challenges, a more holistic approach in each of our fields and a better integration of the fields is needed. This will not only improve the production of scientific evidence, it will also render policy advice for system transformation and innovation-based competition more meaningful and relevant.

We do hope in analysing SIR in its historical development this volume is a contribution towards such a more holistic—and more impactful—analysis in the future.

Jakob Edler is Managing Director of the Fraunhofer Institute for Systems and Innovation Research ISI since October 2018. The Anglo-German is also Professor of Innovation Policy and Strategy at the Manchester Institute of Innovation Research (MIoIR), which he led from 2011 to September 2018. His research focuses on analysing and conceptualising governance, policy and management of innovation. J. Edler is Speaker of the Fraunhofer Group Innovation Research, member of the German Science Council, the German Academy of Science and Engineering (Acatech) and the Royal Society for the Encouragement of Arts, Manufactures and Commerce (UK).

Rainer Walz is Deputy Director of the Institute and Head of the Competence Centre Sustainability and Infrastructure Systems at the Fraunhofer Institute for Systems and Innovation Research ISI in Karlsruhe. He obtained his PhD and "habilitation" in economics. In 1991, Prof. Dr. Walz joined the Fraunhofer ISI. Prior employment includes the University of Wisconsin and the Enquête Commission "Protecting the Earth's Atmosphere" of the German Bundestag. He is also an adjunct professor of economics at the Karlsruhe Institute of Technology (KIT) and member of the EU expert group "Economic and Social Impact of Research".

Understanding Paradigm Change in Science, Technology, and Innovation Policy: Between Science Push and Policy Pull

Ralf Lindner, Jakob Edler, and Stephanie Daimer

Abstract Science, technology, and innovation (STI) policy experienced far-reaching changes with regard to both political aims and the underlying rationales. Drawing on the concept of policy paradigms, we analyse the factors that shaped the dynamics of STI policy since the first post-war decades in the OECD world. Next to changes in the context conditions, the chapter focuses primarily on endogenous phenomena related to changes of the conceptual understanding of knowledge generation and innovation, of the main STI policy objectives, and of preferred policy solutions. Of particular interest is the role of scientific expertise in these processes of policy change. The first of the three STI policy paradigms identified is characterised by its emphasis on addressing market failures in processes of knowledge generation. The second paradigm shares key objectives of the first, but is based on the systems of innovation heuristic, aiming to improve system performance. The third paradigm supplements the primarily economic rationales of the previous paradigms with the objective of addressing societal challenges. In view of the conclusion that scientific contributions and policy advice were less influential in the second paradigm shift than during the first, we develop suggestions for a future-oriented research agenda for STI policy research.

R. Lindner (✉) · S. Daimer
Fraunhofer Institute for Systems and Innovation Research ISI, Karlsruhe, Germany
e-mail: Ralf.Lindner@isi.fraunhofer.de; Stephanie.Daimer@isi.fraunhofer.de

J. Edler
Fraunhofer Institute for Systems and Innovation Research ISI, Karlsruhe, Germany

Manchester Institute of Innovation Research (MIoIR), University of Manchester, Manchester, United Kingdom
e-mail: Jakob.Edler@isi.fraunhofer.de

J. Edler, R. Walz (eds.), *Systems and Innovation Research in Transition*, Sustainability and Innovation, https://doi.org/10.1007/978-3-031-66100-6_2

1 Introduction

Science, technology, and innovation policy (STI policy) has evolved extensively over the past decades and undergone fundamental changes in the process. These changes were related to the respective political objectives of STI policy, the underlying rationales for government action, basic conceptual and theoretical assumptions as well as the instruments employed and the governance arrangements. Since the Second World War, three distinct phases of STI policy can be distinguished in the OECD world, each of which is characterised by specific rationales and fundamental concepts (Gassler et al. 2006; Kuittinen et al. 2018; Breitinger et al. 2021); see Fig. 1.

The focus during the first post-war decades was on building or expanding the science systems and supporting the generation of fundamental knowledge. The intellectual and conceptual foundation of STI policy was primarily shaped by the mainstream neoclassical economics prevalent at that time and a largely linear understanding of the innovation process. State interventions in this linear model were justified by the need to compensate for market failures during the generation of knowledge as a public good. The 1970s saw the start of a reorientation of STI policy. The end of the long phase of economic growth and the increasing intensity of international competition led to a revision of the basic assumptions about the nature of the innovation process, which finally resulted in the rise of the systems approach in STI policy. The development of a more complex, non-linear understanding of innovation and of the crucial role of interactions between heterogeneous actors of the innovation system was particularly influential. Consequently, according to this understanding, the main task of STI policy was to improve the performance of the innovation system. While economic goals such as competitiveness and growth were

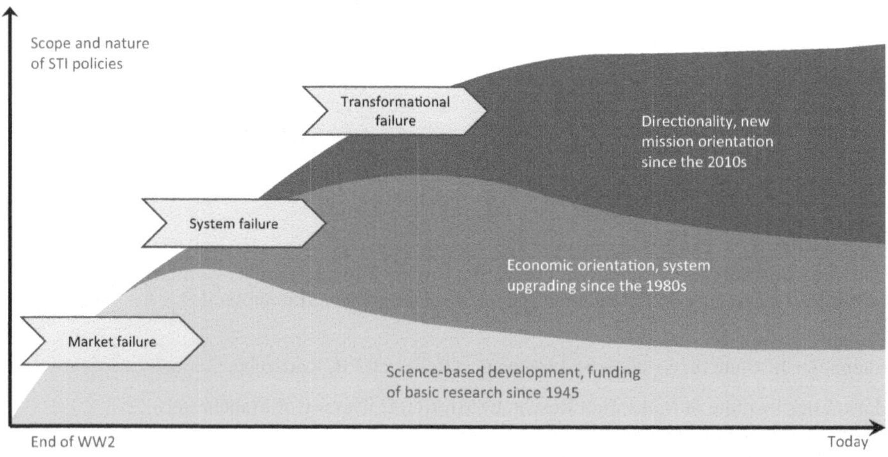

Fig. 1 Phases of STI policy. Sources: Breitinger et al. (2021); based on Daimer et al. (2012); Gassler et al. (2006)

the main reasons for state interventions in the first two phases of STI policy, from 2000 onward, these primarily economic rationales were increasingly supplemented and overlaid by the goal of mobilising knowledge, technology, and innovation to address societal challenges such as climate change. Questions about the directionality of research and innovation and the growing importance of gearing these towards societal needs are reflected, for example, in new policy approaches such as mission-oriented or transformative innovation policy. At the same time, it cannot be claimed that there was no directionality in STI policy prior to this third phase. Research programmes based on the needs of various sectors have been around for decades, such as those in the energy sector, the health sector, environmental protection, or in agriculture. The responsibility for these programmes lies partly with STI policy and partly with the respective sectoral policies. The research programmes do not primarily pursue economic goals, but rather sector-specific (policy) objectives. However, there was no overarching strategic approach in STI policy that aimed at directing a large part of applied research at specific targets or at sparking transformative change. Nor were sectoral programmes or sectoral policies necessarily oriented towards transformation.

While tracing and characterising the historical sequence of different phases of STI policy is valuable in and of itself, the main interest of this chapter is to improve our understanding of the conditions and factors that shape and drive the empirically observable dynamics of STI policy. What were the crucial influencing factors in processes of policy change, how did they interact and what effect did the changing context conditions have? There is a particular focus on the role played by scientific expertise, ideas, and conceptualisations in these processes of change. The concept of policy paradigms is useful to examine these questions more closely. It assigns strong explanatory power to ideas-based and cognitive aspects in processes of policymaking and policy change.

This chapter is structured as follows: First, the analytical framework is developed that is used to examine the developments in STI policy in recent decades. The main part of the chapter then applies this analytical framework to describe the key characteristics and development dynamics of the three major STI policy paradigms since the 1950s. The final section summarises the key findings on the processes of STI policy change and draws conclusions both for the future development of scientific policy advice in STI policy and for the applied policy-analytical tools.

2 Policy Paradigms as Framework for Analysis

Explaining policy change is an important research perspective for policy analysis. However, identifying the causes, drivers, and relevant contextual conditions of policy change is an analytical challenge due to the multidimensional factors influencing change processes. For a long time, policy change processes were predominantly attributed to the actors involved being able to assert their interests as well as to institutional conditions and exogenous events. However, since the 1990s, ideas, knowledge, interpretations, and beliefs have also been included as key explanatory factors and have undoubtedly contributed to improving the conceptualisation of the phenomenon of policy change. The growing recognition of the role of ideas as factors in their own right in processes of policy change has made the key dimensions of policy content accessible to systemic analysis. In the literature on policy analysis, the heading of "ideas" usually refers to conceptual models, assumptions about causalities, theories but also world views, beliefs, values, and norms (Campbell 2002), with particular policy relevance given to complex, structured ideas (Carson 2004; Edler 2003). Ideas are important in the policy process because they provide interpretative frameworks that can be used to determine values and preferences and enable political and economic interests to become actionable (Carstensen and Schmidt 2016). Important contributions to this "interpretative change" (Münch 2016) or "ideational turn" of policy analysis (Daigneault 2014b) were made by approaches such as multiple streams (Kingdon 1984), advocacy coalitions framework (Sabatier 1998), the epistemic communities framework (Haas 2001), as well as approaches rooted more strongly in neoinstitutionalism, such as the punctuated equilibrium theory (True et al. 2007) and discursive institutionalism (Schmidt 2010; Edler 2003). Hall's work on policy paradigms and policy change (1990, 1993) is considered particularly influential.

While ideas represent important factors in policy change processes, they are not the only driver of transformation. On the one hand, the specific institutional and political conditions influence policymaking and thus both stability and change in policy. In particular, variants of historical institutionalism explain policy stability through the long-term effects of previous policy measures (path dependencies, policy legacies) (Béland 2009). From this perspective, policy change is predominantly incremental and is explained by institutional opportunity structures used by political actors within the framework of the given conditions (Skocpol 1992; Streeck and Thelen 2005). On the other hand, the rarely occurring, far-reaching changes are primarily explained by exogenous shocks and crises that break up established and ingrained policy pathways and open up new ones (Hogan and Howlett 2015). However, both approaches have significant blind spots when it comes to uncovering and explaining the concrete content and direction of policy—whether in the mode of incremental policy development or in disruptive, path-breaking phases. This is where the above-mentioned ideas-based, interpretative approaches come in and help to illuminate how beliefs, values, and the understanding of problems change and become effective in the complex processes of constructing meaning, agenda setting, and problem-solving (Béland and Cox 2013). In this context, it is worth recalling Heclo's famous dictum of 1974 that policymaking is not only determined by power and interests, but is shaped to a significant extent by the search for solutions to problems (Heclo 2010).

Policy paradigms can be understood as a cognitive model or a coherent set of ideas and beliefs concerning a policy problem and suitable solutions to it, which is shared by a specific group of actors and provides orientation in the relevant policy field (Carson 2004; Daigneault 2014a). Hall's analysis of monetarism displacing the macroeconomic policy paradigm of Keynesianism was groundbreaking for the link between paradigms and policy change. The essential starting point of this approach is the observation that processes of policy formulation and design take place within the context of discursive relationships (Hall 1993). Such discourses are powerful and effective because they occur within a framework—consisting of ideas, standards, and assumptions about the relevant policy field. Hall refers to this reference framework as a policy paradigm, defined as an "overarching framework of ideas that structures policy-making in a particular field" (Hall 1990). These interpretative frameworks affect policy by determining fundamental relationships, structuring policy discourses according to the prevalent parameters and thus influencing policy goals. Once a policy paradigm has become established, fundamental disputes about its key elements usually only take place to a very limited extent and transaction costs between stakeholders in the policy field are reduced due to shared patterns of interpretation. To explain policy change that goes beyond incremental shifts, Hall draws on Kuhn's concept of scientific paradigms (1962). Following Kuhn, Hall distinguishes three modes of policy formulation (Hall 1990): While in phases of (1) first-order change, changes in a policy field are manifested as continuous further developments of existing policy instruments, in phases of (2) second-order change, revisions, or even the exchange of instruments take place in order to still achieve the goals pursued within the framework of the current paradigm despite emerging policy problems. Finally, (3) third-order change involves the fundamental departure from the previously dominant system of ideas and is accompanied by profound changes in the overarching discourse. First- and second-order changes are thus "normal", incremental phases of policy change, whereas third-order change signals a policy paradigm shift.

While Hall's approach has received a lot of attention and has been widely used to analyse processes of fundamental policy change, the understanding of policy paradigms has continued to evolve. Today, a generally more fluid and differentiated understanding of policy paradigms is prevalent, in which some of the central postulates of the original concept are no longer shared or shared only in a modified form. Above all, this concerns the assumption that competing paradigms are strictly incommensurable. In contrast to the arguments by Kuhn (1962) and Hall (1993), linkages and combinations of different paradigms are now regarded as possibilities (Hogan and Howlett 2015), these are sometimes referred to as "synthetic" and "hybrid" paradigms (Béland 2007; Kay 2007; Wilder 2015). The strict assumption that a new paradigm always completely supersedes and replaces its predecessor has also been watered down. Instead, different paradigms can co-exist, although often not without conflict, and a type of "policy layering" results. Sometimes there are recombinations and novel combinations of ideas that can lead to paradigms that incorporate partially contradictory ideas (Diercks et al. 2019). Additionally, with regard to the dimension of time, paradigm change is no longer understood as an exclusively revolutionary process that occurs in phases, but also as a gradual process, that takes place over longer periods of time (Carstensen 2011; Mahoney and Thelen 2010).

Based on the conceptualisation of policy paradigm change and its further developments, this chapter focuses on the following perspectives to analyse the change in the fundamental orientations of STI policy since the middle of the twentieth century in the OECD world. A distinction is made between phenomena that are predominantly external to the policy field and those that are internal to it.

Primarily exogenous phenomena:

- What changes in the context conditions (economic, societal, political) can be identified that have played a role in fundamental change processes in STI policy?

Primarily endogenous phenomena:

- How has the main theoretical-conceptual basic understanding of the origins of knowledge and innovation changed?
- How have the STI policy objectives changed?
- How have problem constructions and the respective preferred solutions and their instrumentation changed in STI policy discourse and in application?

In all three main endogenous dimensions of analysis, it is of particular interest what role scientific policy consultation played in the processes of change in each case, and which questions research providing policy advice was confronted with in the different phases.

3 The Paradigmatic Development of STI Policy

3.1 The Point of Departure: STI Policy Centred on Knowledge Generation and Addressing Market Failures

In OECD countries, STI policy in the first two to three decades after the Second World War was strongly focused on funding, strengthening, and differentiating the science system (OECD 1972). It was considered a genuine task of the state to support the development of scientific capacities as a foundation for progress and prosperity. OECD countries generally adhered to the logic that it should be a task of the state to promote basic knowledge, which would then make its way into innovation via market processes or direct state (especially military) demand (Bush 1945). This idea was based on experiences during the Second World War, when massive government scientific programmes had laid the groundwork for military innovations (Gassler et al. 2006).

The guiding principle of the state funding basic knowledge, especially for non-state utilisation, was scientifically underpinned in the following years, primarily by economists. Science and technology were increasingly perceived as the engines of economic growth and social progress in neoclassical welfare economics (especially Solow (1957)). However, it was stated that the necessary scientific foundations could only be secured through state funding. The reason for this is the neoclassical concept of market failure as an underlying problem: The economic benefit of basic knowledge is, by its very nature, not foreseeable ex ante, and its utilisation is therefore highly uncertain. At the same time, the potential economic benefit can usually not be limited to the actor producing the required basic knowledge due to spill-over effects. Due to these positive externalities and the associated free rider behaviour, there is insufficient incentive for private actors to invest in the creation of basic knowledge (Arrow 1962; Nelson 1959). Following this logic, the public good of basic knowledge needs to be largely financed by the state and then progresses along a linear process towards economic utilisation as innovation via the development of technologies (among others Pavitt 1976; Tidd et al. 2005; Klodt 1987).

Consequently, the state has the task of creating and maintaining the financing and research institutions relevant for the respective disciplines and state or private utilisation contexts, promoting the transfer from publicly financed research to industry, and subsidising strategic research conducted in companies (OECD 1972). Additionally, intellectual property rights have to be strengthened in this logic, especially via patents, in order to create incentives for investment in uncertain innovations (Edler and Fagerberg 2017). This also explicitly meant that the state set certain priorities in science and technologies that were viewed as particularly worth supporting in terms of state or private utilisation interests (Soete and Arundel 1993). This phase of public governance of science and technology was thus in no way devoid of setting priorities or supporting selected technology trajectories. Science and technology policy was institutionalised and differentiated in all OECD countries in the post-war period based on this understanding (OECD 1972). The first generation of large, mission-oriented research and development programmes also fall into this context, especially those in the USA such as the Apollo projects of the 1960s. These so-called "old missions" were characterised by their almost exclusive public funding and their focus on developing novel, hitherto non-existent technologies (Foray et al. 2012).

The dominant research questions of analysts in the context of this differentiation were the empirical documentation of expenditures with regard to research, development, and innovation activities (see Frietsch et al. 2024 in this anthology), in order to determine empirically what contribution science and research make to productivity, economic growth, and social progress. This included the institutional analysis and international comparison of "science systems" (OECD 1972) as well as the importance of state demand for technological innovations (Mowery and Rosenberg 1979).

The conceptual development within this underlying paradigm can be illustrated using the example of German policy advice. In a study for the then German Federal Ministry for Research and Technology (BMFT), the founding director of Fraunhofer ISI, Helmar Krupp, analysed the institutional conditions of science and its governance in Germany (Krupp 1972; see also Stucke 1993). To start with, Krupp still followed the linear model outlined above and the idea of market failure as the rationale for political steering and stated the need to particularly focus on changing the knowledge transfer and utilisation conditions. Implicitly, however, Krupp went beyond the linear model and emphasised the huge variety and heterogeneity of all the actors involved in the creation, transfer, and exploitation of scientific knowledge in innovations and who therefore contributed to the diffusion of innovations in the economy and society. His use of the term “innovation system” already pointed the way to concepts that only experienced a breakthrough in the 1980s (see below). The early 1970s also marked the beginning of a comprehensive but prolonged revision of the understanding of the nature of the innovation process throughout the OECD world. This revision culminated in the development towards a more differentiated technology and later innovation policy (Sweeney 1985; Dodgson and Rothwell 1994; Rothwell and Zegveld 1981), see also Fig. 2.

3.2 *Systems of Innovation as the Dominant Heuristic for Analysis and Policymaking*

This reorientation towards a systems approach during the 1970s and early 1980s was largely driven by a generation of economists inspired by evolutionary theory. This group exchanged ideas through various expert groups and conferences of the OECD, among others, and this helped their ideas to quickly disseminate internationally and gain legitimacy (OECD 1981; Edler 1999). There were different intellectual sources for this reorientation. The broader context, founded on political science, was the fundamental optimism about steering by means of state intervention in the early 1970s. In a sort of “planning euphoria” (Mayntz 1996; Mayntz and Scharpf 1973) in the late 1960s and 1970s, there was the notion that state instruments, regulations, and incentives could be used to comprehensively steer social dynamics. The optimistic basic notion of state steering capacities came to an abrupt end in the 1980s, which also meant the steering theory of political science became less important (Mayntz 1996). Nevertheless, this fundamental belief in the state during the social democratic decade of the 1970s was an important cognitive and

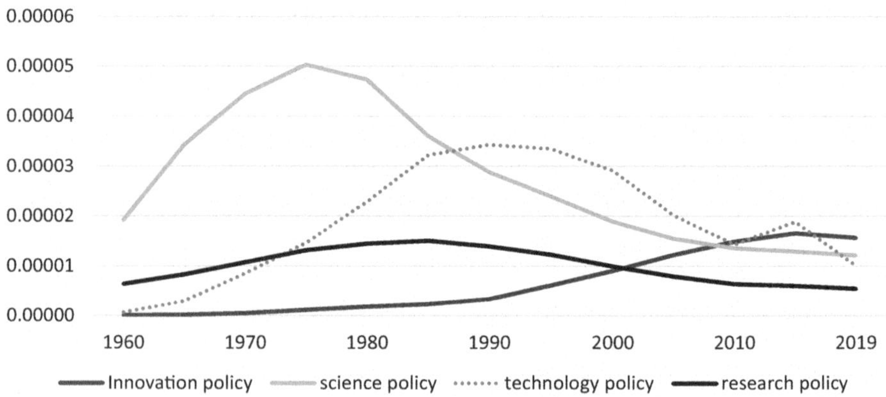

Fig. 2 Differentiation of the field from science to STI policy. Source: own representation. Based on Google Books Ngram Viewer, accessed 13.09.2023

normative basis for the academic and political discussion that then began concerning a more proactive state, which was also reflected in research and innovation policy.

On the macroeconomic level, against the backdrop of stagflation in the 1970s, the insight from welfare theory became increasingly widespread that research and development (R&D) and technological innovation were major drivers and conditioning factors for the economic and social development of national economies (Pavitt 1979; OECD 1980: 91 ff.). Entrepreneurial R&D became more central to explaining economic dynamics, but at the same time, there was an increasing emphasis on the scientific knowledge of technological innovations (Freeman 1973), with the result that the link between state-funded scientific research and private R&D became even more important. Targeted R&D directed at market innovations became the core of economic policy considerations. According to this logic, the crisis in the 1970s had to be addressed by the targeted support of R&D activities, especially in view of the stagnation in productivity development (Edler 1999).

A vital insight from the systemic view of innovation was that it is the increasingly generic and systemic character of technologies that determines these dynamics:

> [O]ne of the most important advances made in the area of the theory of technology during the last five years or so has been the move made from several sides towards an understanding of major technologies as having systemic traits (Chesnais 1982).

In concrete terms, microelectronics was the most important technological driver on which the conceptual developments were based. This was characterised as a "technical revolution" (Krupp 1975) that was going to fundamentally alter various sectors and social practices. The changes that were associated with this pervasive technology led to reflections on how technological innovations and socio-institutional configurations interacted. This was new in this form (see Heyen et al. 2024 in this anthology).

This also involved rethinking the innovation process itself. The linear model was replaced—or rather supplemented—by a model of different phases and stages which interact with one another. The chain-linked model of Kline and Rosenberg (1986) is one prominent example among several at the time, or the coupling model of Keck (1986). Innovation was thought of as resulting from complex interactions of different actors, with feedback loops between different stages of the process (see also Mowery and Rosenberg 1979) and between different sectors and scientific disciplines. Attention was also increasingly paid to innovations not only being the result of implementing technological inventions but also emerging in response to the needs of the market or as recombinations of existing technologies. This more complex view of the innovation process and the understanding of the need for multiple horizontal and vertical interactions and collaborations had become widespread by the early 1980s (European Industrial Research Management Association 1982; Teece 1986; Rosenberg 1991; Wissenschaftsrat 1992; Edler 1999).

This new conceptualisation of the innovation process was accompanied by the development of a specifically systemic view of innovation dynamics. R&D and the generation of innovations were then conceptualised as the interplay of the economy, science, and society (Freeman 1982). This insight was also the result of empirical analyses that were no longer able to explain productivity differences only in terms of different sectoral production functions. Instead, different socio-institutional configurations were considered important complementary explanations (Nelson and Winter 1977; Gaudin 1985; Edler 1999). Consequently, the understanding became increasingly widespread among researchers and analysts that technological innovations and socio-institutional configurations are influenced by their close interdependency.

The various academic developments then led to an explicit and broadly conceptualised innovation systems approach in the 1980s. While this term had already appeared in the early 1970s (Krupp 1972), it was conceptually elaborated by a number of leading innovation economists, i.e., Freeman et al. (1982); Lundvall (1985, 1992); Nelson (1993) as well as Edquist (1997). The different variants of the innovation systems approach place a different emphasis on cumulative learning processes and on institutional configurations, especially with regard to user–producer relationships and the relative importance of the public sector vs. the market. In addition, the different approaches are influenced to varying degrees by insights and assumptions from evolutionary economics, according to which actors behave with bounded rationality under conditions of high uncertainty and therefore develop adaptive strategies instead of striving for optimal ones (Nelson and Winter 1982; Fagerberg 2003), as postulated in a neoclassical world with perfect information and technology as external variables. These different variants were widely adopted in practical policy strategies not only in the OECD world (Lundvall 1992), but also in developing and newly industrialising countries (Chaminade et al. 2018).

From the end of the 1990s onward, the research agenda included not only actor relationships and dynamics in the innovation system, but also more holistic system considerations as objects of investigation. Alongside the evaluation of individual funding programmes, institutional evaluations became more important, i.e., the role and performance of research or research funding organisations in the innovation system were also taken into account. The evaluation of the Research Council of Norway, which was co-developed and conducted at Fraunhofer ISI, was the first to work with a graphical representation of the innovation system (Arnold et al. 2001; Kuhlmann and Arnold 2001; see also Bührer et al. 2024 in this anthology). The diffusion of the innovation system approach, which was formulated by different scientific communities, had far-reaching consequences for the role and instruments of STI policy. If innovation is the driver of economic growth and social development, if innovation is the result of a complex, recursive interplay of different actors, scientific-technological disciplines and sectors, if recursive learning processes must be enabled and enhanced, if publicly funded research remains a mandatory foundation for market-based, radical innovations, then it is incumbent upon the state to actively support and help shape the necessary socio-institutional configurations and interactions in the system. Not only the financial subsidising of existing constellations is important in this perspective, but also support for reconfigurations, for new linkages, for connections that do not result on a sufficient scale from a purely market-based logic. It is therefore a task of the state to adapt socio-institutional configurations over time or to enable this adaptation to come about in an evolutionary way. This does not question the role of the market in generating and selecting innovations, but greater attention must be paid to the conditions under which innovations are generated and diffuse in market economies. According to this perspective, it is no longer sufficient for the state to fund science or long-term oriented R&D in companies. In terms of research and innovation policy, it was now increasingly important to think of R&D in broader contexts, as more cooperative and more systemic.

In the early 1980s, the term innovation policy became increasingly dominant for all the policies that actively addressed these systemic challenges and linked them to economic and social goals (Rothwell and Zegveld 1981; Sweeney 1985). In view of the stated complexity of the system, the relevant innovation economists were quick to point out that a number of different ministries and policy fields would have to be an active part of such an innovation policy, and at the same time, that they would be overwhelmed by the complexity of the interrelationships. Thus, although the fundamental steering optimism of the 1970s was largely obsolete and the neo-liberal or neo-conservative turn in the late 1970s and early 1980s assigned the state a more reticent role in the economy, the role of the state became de facto increasingly important for the "new" innovation policy in light of the complementary insights into the systemic nature of innovation outlined above. This was accompanied by a greater differentiation of instruments that took different forms in different countries (Rothwell and Zegveld 1981, 55 ff.; Gaudin 1985; Stuart and Kuntze 1982).

In view of the many systemic points of connection in the innovation systems approach, the concrete set of instruments continued to become increasingly differentiated over the course of the 1980s and 1990s (Edler and Georghiou 2007; Edler et al. 2016). The focus was on improving the framework conditions, and especially on incentives to support collaboration or networking among different actors. Examples include Alvey in the United Kingdom, ESPRIT and BREITE/EURAM at EC level, as well as the "Verbundprogramme" in Germany. In addition, there were cluster and network programmes, mobility programmes, knowledge and technology transfer programmes, further education and training on innovation management and entrepreneurship, as well as awareness-raising measures and the first large-scale foresight processes (see Cuhls et al. 2024 in this anthology). These were intended to provide actors in the system with more structured opportunities to jointly reflect on the orientation of innovation activities. The latter has been part of a broader understanding of policy advisory studies associated with the catchphrase of strategic intelligence (cf. Kuhlmann 2003 as well as Bührer et al. 2024 in this anthology). The differentiation and policy design of sectoral (among others, Malerba 2004), regional (e.g., Braczyk et al. 1998; Koschatzky 2001), and technological innovation systems (e.g., Carlsson and Stankiewicz 1995) became significantly more important during this phase. Accordingly, the epistemic community also broadened to include, for instance, economic geography and various forms of sustainability research.

Since both the European and, in federal countries, the regional level developed policymaking ambitions, and since it made sense to apply the innovation systems approach at different levels, this differentiation took place on all three political system levels. Simultaneously, on all three levels, the institutions of innovation policy also became increasingly more differentiated, i.e., innovation and funding agencies, ministerial departments, advisory bodies, etc. From around the late 1980s, this led in turn to a growing need for vertical and horizontal coordination.

Similar to the institutional expansion and differentiation of the innovation systems, the field of STI policy research also experienced major steps of institutionalisation in the early 2000s. Taking advantage of an EU sponsored scheme for European Network of Excellence (NoE), a number of senior researchers, led by Philippe Larédo of Paris, initiated the NoE "Policies for Research and Innovation in the Move towards the European Research Area" (PRIME). The founders were motivated by the recognition of "major transformations in the locus, formulation and implementation of science, technology, higher education and innovation policies, which call for new theoretical and empirical research in research and innovation policy studies" (see Larédo 2003). Furthermore, as the STI policy field at the time was highly interdisciplinary and consisted of researchers dispersed in universities and non-university research organisations, it was poorly institutionalised. The intention thus was to create a pan-European intellectual homebase for researchers, in particular the young generation (Luukkonen et al. 2006). In 2010, building on the experiences of PRIME, the "European Forum for Studies of Policies for Research and Innovation" (EU-SPRI) was founded as a European Association which has meanwhile grown to include 19 member organisations (https://euspri-forum.eu/).

3.3 The "Normative Turn" and the Orientation Towards an Innovation Policy Aimed at Addressing Societal Challenges

In the early 2000s, there was a first wave of criticism of the dominant policy approach of the time, heralding, as it were, the policy shift towards addressing societal challenges. Although innovation policy as a policy of improving the system performance also set certain content-related priorities or directionality in technology or sector-based programmes, it was generally not oriented towards specifically applying innovations to solve societal problems. Aligning innovation with societal needs was left to the state or other policy areas, not innovation policy. In terms of competencies of departments and ministries, negative coordination was the dominant mode of governance, in essence leading to a demarcation of responsibilities. In the area of research and innovation funding, technology or sector programmes focused primarily on basic research and the early phases of the innovation process, while research and innovation with a stronger focus on strategic policy or societal goals such as energy efficiency or environmental protection were (and still are) the responsibility of the respective sectoral ministries. Although these disconnected responsibilities are understandable in their political-institutional context, it can be stated that the epistemic communities that were engaged in analysing and developing approaches in the respective policy areas were not able to sufficiently bridge the gap between the different rationales. There was no overarching discussion of the different objectives and possible implications of an increased focus on sustainability issues, e.g., in the areas of energy, the environment, or water innovation (cf. the contributions in this anthology). The lack of an integrated approach has been noted over time by several authors, such as Walz et al. (2019), who criticised the lack of integration between environmental and innovation policies. Weber and Rohracher (2012) spoke in this context of policy coordination failure. So, although other policy areas were engaged in directional research and innovation funding earlier than STI policy in the stricter sense, this does not mean that this was done under the paradigm of initiating transformative change and the socio-technical system innovations necessary for it. The lack of a transformative paradigm has been noted to this day, for example, in environmental policy (Jacob and Ekins 2020).

At the same time, orienting policy to real needs was already part of the concept of innovation policy as it had been developed in the 1980s. Even in the early days of a more explicit innovation policy, innovation researchers called for innovation policy to be more strongly oriented towards societal needs and for innovation policy to be placed at the centre of government policy, as it were, in order to bring about positive change in the economy and society (Rothwell and Zegveld 1982). An international example of this is the summary of an Innovation Policy Workshop of the so-called Six Countries Programme, which was organised by Fraunhofer ISI:

> In fact, with a little thought it is possible to extend the list of things that influence innovation to such an extent that if the concept of an innovation policy is expected to accommodate all of them, then such a policy becomes almost all encompassing in relation to the wide range of already existing policies. This should make clear that by its horizontal nature the goals of innovation policy may interfere with those of other policies. In a rational innovation policy, one would expect these conflicts to be recognized and taken into account in the formulation of policies and measures, i.e. a rational and systematic approach towards moving in desired and predetermined directions (Stuart and Kuntze 1982).

As early as 1985, Helmar Krupp also called for a decidedly demand-oriented approach (Krupp 1985), especially with regard to the necessary adaptation of new technologies to increase efficiency in the energy system. Moreover, Soete and Arundel (1993) developed a conceptual differentiation of innovation policy that included an explicit move towards an approach that was also "mission-oriented".

However, these notions were ahead of their time. In practical policy terms, this only changed with the development of dedicated concepts for demand or needs-based policies (Edler 2007; Edler and Georghiou 2007). The starting point for this was, on the one hand, the attempt to make greater use of government demand to generate societally important innovations (Edquist et al. 2000; Edler et al. 2005). On the other hand, political actors from other policy areas (environment, energy) also began to look more closely for opportunities to use research and innovation policy to address complex problems in their areas of responsibility. At the same time, approaches in other policy areas that had traditionally applied demand-side instruments inspired the innovation policy debate (Edler 2007). Consequently, demand-side instruments were also integrated (OECD 2011) into innovation policy with a focus on innovative procurement in the public sector. More far-reaching demands for tools that not only integrate the demand side but also cover all the functions attributed to innovation systems in a holistic and coordinated manner (Edler et al. 2007; Edler 2007; Smits and Kuhlmann 2004) went largely unheard.

With the shift to the demand side and thus to society's needs, the bridge to other policy areas was finally established. However, the corresponding opening up of innovation policy only really began with the explicit turn towards the "Grand Challenges" and transformation-oriented innovation policy.

In addition to the early opening up of innovation policy research to include a stronger focus on societal challenges, there were simultaneous policy developments in the early 2000s that supported the existing paradigms (and they continued to dominate the majority of research articles published on the topic).

At the EU level, the Lisbon Strategy identified shortfalls in public investment in R&D and it was agreed that EU Member States would make a voluntary commitment to increase their overall economic spending on R&D to three percent of their GDP. At the same time, the lead market idea was emphasised with the aim of bundling and streamlining programmes in terms of topic and technology at both European and national level in order to achieve a stronger focus on promoting fields of strength or key technologies that had been identified. The newly created mission statement of the European Research Area was intended to support these ambitions and create the necessary systemic conditions at the same time.

The so-called Aho Report (Aho 2006) on the "Grand Challenges" was a first impulse in the direction of the emerging transformation in STI policy into the third paradigm on the policy level. In this report, innovation researchers and high-ranking representatives from politics and industry urged that innovation policy should be mobilised to solve societal problems in the EU. This was in no small part a response to the overall disappointing innovation dynamics in the context of the Lisbon Strategy. A few years after publication of the Aho Report, in the Lund Declaration in 2009 (Swedish Presidency 2009), 350 participants from science, politics, industry, and research funding called on European policymakers to focus European-funded research on the grand societal challenges of the day and also to align national research funding accordingly in a conference organised by the Swedish EU Council Presidency in preparation for the 8th EU Research Framework Programme (later Horizon 2020). The statement already addresses the necessary political processes and course-setting that were to determine the debates for more directionality in innovation policy a few years later. In this context, the Lund Declaration (2009) emphasises that the "Grand Challenges" must be identified in a joint process involving politics, business, administration, NGOs, and the research community.

However, the financial crisis of 2008/2009 shifted the focus of many European governments back to stabilising economic strength, with particular attention paid to strengthening the industrial base. Although research and innovation had been considered important drivers of economic strength under the first two paradigms, many governments actually decreased their investments in R&D. The legal basis adopted in 2011 for the EU's 8th Research Framework Programme Horizon 2020 (European Commission (2014)), for example, was dominated by the crisis and emphasised the growth and competitiveness of European industry as its primary goals. Although, for the first time, the programme did not contain any theme-based sub-programmes, it did organise funding along seven societal challenges. However, the instruments used were not changed in any way that would have been conducive to the new paradigm that was emerging. The societal challenges were merely included in the broad research agenda, and mostly only in the form of expected "impacts", whereas the instruments continued to address primarily science and industry as the traditional R&D actors. Further developments of these instruments were primarily focused on strengthening SMEs and the innovativeness of companies (new instruments for SMEs and the subsequent piloting of the European Innovation Council, EIC). Similar developments occurred at the national level. In Germany, for example, the so-called High-Tech Strategy was launched in 2006 (Bundesministerium für Bildung und Forschung 2006), in which societal goals were increasingly proclaimed over successive generations of this strategy (Bundesministerium für Bildung und Forschung 2010, 2014, 2018). However, a consistent shift to the new paradigm had not yet taken place.

The adoption of the United Nations Sustainable Development Goals (SDGs) in 2015, as well as the rapidly worsening climate crisis and internationally agreed climate targets once again increased the pressure on STI policy to focus its approaches more strongly and systematically on addressing these overarching issues. These developments at policy level were accompanied by and interrelated with conceptual developments in innovation research.

After Christopher Freeman, Bengt-Ake Lundvall, Charles Edquist, and Luc Soete, a second generation of innovation economists and researchers played a pivotal role. These included, for example, the economist Mariana Mazzucato, who gained popularity with her call for the state to take a more active role and to shape markets in response to important societal challenges (Mazzucato 2013, 2018). The innovation policy debate in Germany also opened up cautiously in the direction of a needs-based and directional approach as a result of the greater focus on evolutionary economics in the Expert Commission for Research and Innovation (EFI) appointed by the German government. However, the broader research community that was specifically concerned with STI policies dealt with the new issues comparatively cautiously and published only with some delay in reaction to the developments, especially at the European level. An early influential paper from 2012 by Weber and Rohracher created a foundation for broadening the legitimacy base for policy intervention by placing a third group of rationales for government intervention alongside the established ones of market and systemic failures, namely: "transformational failures". Fraunhofer ISI also addressed the issue in its commemorative publication as part of its 40th anniversary celebrations (Daimer et al. 2012) and by organising the European Forum for Studies of Policies for Research and Innovation (EU-SPRI) conference under the heading "Towards Transformative Governance? Responses to mission-oriented innovation policy paradigms" (Fraunhofer Institut für System- und Innovationsforschung ISI 2012). The call for the conference postulated the emergence of a third paradigm for STI policy:

> This (author's note: The Lund) declaration has taken up and reinforced a development in the past few years in which governments and the European Union have adopted a new strategic rhetoric for their research and innovation policy priorities which addresses the major societal challenges of our time. This is evolving into the third major policy rationale besides economic growth and competitiveness.

The third paradigm essentially assumes that science, technological development, and innovation produce relevant contributions to solving the societal challenges of our time and, in particular, trigger transformations in the direction of sustainable development.

The STI policy science community is now intensively addressing issues related to this latest policy paradigm and is increasingly interacting with communities researching sustainability transitions. This has further substantiated the impetus of the third paradigm in normative and conceptual terms. Thus, approaches such as the multi-level perspective were taken up, from which the need for policy interventions aimed at supporting new market niches (technologies or applications) can be derived (Geels 2002; Smith et al. 2010). A conceptual combination of approaches from the two research traditions, i.e., approaches from STI policy and sustainability transition research, and thus an explicit derivation of the third paradigm was provided by another influential publication by Schot and Steinmueller (2018), which popularised the term transformative innovation policy.

Although the new paradigm of innovation policy, which aimed at directionality, problem-solving, and system transformation, quickly became widely accepted (e.g., OECD 2015), little has changed so far at the instrumentation level. Policymakers may have accepted the new objectives ("policy agenda"), but they have not yet fully adapted a broader concept of innovation, which would ultimately have far-reaching implications for innovation policy instrumentation (Diercks et al. 2019).

A broad conception of innovation goes beyond a purely science-based and technological understanding of it to include forms such as "doing, using, and interacting", which places a great deal of emphasis on social or organisational innovation. This also implies that the group of innovating actors is defined more broadly, encompassing not only research and industry, but also a wider range of actors in the public and, above all, the civic sectors (cf. also Warnke et al. 2016). A third crucial point is that, in addition to the supply side ("technology supply"), the demand side (end users, so-called "need owners") is also an important starting point for policy design.

Many of the policies established in recent years address the new objectives, but in effect still remain constrained by a narrow definition of innovation. The above-mentioned example of Horizon 2020 is one of them, as are some of the national innovation strategies initiated in the past decade, such as the German High-Tech Strategy (Bundesministerium für Bildung und Forschung 2018) or the Dutch Top-Sector Strategy (Ministry of EZK 2019). Despite their mission-oriented focus, both these national strategies rely primarily on research and technology and allude, mainly rhetorically, to wider definitions of actors and innovation, but do not actually apply them in the instruments used.

There are systemic instruments that are suitable for addressing the new paradigm because they deal with diagnosed systemic failures in a differentiated way and are suited to integrating a wider definition of innovation and directionality (orientation towards societal challenges). These include, above all, approaches that strengthen the ability of actors to learn and reflect, such as adaptive, supportive, and interactive evaluations of funding programmes, or foresight processes that are designed to be inclusive and systemic (Daimer et al. 2012; Lindner et al. 2016).

Whereas at the strategic level, the alignment of STI policy with the Grand Challenges can now be observed on a broad scale in the OECD world, the understanding of innovation and its instrumentation has by and large not followed suit. Against this background, STI policy in a number of OECD countries as well as at the EU level has increasingly adopted mission-oriented approaches in recent years (Larrue et al. 2019; Larrue 2021), in order to make the unspecific Grand Challenges politically manageable and to operationalise them by defining concrete targets and time horizons (Lindner et al. 2021). Accordingly, current research is concentrating on the new mission-oriented approaches (Janssen et al. 2021; JIIP 2018; Kuittinen et al. 2018; Wanzenböck et al. 2020; Wittmann et al. 2021a, b). It can be observed that mission-oriented approaches are becoming increasingly differentiated and a number of challenges related to governance have been identified, which still stand in the way of effectively implementing mission approaches (Edler and Boon 2018; Lindner et al. 2021). Simultaneously, the first attempts are being made to evaluate these new policy approaches (Ghosh et al. 2021; Haddad and Bergek 2023; Wittmann et al. 2022), which are primarily concerned with the question of how to measure political and societal "impacts" (cf. Bührer et al. 2024 in this anthology on this and on the generally increasing impact orientation). Another major issue that has so far been discussed mainly in research is the need for the transformation of entire systems, understood as a reconfiguration of the provision of important societal functions such as mobility or energy supply. This requires a complex process of transformation involving not only technological but also non-technological innovations and changes in actor behaviour, as well as the reconfiguration of actor networks (Borrás and Edler 2020; Grillitsch et al. 2019).

4 Conclusions and Outlook

4.1 Making Sense of Paradigmatic Changes

The development of STI policy in the OECD world can be divided into three distinct phases that differ in terms of the underlying theoretical basic assumptions about how knowledge and innovations emerge, the policy goals, the dominant approaches to solutions, policy instruments and governance structures as well as the contributions of (policy advising) innovation research.

1. In the first post-war decades, the theoretical foundations for early STI policy were clearly dominated by the key postulates of neoclassical economics. State intervention in processes of knowledge generation was justified by the need to address phenomena of market failure, especially in the funding of basic research. However, the policy paradigm was not without contradictions at the level of instrumentation, as priorities and steering of technology policy already existed, albeit in a comparatively weak form that were in conflict with neoclassical orthodoxy. Over the course of the 1970s, there were the first cautious conceptual developments from science that challenged the previously dominant basic assumptions and problem constructs. These were based on a gradually improved understanding of knowledge generation processes as well as initial approaches of a systemic understanding of innovation processes.
2. From the end of the1970s, the second STI policy paradigm gradually asserted itself as the dominant perspective. As with the first paradigm, the key policy goals of STI policy focused mainly on economic objectives such as growth and competitiveness. The main drivers behind this paradigm shift were the end of the long period of economic growth in the first post-war decades and intensified international competition, which significantly increased the importance of national (and later also regional and sectoral) innovation performance as an essential prerequisite for economic prosperity in the political and scientific debates. In this context, these were the expansion of the theoretical and conceptual foundations of the STI policy paradigm to also include concepts from evolutionary economics and other academic disciplines (such as political science) on the one hand, and new insights into innovation processes and their factors of influence on the other. When combined, at the time, these provided convincing epistemic explanations to challenge STI policy. The goals of STI policy remained basically the same compared to the first paradigm (economic growth), but with a much stronger emphasis on company-based innovation activities. At the level of instruments, the new paradigm was accompanied by numerous innovations in the form of "system-strengthening" instruments. However—and this is an indication of the phenomenon of policy layering—key instruments from the first policy paradigm were largely continued without any changes. Criticism of the goals of the system-strengthening paradigm started at the beginning of the 2000s, mainly due to the insufficient orientation of STI policy towards society's needs.
3. The rise of the third STI policy paradigm became apparent around the mid-2000s. The main characteristic of this most recent paradigm is the targeted orientation of STI policy towards addressing urgent societal challenges. Within this paradigm, STI policy measures are justified by the potential contributions that STI can make to addressing the Grand Challenges. This new orientation at the level of objectives was mainly driven by the growing pressure of problems related to worsening of the climate crisis and the crossing of planetary boundaries as well as the growing realisation that previous STI policy approaches were unable to effectively mobilise science, technology, and innovation and direct them to tackle current problems. However, economic crises have slowed down the transformation process and reinforced the strong influence of established STI policy

paradigms. Overall, the fundamental objective of the third policy paradigm has become established over the last 10–15 years and it is now widely recognised that this new generation of innovation policy should not only address market and system failures, but also the failure to transform (Weber and Rohracher 2012). However, the debate in STI policy research on the conceptual foundations and implications of the new paradigm has largely only taken place as a follow-up to the strategic reorientation of STI policy, which, as has been shown, was strongly influenced by the course set at EU level. Important discursive contributions to the paradigm shift from research are related, for example, to re-assessing the function of the state in research and innovation processes, which is in contrast to the main, primarily neoclassically influenced axioms of previous policy paradigms, as well as to systematically deriving the policy intervention rationale for the new paradigm. The fact that a significant mismatch between paradigm objectives, on the one hand, and the instruments and governance structures, on the other, can still be observed is also an indication that the process of paradigm change is still ongoing.

Table 1 provides an overview of the three STI policy paradigms based on the key analytical dimensions of this chapter.

In order to further develop the policy-analytical understanding of the processes of paradigm shifts, some conclusions can be drawn, at least with regard to the STI policy field. That policy paradigms are not subject to strict incommensurability and that one paradigm does not have to be completely replaced by the next is supported by the analysis of the three STI paradigms. The fact that the dominant paradigm in each case has incorporated theoretical and conceptual elements as well as partial goals of the previous paradigms clearly confirms empirically the possibility of recombinations and hybrid forms. This rather "soft" form of paradigm shift is coupled with predominantly incremental processes of paradigm shift that take place over longer periods of time. One explanation for the fact that the process of paradigm shift in STI policy has not been characterised by a clear replacement of one paradigm by another, as conceptualised by Hall (1990, 1993), could be that he based his analyses on the replacement of an economic policy paradigm, whose basic assumptions were developed and discussed exclusively by one scientific discipline. In STI policy, on the other hand, a number of different disciplines have played a role since the 1970s at the latest, with the consequence that academic discourses are more diversified and thus the processes of paradigm change could also be more multi-layered. We have seen that, at least in the two most recent STI paradigms, the degree of coherence between the systems of ideas is low, given the coexistence of sometimes contradictory intervention logics, approaches, and goals.

The three STI policy paradigms have thus been superimposed on one another in an additive manner in the sense of policy layering, reducing the effectiveness of the preceding, older paradigm in each case, but not displacing it completely. This sometimes results in considerable tensions both between and within the paradigms. This last point applies, for example, to certain measures of targeted technology support that took place within the framework of the first and second paradigms, but which

Table 1 Overview of the three STI policy paradigms

Analytical dimensions	Paradigm 1: STI policy to generate knowledge and address market failures	Paradigm 2: Innovation systems approach	Paradigm 3: Directionality and problem orientation, transformative change
Context conditions	Post-war years with a specific focus on reconstructing science systems	End of the long phase of economic growth in the post-war decades; stagflation in Western industrial nations	Growing political urgency of the "Grand Challenges" and the need for system transformations
Basic understanding	Development and expansion of scientific capacities, public funding, especially for the generation of basic knowledge; Neoclassical welfare theory as a guiding theoretical and conceptual idea	Ideas based on evolution theory; Research and technology as particularly important drivers of economic growth	STI as an essential prerequisite for and contributor to solving societal challenges; Emerging change: increasingly broad understanding of innovation, which also sees non-research-based forms of innovation as key to addressing societal challenges
Policy objectives	Promoting growth and competitiveness through research	Growth, economic dynamism, and competitiveness through innovation	Focus increasingly on addressing societal challenges, complemented by economic goals such as competitiveness
Problem constructs and solutions	Since knowledge is conceptualised as a public commodity, government funding is required (basic research, research institutions); Utilisation of linear understanding predominantly through spill-over effects Support through IP governance; To some extent technology policy priorities	Systems of innovation and interaction between actors as dominant heuristic; Understanding the innovation process in a recursive way; Government interventions aimed at improving the performance of the innovation system (systemic instruments); Supply orientation; Innovative companies as the key to globally competitive economies	Systems of innovation remain key; Sustainability transformation concepts find their way into STI policy discourses; At the level of instruments, largely as in paradigm 2, quantitative growth of thematic research funding; First approaches of MOIP and transformative innovation policy; Growing impact orientation; Need for system transformations

(continued)

Table 1 (continued)

Analytical dimensions	Paradigm 1: STI policy to generate knowledge and address market failures	Paradigm 2: Innovation systems approach	Paradigm 3: Directionality and problem orientation, transformative change
Main focus of (policy advising) research	Requirements of knowledge generation in publicly funded research and in industry; Requirements for the transfer of basic knowledge to its application	Further development of the understanding of innovation processes and the implications for innovation support; Identification and analysis of requirements and effects of the innovation system and its deficits (also comparatively); Instruments for system upgrading and their effects	Further development of STI policy instruments and their governance to achieve directionality; Interplay between innovation and system transformation; Identification of innovation paths that contribute to solving problems; Mobilising strategic intelligence, including impact analysis, to support transformations

were at odds with the dominant economic assumptions of the time. With regard to the third paradigm, there is a potential conflict of goals with the earlier paradigms, both of which, in different ways, aim at the economic growth of separate, mainly national, innovation systems. As already stated in the Club of Rome report "The Limits to Growth" (Meadows et al. 1972), planetary limitations such as resource availability necessitate a move away from a definition of welfare that is based exclusively on economic growth. In addition, sustainable innovation needs to diffuse as quickly as possible. This implies that the necessary demand-side support can become support for innovation imports and thus contribute to positive economic benefits in other innovation systems (Edler 2010).

The overall picture also reveals differences in the contributions of the STI policy research community and scientific policy advice community to the processes of paradigm change. This becomes clear when looking at the recent STI policy paradigm shift, which differs from the previous one primarily in terms of changes at the level of objectives, while the shift from the first to the second paradigm was significantly driven by changes at the level of basic theoretical assumptions and conceptions of the problem. To put it bluntly, the first case was therefore primarily science-driven, while the second paradigm shift was and is more strongly influenced by shifts at political and normative levels. In both paradigm shift processes, however, changes to the context conditions—in brief: end of the growth phase of the post-war decades and stagflation crisis, urgency of the climate and environmental crises—created important prerequisites for paradigm reorientation. It is a perplexing finding that, apart from a few exceptions, the scientific community in the field of STI policy has not actively promoted and prepared the conceptual basis for the most recent paradigm shift, but has delayed incorporating it into the research agendas. One reason for this omission may be the process of professionalising the discipline of innovation research in recent decades. This has led to establishing and formalising the discipline of innovation research at universities and non-university institutions, but at the same time might have been accompanied by a tendency to close off innovation research from related disciplinary discourses, such as sustainability transition research. A perspective that was strongly influenced by economic and industrial policy also prevented a discourse being established with the epistemic communities of important sectoral policies, where the directionality towards sustainability became apparent early on, which was also reflected in the corresponding sectoral research and innovation funding programmes. It was not until the rise of the third STI policy paradigm that efforts were actively made to correct these past omissions.

4.2 Challenges for the STI Policy Community

Looking at the changing paradigm of STI policy since the post-war period has clearly revealed the profound interplay between changing environmental conditions, new normative policy objectives, and the conceptual work of the scientific community. Looking at the future development of STI policy poses the question with which research agenda, but also with which understanding of its role STI policy research can make the most constructive, future-oriented, and socio-politically relevant contributions. This seems to require a three-pronged approach:

1. Coherence: In the course of the policy layering of the three distinct phases of STI policy to date, tensions and contradictions are evident in some of the central basic assumptions as well as in the orientation of STI policy actions derived from them. Contributing to a more coherent theoretical-conceptual framework of STI policy and its further development seems to be an important task for the relevant STI policy researchers and their scientific policy advice services. According to Hogan and Howlett (2015) and Daigneault (2014a), the coherence of idea systems or policy paradigms plays a major role in determining how influential and "policy effective" they are. As effective contributions of science, technology, and innovation to the complex processes of system transformations are increasingly called for, greater urgency is attached to questions concerning the right balance and effective interplay between market-oriented mechanisms and primarily state-orchestrated directionality. So far, the tensions between the market-oriented basic assumptions of the first two paradigms and the problem- and/or solution-oriented approach of the third paradigm have hardly been addressed. One exception is the attempt by Breitinger et al. (2021) to productively combine findings from classic innovation research with the requirements of transformation processes.

   Of even more fundamental importance is the conflict between the goals of economic growth and the need to respect planetary boundaries, which has so far only been touched upon marginally in the STI policy community. The discourses on green growth, de-growth, or post-growth have so far largely taken place without any significant participation of innovation researchers. A systematic examination of the question of how innovation processes can be shaped under a different growth policy premise appears increasingly necessary.

2. Positioning: It is obvious that findings from STI policy research can make significant contributions to the effective design of transformation processes. Particularly relevant STI policy perspectives include questions of improving the performance of innovation systems, the diffusion speed of new applications and innovative processes, or the analysis of policy and governance capacities. While the need to mobilise STI for transformation processes is indisputable, there has been no approach to date that effectively links STI policy with sectoral policies. The problem is illustrated by the example of mission-oriented innovation policy (MOIP), which often claims to be transformative, but in fact remains mostly within the traditional confines of the STI policy field in terms of its main focus. The fact that effectively integrating STI policies into the broader policy mixes of sectoral transformation processes might also be accompanied by a relative weakening of the role of STI policy in relation to the respective sectoral policies has hardly been openly discussed so far, although the field of sectoral policy research has touched upon this. Here, STI policy is conceived as part of an overarching policy mix that complements sectoral policies which play a role primarily in innovation diffusion and the exnovation of established solutions (e.g., Kivimaa and Kern 2016; Rogge and Stadler 2023).

   The growing importance of transformation processes raises further questions of identity and positioning for the STI policy community. Since successful transformations aim at comprehensive behavioural and structural changes, generating winners and losers and often intervening to a great extent in people's everyday lives, transformative policies generally receive more political and media attention than conventional innovation policies. The question of how to deal with the potential politicisation of transformation processes also arises indirectly for (policy advising) academia. So far, it seems that STI policy researchers have not paid much attention to these issues.

3. Responsiveness: The last few years have been repeatedly marked by turbulent developments such as the COVID 19 pandemic or the deep geopolitical upheavals resulting from Russia's military aggression. In general, the STI policy community has responded quickly to each crisis. For example, in the case of the pandemic, methods were developed to assess the effectiveness of crisis response (Weber et al. 2021). Fundamental questions of systemic resilience (cf. Roth et al. 2021) in response to increased crisis probability or technology sovereignty (Edler et al. 2023) in the face of growing geopolitical uncertainties and global trade conflicts were also incorporated into the research agenda. In view of the growing frequency and intensity of crises, the question also arises as to the implications for STI policy research. Two things seem obvious: in the short to medium term, the aim is to improve the responsiveness of research and innovation actors without compromising quality and content; in the medium to long term, it will be a matter of further developing research and innovation systems in

the direction of resilient yet transformation-capable structures. The idea behind this is that policymaking based on a concept of the reasons for failure will always remain reactive. This should be accompanied by a more forward-looking, proactive approach to policymaking that enables policy interventions based on a potential future failure to address crisis-related developments (Kubeczko and Weber 2009). Especially for the latter, the STI policy community seems rather well equipped due to its wealth of knowledge and experience with system analyses.

References

Aho E (2006) Creating an innovative Europe: Report of the Independent Expert Group on R & D and Innovation appointed following the Hampton Court Summit, 2006th edn. EUR, vol 22005. Office for Official Publications of the European Communities, Luxembourg

Arnold E, Kuhlman S, van der Meulen B (2001) A singular council: evaluation of the Research Council of Norway. Technopolis, Brighton, England

Arrow KJ (1962) Economic welfare and the allocation of resources for invention. In: Universities-National Bureau Committee for Economic Research, Committee on Economic Growth of the Social Science Research Council (ed) The rate and direction of inventive activity: economic and social factors. Princeton University Press, Princeton, NJ, pp 609–626

Béland D (2007) Ideas and institutional change in social security: conversion, layering, and policy drift. Soc Sci Q 20–38

Béland D (2009) Ideas, institutions, and policy change. J Eur Public Policy 16:701–718. https://doi.org/10.1080/13501760902983382

Béland D, Cox RH (2013) The politics of policy paradigms. Governance 26:193–195. https://doi.org/10.1111/gove.12034

Borrás S, Edler J (2020) The roles of the state in the governance of socio-technical systems' transformation. Res Policy 49:103971. https://doi.org/10.1016/j.respol.2020.103971

Braczyk H-J, Cooke P, Heidenreich M (1998) Regional innovation systems: The role of governance in a globalized world. UCL Press, London

Breitinger JC, Edler J, Jackwerth T, Lindner R, Schraad-Tischler D (2021) Good-Practice-Beispiele für missionsorientierte Innovationsstrategien und ihre Umsetzung. Innovation for transformation / Bertelsmann Stiftung, Ergebnispapier 1. Bertelsmann Stiftung, Gütersloh

Bührer S, Edler J, Kuhlmann S, Seus S (2024) Evaluating public research and innovation policies: a short history of co-evolution. In: Edler J, Walz R (eds) Systems and innovation research in transition: Research questions and trends in historical perspective. Springer, Heidelberg

Bundesministerium für Bildung und Forschung (2006) Die Hightech-Strategie für Deutschland. Kommunikation und publizitisk GmbH, Bonn

Bundesministerium für Bildung und Forschung (2010) Ideen. Innovation. Wachstum.: Hightech-Strategie 2020 für Deutschland, Berlin

Bundesministerium für Bildung und Forschung (2014) Die neue Hightech-Strategie: Innovationen für Deutschland, Berlin

Bundesministerium für Bildung und Forschung (2018) Forschung und Innovation für die Menschen: Die Hightech-Strategie 2025, Berlin

Bush V (1945) Science, the endless frontier: A report to the President. United States Government Printing Office, Washington

Campbell JL (2002) Ideas, politics, and public policy. Annu Rev Sociol 28:21–38

Carlsson B, Stankiewicz R (1995) On the nature, function and composition of technological systems. In: Carlsson B (ed) Technological systems and economic performance: the case of factory automation, vol 5. Springer, Dordrecht, pp 21–56

Carson M (2004) From common market to social Europe?: Paradigm shift and institutional change in European Union Policy on Food, Asbestos & Chemicals, and Gender Equality. Stockholm studies in sociology, N.S, vol 22. Acta Universitatis Stockholmiensis

Carstensen MB (2011) Ideas are not as stable as political scientists want them to be: a theory of incremental ideational change. Polit Stud 59:596–615. https://doi.org/10.1111/j.1467-9248.2010.00868.x

Carstensen MB, Schmidt VA (2016) Power through, over and in ideas: conceptualizing ideational power in discursive institutionalism. J Eur Public Policy 23:318–337. https://doi.org/10.1080/13501763.2015.1115534

Chaminade C, Lundvall B-Å, Haneef S (2018) Advanced introduction to national innovation systems. Elgar advanced introductions. Edward Elgar, Cheltenham, UK

Chesnais F (1982) Schumpeterian recovery and the Schumpeterian perspective—some unsettled issues and alternative interpretations. Emerging technologies. JCB Mohr, Tübingen

Cuhls K, Dönitz E, Erdmann L, Gransche B, Kimpeler S, Schirrmeister E, Warnke P (2024) Foresight: Fifty years to think your futures. In: Edler J, Walz R (eds) Systems and innovation research in transition: Research questions and trends in historical perspective. Springer, Heidelberg

Daigneault P-M (2014a) Puzzling about policy paradigms: precision and progress. J Eur Public Policy 21:481–484. https://doi.org/10.1080/13501763.2014.893080

Daigneault P-M (2014b) Reassessing the concept of policy paradigm: aligning ontology and methodology in policy studies. J Eur Public Policy 21:453–469. https://doi.org/10.1080/13501763.2013.834071

Daimer S, Hufnagl M, Warnke P (2012) Challenge-oriented policy-making and innovation systems theory: reconsidering systemic instruments. In: Fraunhofer ISI (ed) Innovation system revisited: Experiences from 40 years of Fraunhofer ISI research. Fraunhofer, Stuttgart, pp 217–234

Diercks G, Larsen H, Steward F (2019) Transformative innovation policy: Addressing variety in an emerging policy paradigm. Res Policy 48:880–894. https://doi.org/10.1016/j.respol.2018.10.028

Dodgson M, Rothwell R (1994) The handbook of industrial innovation. Edward Elgar, Cheltenham

Edler J (1999) Die Bedeutung von Leitideen bei der Genese und Implementierung europäischer Forschungspolitik am Beispiel BRITE. PhD dissertation, Universität Mannheim

Edler J (2003) Change in European R&D policy as a complex consensus-building process: Experiences from the past and what they can teach us for the present. In: Edler J, Kuhlmann S, Behrens M (eds) Changing governance of research and technology policy: The European research area. Edward Elgar, Cheltenham

Edler J, Ruhland S, Hafner S, Rigby J, Georghiou L, Hommen L, Rolfstam M, Edquist C, Tsipouri L, Papadakou M (2005) Innovation and public procurement. Review of Issues at Stake. Fraunhofer ISI, Karlsruhe

Edler J (ed) (2007) Bedürfnisse als Innovationsmotor: Konzepte und Instrumente nachfrageorientierter Innovationspolitik. Studien des Büros für Technikfolgen-Abschätzung beim Deutschen Bundestag, vol 21. ed. sigma, Berlin

Edler J (2010) Demand-based innovation policy. In: The theory and practice of innovation policy: an international research handbook. Elgar, Cheltenham, pp 275–301

Edler J, Boon WP (2018) The next generation of innovation policy: Directionality and the role of demand-oriented instruments—Introduction to the special section. Sci Public Policy 45:433–434. https://doi.org/10.1093/scipol/scy026

Edler J, Fagerberg J (2017) Innovation policy: What, why, and how. Oxf Rev Econ Policy 33:2–23

Edler J, Georghiou L (2007) Public procurement and innovation—Resurrecting the demand side. Res Policy 36:949–963

Edler J, Beckert B, Betz R, Blind K, Bührer S, Bührlen B, Ebersberger B et al (2007) Bedürfnisse als Innovationsmotor: Konzepte und Instrumente nachfrageorientierter Innovationspolitik. Edition sigma, Berlin
Edler J, Gök A, Cunningham P, Shapira P (2016) Introduction: Making sense of innovation policy. In: Edler J, Cunningham P, Gök A, Shapira P (eds) Handbook of innovation policy impact. Edward Elgar, Cheltenham
Edler J, Blind K, Kroll H, Schubert T (2023) Technology sovereignty as an emerging frame for innovation policy. Defining rationales, ends and means. Res Policy 2021–11 50(9):104341, 52:104765. https://doi.org/10.1016/j.respol.2023.104765
Edquist C (ed) (1997) Systems of innovation: technologies, institutions and organisations. Pinter, London
Edquist C, Hommen L, Tsipouri L (2000) Public technology procurement and innovation. Kluwer, Dordrecht
European Commission (2014) Horizon 2020 in brief. The EU framework programme for research & innovation. Luxembourg (KI-02-13-413-EN-C)
European Industrial Research Management Association (EIRMA) (ed) (1982) The role of R&D in the innovation process. Working group report, 27, Paris
Fagerberg J (2003) Schumpeter and the revival of evolutionary economics: an appraisal of the literature. J Evol Econ 13:125–159
Foray D, Mowery DC, Nelson RR (2012) Public R&D and social challenges: What lessons from mission R&D programs? Res Policy 41:1697–1702. https://doi.org/10.1016/j.respol.2012.07.011
Fraunhofer Institut für System- und Innovationsforschung ISI (ed) (2012) Towards transformative governance? Responses to mission-oriented innovation policy paradigms
Freeman C (1973) A study of success and failure in industrial innovation. In: Williams BR (ed) Science and technology in economic growth. Springer, Cham
Freeman C (1982) The economics of industrial innovation, 2nd edn. Pinter, London
Freeman C, Clark J, Soete L (1982) Unemployment and technical innovation: A study of long waves and economic development. Contributions in economics and economic history, vol 50. Greenwood, Westport, CT
Frietsch R, Reiß T, Schmoch U (2024) Development of innovation monitoring and innovation indicators in the past 50 years. In: Edler J, Walz R (eds) Systems and innovation research in transition: Research questions and trends in historical perspective. Springer, Heidelberg
Gassler H, Polt W, Rammer C (2006) Schwerpunktsetzungen in der Forschungs-und Technologiepolitik: eine Analyse der Paradigmenwechsel seit 1945. Österr Z Polit 35:7–23
Gaudin T (1985) Definition on innovation policies. In: Sweeney G (ed) Innovation policies: an international perspective. Frances Pinter, London
Geels FW (2002) Technological transitions as evolutionary reconfiguration processes: a multi-level perspective and a case-study. Res Policy 31:1257–1274. https://doi.org/10.1016/S0048-7333(02)00062-8
Ghosh B, Kivimaa P, Ramirez M, Schot J, Torrens J (2021) Transformative outcomes: assessing and reorienting experimentation with transformative innovation policy. Sci Public Policy 48:739–756. https://doi.org/10.1093/scipol/scab045
Grillitsch M, Hansen T, Coenen L, Miörner J, Moodysson J (2019) Innovation policy for system-wide transformation: The case of strategic innovation programmes (SIPs) in Sweden. Res Policy 48:1048–1061. https://doi.org/10.1016/j.respol.2018.10.004
Haas PM (2001) Epistemic communities and policy knowledge. In: Smelser NJ, Baltes PB (eds) International encyclopedia of social and behavioral sciences. Elsevier, Oxford, pp 11578–11586
Haddad CR, Bergek A (2023) Towards an integrated framework for evaluating transformative innovation policy. Res Policy 52:104676. https://doi.org/10.1016/j.respol.2022.104676
Hall PA (1990) Policy paradigms, experts and the state. In: Brooks S, Gagnon A (eds) Social scientists, policy and the state. Prager, New York
Hall PA (1993) Policy paradigms, social learning, and the state: the case of economic policymaking in Britain. Comp Polit 25:275. https://doi.org/10.2307/422246

Heclo H (2010) Modern social politics in Britain and Sweden: From relief to income maintenance, 2nd edn. ECPR, Colchester

Heyen NB, Friedewald M, Jochem E, Reiß T, Thielmann A (2024) Assessing technological innovations: From early warning to the governance of socio-technical transformations. In: Edler J, Walz R (eds) Systems and innovation research in transition: Research questions and trends in historical perspective. Springer, Heidelberg

Hogan J, Howlett M (2015) Reflections on our understanding of policy paradigms and policy change. In: Hogan J, Howlett M (eds) Policy paradigms in theory and practice. Palgrave Macmillan, London, pp 3–18

Jacob K, Ekins P (2020) Environmental policy, innovation and transformation: affirmative or disruptive? J Environ Policy Plan 22:709–723. https://doi.org/10.1080/1523908X.2020.1793745

Janssen MJ, Torrens J, Wesseling JH, Wanzenböck I (2021) The promises and premises of mission-oriented innovation policy—A reflection and ways forward. Sci Public Policy. https://doi.org/10.1093/scipol/scaa072

JIIP (2018) Mission-oriented research and innovation : assessing the impact of a mission-oriented research and innovation approach: Final report. Publications Office, Brussels

Kay A (2007) Tense layering and synthetic policy paradigms: the politics of health insurance in Australia. Aust J Polit Sci 42:579–591. https://doi.org/10.1080/10361140701595775

Keck O (1986) Die gesellschaftliche Steuerung der Technik: Ein institutioneller Ansatz. In: Mayer-Krahmer F, Beckmann G (eds) Technologiepolitik und Sozialwissenschaft, Frankfurt a.M., pp 17–42

Kingdon JW (1984) Agendas, alternatives and public policies. Little Brown, Boston

Kivimaa P, Kern F (2016) Creative destruction or mere niche support? Innovation policy mixes for sustainability transitions. Res Policy 45:205–217. https://doi.org/10.1016/j.respol.2015.09.008

Kline SJ, Rosenberg N (1986) An overview of innovation. In: Landau R, Rosenberg N (eds) The positive sum strategy: harnessing technology for economic growth. National Academy Press, Washington, DC, pp 275–304

Klodt H (1987) Wettlauf um die Zukunft: Technologiepolitik im internationalen Vergleich. Kieler Studien, vol 206. Mohr, Tübingen

Koschatzky K (2001) Räumliche Aspekte im Innovationsprozess: Ein Beitrag zur neuen Wirtschaftsgeographie aus Sicht der regionalen Innovationsforschung. Wirtschaftsgeographie, vol 19. LIT, Münster

Krupp H (1972) Die Funktion der Fraunhofer-Gesellschaft im Innovationssystem der Bundesrepublik Deutschland: Studie für das Bundesministerium für Bildung und Wissenschaft, Karlsruhe

Krupp H (1975) Strukturpolitisch orientierte Forschungspolitik: ihre Möglichkeiten und Grenzen. Wirtschaft und Wissenschaft

Krupp H (1985) Public promotion of innovation: disappointments and hopes. In: Sweeney GP (ed) Innovation policies: an international perspective. St. Martin's Press, New York, pp 48–79

Kubeczko K, Weber KM (2009) Proaktive legitimation FTI-politischer interventionen: Jenseits von Markt-und systemversagen. In: Innovationsforschung und Technologiepolitik in Österreich. Neue Perspektiven und Gestaltungsmöglichkeiten. StudienVerlag, pp 127–153

Kuhlmann S (2003) Evaluation as a source of "strategic intelligence". In: Shapira P, Kuhlmann S (eds) Learning from science and technology policy evaluation: Experiences from the United States and Europe. Edward Elgar, Cheltenham, pp 352–379

Kuhlmann S, Arnold E (2001) RCN in the Norwegian research and innovation system: background report no 12 in the evaluation of the Research Council of Norway

Kuhn TS (1962) The structure of scientific revolutions. University of Chicago Press, Chicago

Kuittinen H, Polt W, Weber M (2018) Mission Europe? A revival of mission-oriented policy in the European Union. In: RFTE—Council for Research and Technology Development (ed) Re: thinking Europe. Positions on shaping an idea. Verlag Holzhausen, Vienna

Larédo P (2003) Six major challenges facing public intervention in higher education, science, technology and innovation. Sci Public Policy 30(1):4–12

Larrue P (2021) The design and implementation of mission-oriented innovation policies: A new systemic policy approach to address societal challenges. OECD science, technology and industry policy papers 100

Larrue P, Machado D, Yoshimoto T (2019) New mission-oriented initiatives as systemic policies to address societal challenges: Analytical framework and typology. In: 2019 EU-SPRI conference—Science technology and innovation policies for sustainable development goals, Rome, 06.06

Lindner R, Daimer S, Beckert B, Heyen N, Koehler J, Teufel B, Warnke P, Wydra S (2016) Addressing directionality: Orientation failure and the systems of innovation heuristic: Towards reflexive governance. Fraunhofer ISI discussion papers innovation systems and policy analysis, no. 52. Fraunhofer ISI, Karlsruhe

Lindner R, Edler J, Hufnagl M, Kimpeler S, Kroll H, Roth F, Wittmann F, Yorulmaz M (2021) Missionsorientierte Innovationspolitik: Von der Ambition zur erfolgreichen Umsetzung. Perspektiven—Policy Brief / Fraunhofer-Institut für System- und Innovationsforschung ISI, 2021, 02. Fraunhofer-Institut für System- und Innovationsforschung ISI, Karlsruhe

Lundvall B-Å (1985) Product innovation and user-producer interaction. Industrial development research series 31

Lundvall B-Å (ed) (1992) National systems of innovation: Towards a theory of innovation and interactive learning, 1st edn. Pinter, London

Luukkonen T, Nedeva M, Barré R (2006) Understanding the dynamics of networks of excellence. Sci Public Policy 33(4):239–252

Mahoney J, Thelen K (2010) A gradual theory of institutional change. In: Mahoney J, Thelen KA (eds) Explaining institutional change: Ambiguity, agency, and power. Cambridge University Press, Cambridge, pp 1–37

Malerba F (ed) (2004) Sectoral systems of innovation: Concepts, issues and analyses of six major sectors in Europe. Cambridge University Press, New York

Mayntz R (1996) Politische Steuerung: Aufstieg, Niedergang und Transformation einer Theorie. VS Verlag für Sozialwissenschaften

Mayntz R, Scharpf FW (1973) Planungsorganisation: die Diskussion um die Reform von Regierung und Verwaltung des Bundes. Piper

Mazzucato M (2013) Financing innovation: creative destruction vs. destructive creation. Ind Corp Chang 22:851–867

Mazzucato M (2018) Mission-oriented research & innovation in the European Union: A problem-solving approach to fuel innovation-led growth. Publications Office of the European Union

Meadows DH, Meadows DL, Randers J, Behrens WW (1972) The limits to growth: a report for the Club of Rome's Project on the Predicament of Mankind. Universe Books, New York

Ministry of EZK (2019) Missies voor het topsectoren- en innovatiebeleid, Den Haag

Mowery DC, Rosenberg N (1979) The influence of market demand upon innovation: A critical review of some recent empirical studies. Res Policy 8:103–153

Münch S (2016) Interpretative Policy-Analyse: Eine Einführung. Springer, Wiesbaden

Nelson RR (1959) The simple economics of basic scientific research. J Polit Econ 67:297–306

Nelson RR (1993) National innovation systems: a comparative analysis. Oxford University Press on Demand, Oxford

Nelson RR, Winter SG (1977) In search of useful theory of innovation. Res Policy 6:36–76. https://doi.org/10.1016/0048-7333(77)90029-4

Nelson RR, Winter SG (1982) An evolutionary theory of economic change. Harvard University Press, Cambridge

OECD (1972) The research system: Vol 1—France, Germany, United Kingdom. OECD, Paris

OECD (1980) Technical change and economic policy. OECD, Paris

OECD (1981) Science and technology policy for the 1980s. OECD, Paris

OECD (2011) Demand-side innovation policies. OECD, Paris

OECD (ed) (2015) System innovation. Synthesis report. Online https://archiwum.pte.pl/pliki/2/1/OECD%20System.pdf

Pavitt K (1976) Governmental support for industrial research and development in France: Theory and practice. Minerva 14:330–354. https://doi.org/10.1007/bf01096276

Pavitt K (1979) Technical innovation and industrial development. Futures 11:458–470. https://doi.org/10.1016/0016-3287(79)90044-2

Rogge K, Stadler M (2023) Applying policy mix thinking to social innovation: from experimentation to socio-technical change. Environ Innov Soc Transit 47. https://doi.org/10.24406/publica-1205

Rosenberg N (1991) Critical issues in science policy research. Sci Public Policy 18:335–346. https://doi.org/10.1093/spp/18.6.335

Roth F, Warnke P, Niessen P, Edler J (2021) Systemische Resilienz: Einsichten aus der Innovationsforschung. Perspektiven—Policy Brief / Fraunhofer-Institut für System- und Innovationsforschung ISI, 2021, 03. Fraunhofer-Institut für System- und Innovationsforschung ISI, Karlsruhe

Rothwell R, Zegveld W (1981) Industrial innovation and public policy: Preparing for the 1980s and the 1990s. Contributions in economics and economic history, no. 42. Greenwood Press, Westport, CT

Rothwell R, Zegveld W (1982) Innovation and the small and medium sized firm. Frances Pinter, London

Sabatier PA (1998) The advocacy coalition framework: revisions and relevance for Europe. J Eur Public Policy 5:98–130. https://doi.org/10.1080/13501768880000051

Schmidt VA (2010) Taking ideas and discourse seriously: explaining change through discursive institutionalism as the fourth 'new institutionalism'. Eur Polit Sci Rev 2:1. https://doi.org/10.1017/S175577390999021X

Schot J, Steinmueller WE (2018) Three frames for innovation policy: R&D, systems of innovation and transformative change. Res Policy 47:1554–1567. https://doi.org/10.1016/j.respol.2018.08.011

Skocpol T (1992) Protecting soldiers and mothers: The political origins of social policy in the United States. Belknap Press of Harvard University Press, Cambridge, MA

Smith A, Voß J-P, Grin J (2010) Innovation studies and sustainability transitions: The allure of the multi-level perspective and its challenges. Res Policy 39:435–448. https://doi.org/10.1016/j.respol.2010.01.023

Smits R, Kuhlmann S (2004) The rise of systemic instruments in innovation policy. Int J Foresight Innov Policy 1:4–32

Soete L, Arundel A (1993) An integrated approach to European innovation and technology diffusion policy (a Maastricht memorandum). Commission of the European Communities, SPRINT Programme, Luxembourg

Solow RM (1957) Technical change and the aggregate production function. Rev Econ Stat 39:312. https://doi.org/10.2307/1926047

Streeck W, Thelen K (2005) Beyond continuity: Institutional change in advanced political economies. Oxford University Press, Oxford

Stuart G, Kuntze U (1982) National innovation policies: the challenges in looking ahead. Monografia, Karlsruhe

Stucke A (1993) Institutionalisierung der Forschungspolitik: Entstehung, Entwicklung und Steuerungsprobleme des Bundesforschungsministeriums. Schriften des Max-Planck-Instituts für Gesellschaftsforschung, Bd. 12. Campus, Frankfurt a.M.

Swedish Presidency (2009) Lund declaration: Europe must focus on the grand challenges of our time. https://era.gv.at/object/document/130

Sweeney GP (ed) (1985) Innovation policies : an international perspective. St. Martin's Press, New York

Teece DJ (1986) Profiting from technological innovation: Implications for integration, collaboration, licensing and public policy. Res Policy 15:285–305. https://doi.org/10.1016/0048-7333(86)90027-2

Tidd J, Bessant J, Pavitt K (2005) Managing Innovation: Integrating technological, market and organizational change. Wiley, Chichester

True JL, Jones BD, Baumgartner FR (2007) Punctuated-equilibrium theory: explaining stability and change in American policymaking. In: Sabatier P (ed) Theories of the policy process, 2nd edn. Routledge, New York, pp 97–115

Walz R, Ostertag K, Eckartz K, Gandenberger C, Bodenheimer M, Peuckert J, Ramel F, Gigli M, Doranova A, Miedzinki M (2019) Ökologische Innovationspolitik in Deutschland: Bestandsaufnahme und Handlungsempfehlungen. Umwelt, Innovation, Beschäftigung, 01/2019. Umweltbundesamt, Dessau-Roßlau

Wanzenböck I, Wesseling JH, Frenken K, Hekkert MP, Weber KM (2020) A framework for mission-oriented innovation policy: Alternative pathways through the problem-solution space. Sci Public Policy 47:474–489

Warnke P, Koschatzky K, Dönitz E, Zenker A, Stahlecker T, Som O, Cuhls K, Güth S (2016) Opening up the innovation system framework towards new actors and institutions. Fraunhofer ISI discussion papers innovation systems and policy analysis, no. 49. Fraunhofer ISI, Karlsruhe

Weber KM, Rohracher H (2012) Legitimizing research, technology and innovation policies for transformative change. Res Policy 41:1037–1047. https://doi.org/10.1016/j.respol.2011.10.015

Weber, KM, Biegelbauer P, Brodnik C, Dachs B, Dreher C, Kovaćc M et al (2021) Agilität in der F&I-Politik: Konzept, Definition, Operationalisierung. Studien zum deutschen Innovationssystem. Berlin (8-2021). Online http://hdl.handle.net/10419/231476

Wilder M (2015) What is a policy paradigm? Overcoming epistemological hurdles in cross-disciplinary conceptual adaptation. In: Hogan J, Howlett M (eds) Policy paradigms in theory and practice. Palgrave Macmillan UK, London, pp 19–42

Wissenschaftsrat (1992) Empfehlungen zur Internationalisierung der Wissenschaftsbeziehungen. Köln

Wittmann F, Hufnagl M, Lindner R, Roth F, Edler J (2021a) Governing varieties of mission-oriented innovation policies: A new typology. Sci Public Policy 48:727–738

Wittmann F, Hufnagl M, Roth F, Yorulmaz M, Lindner R (2021b) From mission definition to implementation: Conceptualizing mission-oriented policies as a multi-stage translation process. Fraunhofer ISI discussion papers innovation systems and policy analysis, no. 71. Fraunhofer ISI, Karlsruhe

Wittmann F, Roth F, Hufnagl M, Lindner R, Yorulmaz M (2022) Towards a framework for impact assessment for mission-oriented innovation policies: A formative toolbox approach. ftval J Res Technol Policy Eval 31–52. https://doi.org/10.22163/fteval.2022.540

Ralf Lindner is the Head of the Department Policy and Society at the Fraunhofer Institute for Systems and Innovation Research ISI. He is particularly interested in policy design and governance approaches for transformative change, responsible research and innovation, and processes of policy learning. In recent years, the main focus of his research has been on the analysis and conceptualisation of mission-oriented innovation policies (e.g., in the context of the German High-Tech Strategy) and the development of practitioner-oriented approaches supporting their design, implementation, and evaluation.

Jakob Edler is Managing Director of the Fraunhofer Institute for Systems and Innovation Research ISI since October 2018. The Anglo-German is also Professor of Innovation Policy and Strategy at the Manchester Institute of Innovation Research (MIoIR), which he led from 2011 to September 2018. His research focuses on analysing and conceptualising governance, policy and management of innovation. J. Edler is Speaker of the Fraunhofer Group Innovation Research, member of the German Science Council, the German Academy of Science and Engineering (Acatech) and the Royal Society for the Encouragement of Arts, Manufactures and Commerce (UK).

Stephanie Daimer is the Coordinator of the Business Unit Policy for Innovation and Transformation in the Competence Centre Policy and Society at the Fraunhofer Institute for Systems and Innovation Research ISI. Her research interests are in the analysis and evaluation of national and European research and innovation (R&I) policies, and more recently in policies for transformative change and the role of public administrations as agents of transformation.

Development of Innovation Monitoring and Innovation Indicators in the Past 50 Years

Rainer Frietsch, Thomas Reiß, and Ulrich Schmoch

Abstract Innovation indicators are instruments to systematically analyse the performance of innovation processes and systems. In this chapter we discuss the evolution of innovation indicators alongside conceptual developments as well as technical and methodological progress. We identify four driving factors, namely (1) new theories/concepts, (2) lower technical thresholds for data analyses and availability of new data, (3) increasing policy demands and (4) technological and economic developments. Our discussion shows that at different stages of the indicator development different factors were the driving forces. The early innovation indicators were mainly R&D-centred with a strong focus on the manufacturing industry and R&D processes in companies as well as the science systems. The innovation system's perspective widened the focus and introduced additional indicators, among them indicators on transfer and collaboration. Data availability and better options for data treatment and analysis gave another push. More recently, information and computer science methods have entered the innovation indicators scene and widened the scope even further. We conclude that indicators are a means to measure and assess constructs which are otherwise not directly measurable. They should not become a means in itself.

1 Introduction

This chapter intends to provide an overview of innovation indicator developments since the early 1970s and against that background to critically discuss current trends and potential future developments. When writing on innovation—and this is essentially the case when writing on innovation indicators—it is always a good idea to start with Schumpeter (1997 [1911, 1934]) and his seminal work on the economic

R. Frietsch (✉) · T. Reiß · U. Schmoch
Fraunhofer Institute for Systems and Innovation Research ISI, Karlsruhe, Germany
e-mail: Rainer.Frietsch@isi.fraunhofer.de; Thomas.Reiss@isi.fraunhofer.de; Ulrich.Schmoch@isi.fraunhofer.de

J. Edler, R. Walz (eds.), *Systems and Innovation Research in Transition*, Sustainability and Innovation, https://doi.org/10.1007/978-3-031-66100-6_3

development. We will come back to his work when defining innovation. However, our journey of the history of innovation indicators and development over time starts after the Second World War when the new discipline of innovation research started off with conceptual and also empirical analysis of the subject in new research units and institutes, among them Fraunhofer ISI, founded in 1972.

Vannevar Bush published his "Science—The Endless Frontier" (Bush 1945) in 1945 that led to the establishment of the National Science Foundation in 1950 where he essentially suggested the linear model that reaches from basic research to innovation (Godin 2006). In the USA a systematic analysis of innovation processes, their management, as well as conceptual considerations on their formation and effects, in particular by economists and sociologists started as early as the 1930s and 1940s (see, for example, Fagerberg et al. 2011; Godin 2006). A regular report by the NSF was first published in the late 1950s (Godin 2003). At the same time Europe—during this period the USA and Europe mainly represented the industrialised world—was lagging behind. Also systematic data collections of innovation processes or outcomes slowly began to emerge on both sides of the Atlantic in the 1950s and 1960s.

Starting in the mid-1960s, the OECD assumed a very important and impactful standard setting role. It thereby changed the development not only of innovation analysis, but even more so the development of innovation indicators that were needed for these analyses (Godin 2003). It was also decisive for the development and diffusion of science and innovation policies on a global scale (Henriques and Larédo 2013). The OECD was following what we call "the R&D paradigm"—focusing on innovation activities in the R&D departments of companies, mainly in large multinational enterprises. In consequence, indicators were developed and employed that allowed a thorough analysis of R&D inputs and outputs in industry. With the availability of large data sources—not called Big Data in the early years—and the capacities to process them in the first decade of the new century, an emancipation from the R&D paradigm resulted in a new paradigm of innovation research (Sundbo 1997, 2001). This accelerated and shifted the focus to the service sector and also to further aspects of the innovation process.

In the following, we will discuss the evolution of innovation indicators in the last 50 years. We will do so in a contextual and conceptual approach. We analyse the development of the generation and use of indicators in the context of the changes in the conceptual understanding of the innovation process. In addition we look at the changes in the political demand to understand innovation dynamics over time. The chapter is structured as follows. We briefly introduce our conceptual framework before we summarise the theoretical foundations of innovations in technologies and firms in the next section of this chapter. Section 3 describes the pioneering activities and first innovation indicators that already broadened the empirical perspective on innovations and continues with indicators that allow to describe innovation systems and their performance in a national and international comparative perspective. New data sources and upcoming target areas are discussed in Sect. 4, while Sect. 5 focuses on recent developments and tries to look what is immediately ahead of us. Section 6 contains the summary and conclusion.

As a conceptual framework we consider four different, but interacting dimensions. They have shaped the demand for and evolution of the indicators in the last 50 years (Fig. 1).

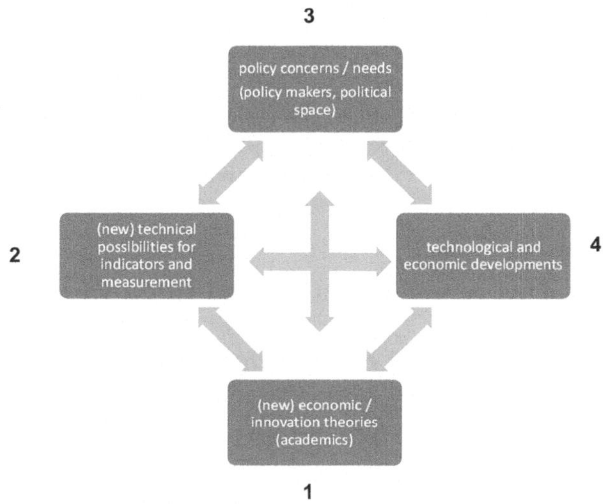

Fig. 1 Framework for the evolution of indicators. Source: Own representation

Economic and innovation theories provide a *first* dimension on the development of innovation indicators. Important examples comprise the replacement of a linear understanding of innovation by a systems perspective, the realisation of different types of innovation systems and their significance, such as national, regional or technological systems (Warnke et al. 2016). Other contributions deal with research on the functions of innovation systems offering new opportunities for innovation analyses (Bergek et al. 2008; Hekkert et al. 2007), or the emergence of transition theories, providing a multi-level perspective on innovation (Geels 2002). For a systematic elaboration of empirical evidence supporting such new concepts, specific indicators are needed.

Second, new technological possibilities to generate data offer an additional perspective for the elaboration of innovation indicators and the empirical analyses. The tremendously increasing computing power or the availability of huge data sets (Big Data) not only in (natural) science, but also on social and economic systems, are important examples illustrating these trends (Glänzel et al. 2019).

The *third* dimension comprises the political perspective. Needs and concerns in the political space call for objective, and if possible, quantitative measures and assessments for informing policy-making. Examples include the discussions on the international competitiveness (Fagerberg 1988; Freeman 2004) of Europe in the light of emerging economies in Asian countries. This trend started in the 1970s with a focus on Japan and for a number of years has been replaced by the development of China as an economic super power. This was complemented by increasing concerns as to the contribution of technologies and innovations to solve societal problems (Edler and Fagerberg 2017). Other recent examples include discussions on technology sovereignty (Edler et al. 2020) of nations or regions considering trade restrictions, breaking down of supply chains, or the dependence on energy supply and raw materials from just a few countries.

The *fourth* dimension concerns technological and economic developments. Examples encompass the rapidly growing biotech sector since the 1980s (Reiss

2001), the ongoing digitisation of basically all industry sectors (Oztemel and Gursev 2018) or recently the emergence of new technological paradigms, such as quantum technologies. Exemplary economic developments embrace, for example, the growing global influence of China (Frietsch et al. 2019), the dominating role of a few global IT companies, or the issue of economic perspectives of countries in the Global South. In order to track such developments, suitable indicators are required.

2 Theoretical Foundations of Innovation Processes

2.1 What Is an Innovation Indicator?

Before we clarify what is an innovation indicator, we first need to define what we see as innovation in the context of our discussion. In other words, we need to define our subject before we can look for means to measure and describe this subject. Schumpeter (1942, 1997 [1911, 1934]) himself did not really define innovations. He only delivered a process definition using five categories of "new combinations", namely new products, new production processes, new markets, new input sources or new positioning in markets.

First globally accepted and codified definitions of innovation at the OECD (1992, p. 31) and also previously used definitions were restricted to technological product or process innovations. These definitions led to (or were inspired by) a focus on the manufacturing industry and a corresponding production of data and indicators. It served well for a long time, but since the second half of the 1990s, in particular, it has repeatedly encountered broad criticism within empirical and theoretical innovation research (Coombs 2003; Gallouj 1997, 2002; Hauknes 1998; Miles 2004; Sundbo and Gallouj 1998; Tidd and Hull 2003), since a superficial focus exclusively on technical innovations no longer seemed appropriate due to the growing importance of the service sector within Western industrialised countries.

In consequence, the OECD adapted the definition of what innovations are. However, similar to Schumpeter the definition is also a process—or in this case an output-oriented—definition. The OECD provided a general definition, describing innovation as the result of different kinds of processes:

> An innovation is the implementation of a new or significantly improved product (good or service), or process, a new marketing method, or a new organisational method in business practices, workplace organisation or external relations (OECD, Eurostat 2005, p. 46).

Following the OECD approach, defining innovation as the process of the generation of new outcomes or the outcome of such a process itself, we continue with the definition of what an innovation indicator is.

Many analytical subjects, especially when they are of an abstract, multidimensional, or in other ways complex form, cannot be directly observed or measured. These are the so-called latent constructs or latent variables in economic and management research. To monitor, assess, compare (benchmark) or measure them, it is therefore necessary to resort to so-called indicators, which—as their name clearly suggests—provide an indication of the underlying concept to be measured (Grupp 1998). In essence, these are factors or variables of which one knows or at least believes to know that they are closely linked to the subject of interest and therefore allow conclusions to be drawn about the actual target variable. These indicators can then be measured, counted or recorded in some other way.

> If 'measurement' is the formal assignment of numbers to circumstances, comparisons will be necessary. [...] The properties which need to be noted when comparing the process investigated against an archetype are termed indicators. If, simultaneously, various indicators are considered plausible (which is standard practice in innovation economics), it can be stated that various operational definitions and hence various measurement processes are available to innovation processes or to their component parts. (Grupp 1998, p. 31)

Besides what Grupp emphasised in this definition of an indicator, namely the link between the measurement and the circumstance itself, the operationalisation of the measurement in form of an indicator and its validation are inevitable. Furthermore, it should always be kept in mind that the indicators are only indications and therefore a means to analyse the end. It should not become an end in itself. In this respect, the measurement itself might have an impact on the validity of the relation between the subject (or circumstance, as Grupp put it) and the indicator. For example, scientific journal publications were seen as an indication of scientific competences, assuming they add to the current scientific knowledge. In addition, their quality as well as scientific perception or visibility was to be reflected by the citations they receive in papers by other researchers. In times when the numbers of publications exploded and when the marginal added value of the majority of these publications is diminishing, the former relation of the subject/circumstance (namely scientific contribution) and the indication by bibliometric data might require a reconsideration.

We leave the answer to these questions to other papers, but would like to stress at this point that indicators are a means and not an end in themselves and should be carefully selected and used based on scientific grounds. Otherwise artefacts and mistakes might be the outcome of indicator studies instead of empirical evidence. The fact that, for example, social media data is available at all does not mean that it is an adequate indication of social impact of scientific or technological activities. Proper conceptualisations and especially validation studies of these conceptualisations are mandatory. A task that nowadays sometimes seems to be neglected, when data availability appears to drive the conceptualisation instead of an operationalisation of the concepts based on proper data and indicators.

2.2 *The Development of a New Discipline: Innovation Studies*

In this part of the chapter we describe the innovation indicators along the dimension one and two of our indicators development scheme (see Fig. 1) as this more or less reflects the chronological order of the historical development. The first dimension is that of innovation research and economic theory. The second dimension is that of data availability and accessibility as well as decreasing technical obstacles due to increasing computing power. To start with, we address the first dimension and especially the paradigm we would like to call the “R&D paradigm” as the indicators at this time are centred on the R&D activities—either the R&D inputs like expenditures or personnel, or the output in form of patents or publications. The “R&D paradigm” bore innovation indicators at the firm level as well as at the macro-economic level to analyse the (linear) innovation process that spans from directed and structured R&D activities to inventions and via new technologies, processes or services to commercialisation and diffusion of these inventions, which then become innovations (Godin 2006; Grupp 1998; Schmoch et al. 2000).

The theoretical basis of innovation economics was elaborated by Joseph Schumpeter as early as 1911 (Schumpeter 1942, 1997 [1911, 1934]). In the 1960s and 1970s, economic theories on innovation were developed by, e.g., Gibbons and Johnston (1974), Gilpin (1975), Mansfield (1968), Merrifield (1979), Nelson and Winter (1977), Price and Bass (1969), Rosenberg (1976), Schmookler (1966), Utterback and Abernathy (1975) or Weingart (1975) who laid the conceptual ground for innovation systems analyses, innovation economics and innovation policy as well as innovation policy analyses. As innovation economics contribute to describing and analysing economic prosperity, the theories induced a strong need for empirical verification and stimulated the conception of innovation indicators and innovation monitoring.

However, the broader establishment of innovation economics as an independent sub-discipline of economics was first initiated by Christopher Freeman with his book “The Economics of Industrial Innovation” (Freeman 1974). Therein, he describes the growing relevance of innovation for economic prosperity by citing various examples such as chemistry, automobiles or electronics. The illustration of the theory by empirical examples was decisive for the broad diffusion of Freeman’s approach that he developed further in his later publications (e.g. Freeman and Soete 1997). For him, research and development (R&D) played a crucial role. For example, he showed the trends for the expenditures on R&D in the 1980s for 50 leading countries. He also provided a table on inputs and outputs in research, invention, development and innovation and listed various items, which could be used to measure these activities. He therefore introduced innovation indicators. For instance, he suggested the working time of researchers or their remuneration, research papers, patents or technological papers as possible measurable quantities.

Freeman had established the Science Policy Research Unit (SPRU) at the University of Sussex as early as 1966 (Fagerberg et al. 2011) and even before that he got involved in the OECD in order to achieve a comparable documentation of the R&D activities of countries that led to the Oslo Manual—setting the standard at that time and continuing to do so today. His work had enormous influence on the field, but also on institutions worldwide, among them Fraunhofer ISI.

While in Solow’s (1956) economic growth model the technological change was external and simply explained by the unexplained component in the model, Mansfield (1968) and later on Romer (1990) endogenised the technological progress and established it as a relevant component in macro-economic modelling, giving especially way to analyses of the technological competitiveness of countries, regions or sectors. At about the same time, analyses at the micro-economic level of the firm began to focus on effects of innovations on firm performance and competitiveness (Crépon et al. 1998; OECD 1996; Teece 1986, 1998). R&D expenditures became the main input indicator to measure and assess the efforts to achieve new scientific and technological knowledge that lays the foundation of this competitiveness, both at the level of countries and at the level of firms. At the throughput level of the innovation process, scientific publications and patents became the core indicators. The main advantage of these indicators is not only the direct comparability of countries (or science and innovation systems, respectively), but also that they enable putting a focus on scientific disciplines and technological areas—even down to individual technologies. In the early years a focus on countries, regions and technologies was mostly taken, while in the 2000s a shift towards analyses of organisations and individual actors in the system became more widespread—among other factors, this trend was definitely influenced by the innovation systems heuristic (Edquist 1997; Lundvall 1992; Nelson and Winter 1982) and its actor orientation.

Box 1: One Example of Organisational Development in the Field of Innovation Indicators: Fraunhofer ISI's Pathway to Indicators

The founder and first director of Fraunhofer ISI, Helmar Krupp, was in close contact with Christopher Freeman. In the first years of the institute's activities, the work was primarily qualitative, e.g., the conception of political measures of initiating R&D activities in small and medium-sized enterprises, or based on limited surveys, e.g. users of energy efficient houses. The general situation at that time is well characterised by a seminar of the German Federal Ministry of Research and Technology (BMFT) with international participants in 1977, where most of the contributions were qualitative (Stroetmann 1977). The only innovation indicator was research and development expenditure (R&D) by countries or by specific industrial sectors. The contributions to patents dealt with the relevance of patent protection for innovations, but not with patent statistics.

Early contributions by Fraunhofer ISI to innovation indicators are Gielow et al. (1982), suggesting improvements of the German survey on industrial R&D, Kuntze et al. (1975) with basic considerations on using patent statistics as an innovation indicator, and Legler (1982b), analysing the German chemical industry based on foreign trade and patent statistics, where the patent data were provided by the US-American and the German Patent Office. The further development of innovation monitoring and innovation indicators was stimulated by a series of small international workshops. For example, at a seminar in Karlsruhe in 1985, researchers from Sweden, Germany, Portugal, Japan, the USA, the Netherlands and Great Britain discussed topics such as bibliometrics, the relation of patents and R&D, patent statistics, technometrics, foreign trade of research-intensive goods or the international comparison of research-intensive technologies such as robotics, genetic engineering or fibre-optics (Grupp and Legler 1987). Similar workshops followed in 1988, 1990, 1991 and 1993 (Grupp 1992; Sigurdson 1990). Due to these workshops, various international co-operations were initiated such as Noyons et al. (1994) or Schmoch et al. (2003).

As regards the output side of the innovation process, the commercialisation of all these efforts is of interest also from an indicators perspective. From a rather macroeconomic view, international trade was the main indicator, while at the microeconomic level the introduction of new products, processes or services was taken as an indication of the innovation output. A differentiation of the analytical models by sectors, technologies or organisation type led to a better understanding of the innovation process as such and provided insights into the complexity of the processes and their effects. Distinctions between R&D-intensive sectors or technologies (see Grupp et al. 2000; Hatzichronoglou 1997; OECD 2003) as well as between knowledge-intensive businesses (Legler and Frietsch 2007) were introduced to categorise the groups and types. These were based on the insight that not all actors need to be or effectively are involved in innovation processes. Furthermore, the particular innovation processes are very different for each technology, company or sector and the categorisation helped to generalise this insight.

2.3 Paradigms of Innovation Research: Shaping Indicators

We referred to Schumpeter's work (Schumpeter 1942, 1997 [1911, 1934]) as the initiation of innovation research, focusing on the entrepreneur and his/her role as a "creative destructor". This perspective on innovation activities was later on named as Schumpeter Mark I (Malerba and Orsenigo 1995; Nelson and Winter 1982). In his later publications, Schumpeter already stressed the role of large companies and their R&D departments for putting forward innovation processes, which was later on named Schumpeter Mark II.

Since the 1960s a systemic perspective has occurred, first focusing on science systems and their competitiveness and then broadening to R&D in general—first also on the systemic level, where most indicators have their origin and later more and more also including company data at the micro-level, which allowed a better understanding of the processes themselves. This is what we referred to as the R&D paradigm (Mark II) and in which the indicator development was closely linked to the work of Christopher Freeman and his endeavours for the OECD (Fagerberg et al. 2011). In this section, following Sundbo (1997, 2001), we want to introduce an additional paradigm (we could also term it era) of innovation research, that is characterised by further differentiation and deepening of the analytical framework. From here on, we would like to rely on this differentiation of three paradigms to link the particular focal points of innovation indicator developments.

Based on Schumpeter's work, Sundbo (1997, 2001) differentiates between three "paradigms" of innovation. In addition to the "founder's paradigm" that mainly addresses the era of Schumpeter and his descriptions of the entrepreneurial innovation processes (Mark I) and the "economic paradigm" that is centred on R&D activities of large companies (Mark II) and which we therefore termed "R&D paradigm" above, he introduces the "strategic paradigm" (Sundbo 1995, p. 400) that widens the innovation definition and perspective especially to services and shifts the attention to demand-side and diffusion activities in addition to the (R&D) input.

The first of the three identified paradigms, the founder's paradigm, falls into the period of the industrial revolution and the subsequent start-up period of companies, often with a patriarchal company owner at the top. This paradigm lasted till about the end of the Second World War (Grupp et al. 2005). According to Sundbo, it is precisely these founders that Schumpeter had in mind with his analysis of the innovation system. In this first phase of innovation research, indicators played a minor role and the research was more of a qualitative and descriptive nature with the intention to understand—in Weber's sense (Weber 1972 [1922])—the process and the success factors. A focus was on the individuals and the inventions they made.

After this start-up period of innovation research, which, depending on the definition, extends into the first half of the twentieth century, the time of technological developments in modern society arose, forming a second paradigm. According to Sundbo, this paradigm is the "economic" one, in which technology and its development are in focus. Investment in research and development (R&D) drives economic growth through the generation of new technologies. These technologies develop new needs or cover existing needs. So new technologies find their buyers. During the period of this paradigm, R&D departments and public research activities are the focus of attention. The innovation process is often associated with a science or technology-push situation, where the "right" and new products will be absorbed by eager consumers/clients in demanding markets. Innovation theory is correspondingly concerned with the kind of "indicators" that can be used to measure these components. Here the focus is on research expenditures, patents or high-tech products.

Therefore, while Schumpeter's theory (1997 [1911, 1934]) clearly aimed at the entrepreneur (Mark I), within the framework of the economic paradigm with its focus on R&D (Mark II), his/her importance has receded somewhat into the background in some parts of the economy. On the one hand, large corporations emerged in which a manager or technical director did not play the same role as the entrepreneur in Schumpeter's model (Mark I). In such corporations, the fate of the individual is generally no longer tied to the fate of the company. On the other hand, it is much more important for the decoupling or the loose connection between the decision-maker and the company that the number and complexity of the tasks and the qualifications involved and competencies have increased significantly over time. A division of labour is essential within companies that want to be economically successful. This leads to specialisation and "expertise" as well as to a systematisation of the process and its organisation (Frietsch 2011; Marengo and Dosi 2003; Teece 2007). The functional division of labour creates the prerequisite for the intensification of knowledge and its importance for all work processes. This also means that innovation indicators are found that are able to describe the different tasks, the different stages of the innovation process. They also describe the role of the different actors relevant for these processes as well as the effects/outcomes of these tasks and stages.

The third paradigm, the "strategic paradigm", emphasises the shift in innovation processes to additional factors, making R&D investments still a necessary, but no longer a sufficient condition for innovation success. Since the technologies have become very complex and the supply of new technologies does not necessarily cover an open need/demand on the part of consumers, other strategies are necessary that lead to the sale of the new products and services. A science-push or technology-push situation occurs less often. A "pull situation" needs to be created, i.e. the technologies attract the consumers'/clients' attention, for example by employing marketing and other sales-promoting measures. From the companies' point of view, which are now moving into the centre of a holistic interest under the aegis of this paradigm, it is crucial to look for the ability to develop corresponding strategies and ideas about products and services that enable growth in the respective markets. The key players in this game are the managers and decision-makers in companies who use their skills and abilities to guide the companies' fortunes. At the same time, however, the employees are also moving further into the centre of the analyses because they generate ideas, contribute their knowledge and thus create the prerequisites for innovations.

In this third (and still ongoing and open-ended) paradigm and the shift of focal points of innovation analyses, new needs and perspectives of innovation indicators arose as well, mainly driven by a differentiation of innovations and innovation processes. Sundbo's (1998, 2001) thesis is that the image of the decision-maker and thus also of the innovation process has changed significantly over time. Neither the image of the classic entrepreneur, who was seen as the "creative destructor", addressing the market needs through "new combinations", nor R&D expenditures or other mainly input-driven factors are the main driving forces in the innovation process any longer. Teece (1986) makes a similar argument when he claims that the innovation process has changed significantly. The increased complexity and the necessary broader knowledge require a different approach to the research and development process and then also imply different forms of organisation. Although the research process is still targeted, it is associated with significantly greater uncertainties and also with significantly larger investments. The latter, in particular, means that not only increasing productivity, but also increasing R&D efficiency is required to ensure competitiveness. In consequence, concepts like open innovation (Chesbrough 2003; von Hippel 1988; von Hippel and Krogh 2013) attract the interest of innovation researchers and managers alike as the insight diffuses that not only effectiveness, but also efficiency in the knowledge-creation processes is of the utmost relevance. Absorption (Cohen and Levinthal 1990; Teece 1986), transfer from public research to industry (Etzkowitz and Leydesdorff 1995; Schmoch et al. 2000), active external knowledge sourcing, for example, from international markets (Arundel et al. 1998; Thursby and Thursby 2006; UNCTAD 1992), as well as knowledge exploitation (Chesbrough 2003) have been accounted for by innovation scholars, managers, and also by policy makers since then.

To sum up, while in the first phase of the "founder's paradigm" hardly any quantitative indicators were used, but rather qualitative analyses of companies and processes, the economic paradigm shifted the attention towards—mostly macro-economic—quantitative indicators. This is where the OECD and its focus on R&D activities played a crucial role. For the description and the analysis of the innovation processes under the "strategic paradigm", additional indicators for collaboration types, knowledge exchange or innovation processes outside the manufacturing sector were needed. Indicators to assess the flow of knowledge between science and industry or industry and industry, international knowledge flows, the use of knowledge by innovators, or the particular knowledge and technology transfer gained attention.

3 Pioneers in Innovation Indicators: Early Activities of the Community

3.1 Main Indicators (R&D, Patents, Foreign Trade)

As outlined above, the early 1970s saw developments in innovation studies to compare the competitiveness and performance of science and innovation systems at the different levels, mainly at the country and later on also at the regional or technological level. Therefore, one focus was a better understanding of industrial R&D as well as innovation systems as a whole. While early indicator developments had mainly focused on the science system and its contributions to innovations in general, the era of the "R&D paradigm" achieved not only a much broader conceptual understanding of innovations and innovation processes, but also a huge differentiation and systematisation of indicators for their empirical underpinning.

The analysis of *industrial R&D* was a major topic already in the analyses of Freeman (1974). Industrial R&D was examined by many authors, such as Majer (1978), Griliches (1979), Caulcutt (1992) or Grenzmann et al. (1991), but already Schmoch et al. (1988) showed that R&D data are only available on the aggregate level of industrial sectors, whereas for specific technologies rarely reports were on hand. Therefore, R&D data are primarily useful at the aggregate level. R&D indicators refer to the first dimension of our indicator scheme (Fig. 1), i.e. policy concerns.

For achieving a finer level of aggregation, *patents* were suggested as innovation indicators quite early on (e.g. in Freeman 1974; Maclaurin 1954). The prospects of patent indicators were discussed in Kuntze et al. (1975), but still on a very basic level with some aggregate data provided by the German Patent Office. Many concerns as to the validity of patent indicators existed. They achieved a broader legitimation by Griliches (1981), whose article was a real turning point for the acceptance of patent indicators. Decisive progress in the use of patent indicators was achieved by using electronic databases. Faust and Schedl (1984) used an in-house version of the international patent database INPADOC, Narin et al. (1987) established an in-house version of the patent database of the USPTO (US Patent and Trademark Office). Narin and his colleagues introduced a variety of new indicators such as patent citations or citations of non-patent literature (NPL) to patents. The Science Policy Research Unit

(SPRU) at the University of Sussex also worked with an in-house version of a USPTO database and contributed basic methodological papers (e.g. Pavitt 1985) and various patent-based economic analyses (e.g. Patel and Pavitt 1991).

In Germany, Fraunhofer ISI began broader patent analyses with the access to the German patent database PATDPA, which the host STN provided as an online database. Schmoch et al. (1988) discussed various topics such as the comparison of national and international patent databases, the conception of patent search strategies, the use of foreign patent applications, the grant rate, the team size of inventors or the citation frequency, and the link of patent indicators for different technologies to R&D data, publications, technometrics or foreign trade. This very basic analysis was largely ignored, as it was published in German. However, by various follow-up publications, such as Schmoch et al. (1991), Grupp et al. (1996) or Grupp and Schmoch (1999), the use of international patent statistics beyond the use of USPTO data was taken up by many research groups and international institutions such as the OECD or the WIPO, among them the Triadic patent approach (see box below) or the classification of technology fields published by the WIPO (Schmoch 2008). Despite some limitations such as the focus on patentable technologies or underrepresentation of small and medium-sized enterprises (SMEs), patent indicators became broadly accepted, as they allow for a very detailed definition of specific technologies, balanced country comparisons, the analysis of enterprises, the transfer activities of academic institutions, etc. Patent indicators offer to retrace technological developments and are often produced in reaction to policy concerns, thus cover dimensions one and two of our indicator development scheme. They are also used for supporting new economic theories (dimension three, Fig. 1).

Patent searches for statistical analysis are mostly based on codes of the International Patent Classification (IPC) or keywords in the title, abstract or claims of the patent documents. Due to more powerful computer systems, it is now possible to define strategies by text mining in the complete text which opens new possibilities of analysis (we will come back to this further below). Thus, in the recent version of the Handbook of Science and Technology Indicators (Glänzel et al. 2019) five chapters on patents deal with text mining approaches.

Box 2: Country Comparisons in Patent Statistics

A major issue of patent statistics is to describe the technological competencies of countries in country comparisons in an appropriate way. Country comparisons at a specific patent office imply a strong advantage for the domestic country linked to that office.

1. A first suggestion to solve this problem was made by Soete and Wyatt with the indicator *RTA (Revealed Patent Advantage)*. However, this indicator only captures relative, but not absolute comparisons.
2. Gerstenberger (1992) used all patent applications which are applied in at least two countries, thus have a *family size of at least two*. This concept was adopted by the World Intellectual Property Organisation (WIPO) (see, e.g., WIPO 2019, p. 123). However, in specific analyses for technologies,

the patent numbers of Japan appeared to be overestimated (Schmoch and Khan 2019, p. 916).

3. A further suggestion were the so-called *triadic patents* (Grupp et al. 1996), which focus on applications filed in the USA–Europe–Japan triad. This approach was taken over by the OECD and used for many years. But it became increasingly obsolete in the late 1990s, as the economic power of China and South Korea grew.
4. *The IP5 concept* was suggested by the OECD. There, patent applications to the five major patent offices in the world are considered: the EPO, the USPTO, the JPO, the SIPO and the KIPO (OECD 2015, p. 20). In this concept, the threshold for Southeast Asian countries is quite low, as e.g., the step from China to Japan is smaller than that from Europe to the USA. In consequence, these countries are overestimated in analyses for specific technologies (Schmoch and Khan 2019, p. 916).
5. *Worldwide Patent Counts* were suggested by the OECD and some universities (de Rassenfosse et al. 2013), which are implemented as counting patent families, including singletons (Maraut and Martínez 2014; Martinez 2011). This needs to be seen very critically as it assumes that any patent at any office has the same technological value. In reality, patent offices differ immensely, for example in the quality or inventive step of the patents they accept, the newness of the inventions (worldwide vs. national prior art) and a number of other dimensions. In addition, it is not possible to analyse emerging technologies (Schmoch and Gehrke 2022). These concepts tend to strongly overestimate the effects of national filings and therefore ignore the structural and legal as well as market differences of patent jurisdictions. The technological competitiveness of countries like South Korea or China is overestimated based on these approaches and on most of the other conceptual approaches.
6. *Transnational Patents* were suggested by Frietsch and Schmoch (2010). The concept is defined by patent families which comprise either European applications and/or international applications (PCT applications), thus families with a substantive size. According to the present state of knowledge, this approach seems to lead to appropriate results (Schmoch and Khan 2019, p. 916). This concept follows the same idea as the Triadic patent approach, which addresses a balanced and internationally comparable market where mainly the technological competitiveness is decisive beyond national or legislative idiosyncrasies. This concept for patent analysis allows to construct the conditions for empirical analysis and comparison of national systems based on technological profiles. The concept makes use of the effective filing routes for most of the companies and technologies, when they are to be filed abroad. Empirically it was shown that in the vast majority of cases, companies tend to use at least one of these filing routes of EPO or PCT. According to the present state of knowledge, this approach seems to lead to internationally comparable results (Schmoch and Khan 2019, p. 916) that are at the same time correlated to other relevant innovation indicators such as R&D expenditures or exports (Frietsch et al. 2014, 2017).

Foreign trade data are standard for a variety of economic analyses, e.g. the annual analysis of the trade balance of countries. The foreign trade data become innovation indicators, once they are linked to R&D. An important activity in the context of R&D was the differentiation of industrial goods by R&D intensity. As early as 1982, the Institute of Economic Research of Lower Saxony (Niedersaechsisches Institut fuer Wirtschaftsforschung, NIW) defined a list of goods according to the foreign trade classification SITC, differentiated by high and medium technology (Legler 1982a). The European Commission published a similar list in 1982 (Kommission der Europäischen Gemeinschaft 1982) and the OECD in 1985 (OECD 1985). Further details can be found in Grupp and Legler (1987). This definition of R&D-intensive goods, high-level technology goods and cutting-edge technology goods was regularly updated (e.g. Legler and Frietsch 2007). An early report based on this classification is Legler et al. (1992). This classification allows for a characterisation of countries by the R&D intensity of goods. In addition, it analyses how economies with specific R&D profiles respond to business cycles.

Based on the classification of goods, it is possible to characterise sectors by R&D intensity. In the case of service sectors, a classification by knowledge intensity is made on the basis of the shares of qualified staff (Legler and Frietsch 2007).

Foreign trade data are documented to fulfil the demand policy and inform decision-makers to allow for analyses of national competitiveness (first dimension of our indicator development scheme, see Fig. 1).

3.2 A Further Broadening of the Scope: Micro-Data Analytics and Additional Indicators for Particular Parts of the Innovation Process

The informational value of patents and R&D as innovation indicators, which essentially reflect the strong focus on inputs and throughputs of innovation processes, was challenged already in the 1980s and 1990s by several authors (e.g. Kleinknecht et al. 1993, 2002; or van der Panne 2007). They suggested to shift the attention to outputs and, for example, to identify new products introduced to the market by statistically evaluating trade journals. The authors achieved quite convincing results, but the challenge of this approach was to classify the products in an appropriate way and to identify comparable trade journals for different countries. Due to these restrictions, this approach did not establish itself to a greater extent. However, the introduction of new products was included in the Oslo Manual (OECD 1992) and incorporated standards for innovation surveys.

Due to the political interest to better understand the competiveness of firms within innovation systems, further innovation indicators for enterprises are generated by regular broad surveys. The Community Innovation Survey (CIS) is the reference survey on innovation in enterprises (OECD, Eurostat 2005). On this basis, it is possible to compare the innovation activity of most countries of the European

Union (Eurostat 2020). For each country the innovation activity in specific sectors and the development of innovation in time can be examined (Rammer et al. 2021). In order to comply with the regulatory requirements and also to respond to the needs of various users, Eurostat together with the member countries develops for each round a standard questionnaire—Harmonised Data Collection (HDC). In addition to core mandatory questions, each survey wave includes varying variables, e.g. environmental benefits of innovation. Thus, many aspects of innovation are covered.

The EU Member States first introduced the survey in 1992 and since then it has become the regular biennial data collection. At present, the survey is carried out in the EU, EFTA and the EU Candidate Countries. Other countries such as South Africa use the CIS as well. Most of these national surveys build on the Oslo Manual of the OECD, which defines the standards for innovation surveys and provides the "guidelines for collecting, reporting and using data on innovation" (OECD 2018). The CIS activities fulfil requirements of the political space (Dimension three of our indicator development scheme, Fig. 1).

A further approach to use production statistics are broad enterprise databases such as ORBIS (Moody's), Crunchbase or Hoovers (Dun & Bradstreet). Therein, the enterprises are classified by economic activities according to classifications such as NACE. Due to improved software and hardware, it is possible to link patent and enterprise databases and to analyse innovation activities by industrial sectors (Neuhäusler et al. 2016; Schmoch et al. 2003. This approach can be associated with dimension two of our indicator development scheme (Fig. 1), namely new technical possibilities for indicators.

Additional data sources of innovation indicators are also available for foreign trade, R&D expenditures, production or labour force. For example, the UN Comtrade database, the Business R&D Expenditure (ANBERD/STAN) databases of the OECD, the production and sales database of the OECD or the ILO Labour Force database (ILOSTAT) provide additional data, mainly for macro-economic analyses of innovation systems.

A further concept for measuring technological performance is technometrics, first described in Grupp et al. This approach aimed at a holistic understanding of the performance of a country (or region) as regards the generation of a technology. For this approach, some technologies such as solar cells, laser beam sources or industrial robots were selected in a first step. Then, characteristic performance features for these technologies were identified based on a literature survey and interviews with enterprises and scientists. In the final step, the level of these performance features was collected for different countries. This way, the performance of a country in a specific technology in comparison with others could be determined and in particular the features with high and low performance. This approach proved to be useful for analysing the situation in a specific technology field. However, the identification of relevant features and the collection of related data are quite laborious, so that a comparison of many technologies and countries is challenging. Therefore, the use of technometrics is quite limited. Technometrics help to describe technological developments, i.e. they can be categorised in dimension four of our indicator development scheme (Fig. 1).

Another innovation indicator for enterprises are trademarks, for which basic conceptual work started at the beginning of the 2000s (e.g. Mendonça et al. 2004; Schmoch 2003). At that time, the innovation process and its outcomes were well described with the indicators at hand, whereas it was not possible to analyse the diffusion and implementation—which makes the crucial difference between an invention and an innovation—in a satisfactory manner. The introduction of trademarks as innovation indicators was intended to close this gap and to push the frontier of innovation indicators further. Whereas patents show the intention to introduce a new product into the market, trademarks indicate that the product already is on the market. The classification for trademarks is quite coarse and comprises only 34 product classes (Nice Classification). In consequence, the analysis of products by patents is generally more differentiated than that by trademarks. Nevertheless, trademarks describe a different aspect of innovations with a closer link to markets and diffusion. In addition, Neuhäusler et al. (2021a) suggested a method to further classify and differentiate the categories addressed by trademarks employing the mostly standardised keywords of the Nice Classification.[1]

In addition, it is also possible to apply trademarks for services with 11 service classes, a dimension which cannot be captured by patents. Therefore, it is feasible to analyse service enterprises such as banks or insurance enterprises by service marks (Schmoch and Gauch 2009). Furthermore, the economic performance depends not only on technology, but also on the quality of services linked to the products. Thus, a combination of patents and service marks offers additional analytical potentials than a simple patent analysis (Mendonça et al. 2019). Again, trademarks are used for describing technological developments, which, in contrast to patents, indicate that the technologies are deemed ready to be introduced into the market.

4 Indicators for Innovation Systems

The discussion of national systems of innovation (NIS) brought about a decisive change in the use of innovation indicators (Freeman 1987; Lundvall 1988; Nelson 1993). The enlarged view on national systems instead of enterprises and individual technologies proved to be necessary, as many observations in the context of innovation could not be explained solely by the activity of enterprises. Rather, the contribution of institutions of education and research, of government bodies, intermediaries, financial institutions, the structure of the socio-economic environment, the legal system, etc. are also relevant for the innovative performance of a country. In consequence, many additional indicators and their interaction have to be taken into account. In the following section, the most important additional indicators are described.

[1] https://www.wipo.int/classifications/nice/en/

4.1 Data Sources for Indicators

For measuring the performance of universities and research institutes, a set of publication indicators was developed. Garfield (1955) built a publication database including the citations on publications very early on. This was the origin of the Web of Science (WoS). The company CHI Research (Philadelphia) owned by Francis Narin used this database already in the 1970s (Frame et al. 1977) for bibliometric analyses in order to compare the performance of institutions and of national science systems. A broader introduction to innovation research was achieved by a series of conferences of the Centre for Science and Technology Studies (CWTS) at the University of Leiden, Netherlands (van Raan 1989). At the beginning, CWTS was confronted with various concerns as to the validity of bibliometrics or the citation analysis of publications. They succeeded in developing a sound methodology of bibliometrics (van Raan 2005). The broad acceptance of bibliometrics was the consequence of the extension of new public management at universities and, linked to that, the need of performance indicators of science.

Around 2004 the database Scopus by Elsevier was established in competition to the Web of Science (WoS) of Garfield, by then provided by the information broker Thomson Reuters and in the year 2017 taken over by Clarivate Analytics. Both databases have a broad coverage of international publications, about 14,000 journals in WoS and 22,000 in Scopus (Bauschmann and Ahnert 2016). Therefore, they are not only useful for citation analysis, but also for the analysis of publication trends of countries and institutions. Since 2018 also the free-of-charge available database Dimensions by Digital Science added to the available and curated bibliometric sources as well as the Open Alex Database that emerged out of Microsoft Academic Graph after it was discontinued. Naturally, many publication databases for special scientific fields are available, for example MedLine, Compendex, and also for particular document types such as arXive in the case of pre-prints. However, most of these latter listed publication databases only provide bibliographic information, while bibliometric data (including citation information) is only covered by a small number of data sources. The steadily developing new ways of bibliometric analysis, documented, e.g., in journals such as Scientometrics, are induced to a large extent by improved technical possibilities to conceive and exploit databases (dimension four of our indicator development scheme, see Fig. 1).

For innovation analyses, not only the performance of the science system of a country is relevant, but also the link between science and technology, which will be discussed in more detail in the following section of this chapter. In the case of science-based technologies, the parallel observation of patents and publications proved to be one insightful approach (e.g. Schmoch 2007). Furthermore, it is possible to analyse which university publications are frequently cited by enterprises (Tijssen 2006) or which were published by authors affiliated to companies (Krieger et al. 2021). The number of citations in patent examination reports is a good indicator for the science-linkage of a technology (Narin et al. 1997; Verbeek et al. 2002). For analysing technology transfer, the patents of universities are frequently employed indicators (e.g. Dornbusch et al. 2013; Neuhäusler et al. 2021a). For this type of analysis, patent and publication databases are linked. More recently, standard-essential patents and standard-relevant publications have gained attention for analysing transfer and market developments (Blind 2004; Blind and Fenton 2022). A further aspect is international collaboration in science (Leydesdorff and Wagner 2008). A central discussion in the context of national systems of innovation is the interaction of enterprises, universities and governments (Leydesdorff and Etzkowitz 1998). Systemic improvements are to be achieved where different actors or sub-systems interact—this is the basic assumption of these analytical approaches. For each of the sub-systems, particular indicators are used—for example, publications in the science system or patents mainly for industrial technological innovations—but at some point these indicators overlap or play a particular role in describing the activities of the sub-systems that go beyond. Examples are co-publications (see above) or co-patents of science and industry. Academic patents (Lissoni et al. 2008; Neuhäusler et al. 2021b)—these are patents invented by staff-members of universities or public research organisations, but not necessarily filed by these organisations—or academic spin-offs (Frietsch et al. 2021) are additional examples where the classical focal tasks and therefore also focal indicators span over the boundaries of the sub-systems.

In the case of publication databases, online versions of the Web of Science (WoS) or Scopus are available, but therein sophisticated citation analysis is not possible on a large scale, as many of the indicators use expectancy rates, field-specific indicators or organisation- or author-specific normalisations (e.g. exclusion of self-citations) as the basis for the calculations. In addition, data cleaning, data treatment and especially data linking with external sources are much harder in web-based applications than in the case of raw data access.

4.2 New Target Areas and Analytical Differentiations

A slightly different perspective arose from science and technology analyses, however, using similar data sources. The relation of science and technology has always been a major topic of analysis in innovation research. Already in the 1960s, various retrospective studies were conducted to assess the impact of basic research on technological innovation. In particular, the US-American studies called "Hindsight and Traces" had a relevant influence on science policy. An early indicator-based study was provided by Narin et al. (1987), which considered references to publications in patent search reports. A further survey-based, important study was performed by Mansfield (1991) looking at the impact of scientific research on industrial innovations. A famous theoretical contribution, why enterprises conduct basic research, was made by Rosenberg (2010). Rappa and Debackere (1992) described the interaction of enterprises and academic institutions in science-based technologies.

This line of research became more precise and tangible by the use of innovation indicators, in particular by specific forms of patent and publication indicators and their combination (Schmoch 1997). In this context, publication indicators are not used for the assessment of universities and research institutes as in bibliometrics, but for the analysis of scientific research, so the subject of research instead of the process or output is the centre of interest. For instance, Schmoch (2007) could describe the parallel development of science and technology over long periods and show the substantial delay between scientific discoveries and their broad implementation into technology (see also Moed 2017).

A phenomenon of the last years is the increasing number of science-based technologies such as nanotechnology, graphene and other two-dimensional materials, artificial intelligence, electrically conducting polymers, plant biotechnology, high temperature superconductivity, light emitting diodes (LEDs), fuel cells, CRISPR-Cas technologies or improved computer storage systems. A typical observation of a parallel analysis of patents and publications is that in the early years of a science-based technology, the academic research is focused on (oriented) basic research and with the intensified search of industry for specific applications, the academic research is increasingly oriented on applications as well. Thus a direct link between science and technology and its development in time can be shown (see, e.g., Schmoch and Thielmann 2012).

Since about the end of the 1990s, a major topic of the relation of science and technology are the university-industry relations. Most of these papers are indicator-based (Perkmann et al. 2013) and aim at assessing the link between academic research and economic impact. A trigger for this broad activity was an article by Meyer-Krahmer and Schmoch (1998) analysing the university-industry relations as to major mechanisms, advantages and disadvantages for universities as well as differences by scientific fields.

In recent years, there has been an increasing interest in knowledge transfer from academic research to society. This is monitored with indicators such as memberships in political advisory councils, consultancy for social institutions, publications in non-scientific journals, etc. and can be observed in many countries. Thus all scientific disciplines, not only the engineering and natural sciences, are involved. For analysing this type of interaction, new indicators have to be conceived. A part of this new approach comprises the so-called altmetrics (Thelwall 2019), but additional tools will be needed for describing the full spectrum of transfer activities.

In the 1960s and 1970s, innovation indicators were primarily oriented on research-intensive technologies. In the middle of the 1990s, innovation in service industries was complemented. However, it was attempted to transfer the methodology for production technologies to services which proved to be problematic. For example, it is possible to compare biotechnology and mechanical engineering on the basis of R&D intensity, but the structures in services such as banking, accommodation, transport logistics or medical treatment are so different that a meaningful comparison based on R&D expenditures is difficult.

One approach to overcome this inadequacy is the concept of knowledge intensity instead of R&D intensity. Knowledge-intensive companies or services are characterised by high shares of highly qualified personnel—most often university or college graduates—whereby these high qualifications play a major role in the value creation of these companies/sectors.

This is even more complicated in the case of digital business models or platform economies, where the platform provider only acts as a broker with a huge market power, but does not provide the service or the product itself. Digital business model indicators to assess their innovativeness and/or their contribution to innovation processes are to a large extent missing.

In the last 15 years or so, the character and meaning of critical technologies in the economy and society have changed. A number of key technologies are much more pervasive across a number of other technologies and domains and thus impact more generally and comprehensively on broad aspects of the economy and society compared to previous critical technologies. For example, information technology has developed into a generic field which is relevant for many other areas such as mechanical engineering, the automotive industry or biotechnology. For the description of this development, new derived indicators are needed. Another topical development is the increasing relevance of biotechnology for practical applications in industrial processes, materials, agriculture or pharmacy. Again, new indicators are needed to encompass the effects more appropriately.

5 New Data Sources, New Data Analytics: Nature, Opportunities and Limits of New Indicators and Measurements

At the beginning of the new century the framework conditions for innovation indicators began to change massively. In about the middle of the first decade a new methodological paradigm began to diffuse that was driven by what is nowadays called Big Data. While up to the middle of the 2000s data access was very expensive—both in terms of fees and in terms of transaction costs—a diffusion of a number of data sources changed the picture completely. First of all, it was the accessibility of patent data that completely changed the landscape—especially the PATSTAT[2] database provided by the EPO, but also the inauguration of the bibliometric database Scopus by Elsevier, which appeared on the scene as a competitive product to the so far—more or less—unique bibliometric data source of the Web of Science, then owned by Thomson Reuters.

Patent data was used by several innovation scholars worldwide, but the exploitation of the analytical potential was seen to be too limited so that several users and scholars asked for better data and broader data access. A first conference[3] was held in Geneva in September 2003 as a start of a series of conferences where user needs were discussed. This was one of the milestones in the direction to the first release of

[2] The official name of PATSTAT is "EPO Worldwide Statistical Patent Database".

[3] https://www.wipo.int/meetings/en/topic.jsp?group_id=230

PATSTAT in 2005. First users started to implement PATSTAT and learned to work with large-scale relational databases. Given the technical restrictions in processing these large amounts of data, several researchers started to work with individual tables only (flat files). By the end of the decade, however, a large number of universities and research institutes in many countries had subscribed to PATSTAT and it offered completely new analytical potentials. Data cleaning, data treatment and data matching moved patent statistics from the macro-economic or technological meso-level to the micro-level of organisations (companies, universities, research institutes) or even the characteristics of individual applications (e.g. legal status, family size, citation rates). A differentiation of patent analytics became possible with micro-data access like PATSTAT.

Other data sources, for example bibliographic data on specific scientific areas like PubMed in the field of medicine or Compendex in engineering, were more and more subscribed by universities for monitoring and publication retrieval purposes and therefore became also accessible for science and innovation analytics. Additional data sources were implemented, for example, on trademarks (see Gotsch and Hipp 2012; Mendonça et al. 2004; Schmoch 2003; Schmoch and Gauch 2009), which became possible at that time as web interfaces by data owners and database providers lowered the thresholds for access to larger groups of researchers.

Most of these data sources are generated as a by-product of otherwise needed processes. In essence, innovation research analytics became possible as a secondary use of the already existing data. For some data providers commercialising the data turned out to be an additional business model. In case of IPR data, it was just an additional way of fulfilling the need for publishing and granting access, which is inherent to the IPR system. Patents are a vested right of exclusive use, but in exchange for this right, those who intend to own the rights are obliged to describe and publish their technology. Innovation statistics benefit from this IPR specific publication requirement.

In addition to those changes in relation to traditional indicators such as patents and publications, a more fundamental development took place that opened up a number of new possibilities to show innovation developments. Contrary to R&D data that requires extra efforts to collect it in large-scale surveys of (suspected) R&D conducting companies, process generated data just existed and was more frequently co-used for statistical purposes. So new business models emerged and innovation research was in demand for this new kind of data. The access to data was not the bottleneck of analyses anymore. Two dimensions gained relevance; on the one hand, an understanding of the data and its particularities; on the other hand, capabilities for data storage, data treatment and linking it to other data sources. These capabilities have recently been supplemented by capabilities to produce and use large amounts of unstructured data, mainly from web sources, from business reports, or from full-texts of patents or publications. Next to the availability of the data and the meeting of supply and demand of new data sources, all this has been possible by two additional trends. The technical evolution of treating large amounts of data with comparatively short computation times, storing them, treating them and developing ever more complex models. In addition, the demand for solid and robust innovation research results as well as evidence-based policy-making increased as more and more countries entered the innovation stage. Accordingly, policy makers all around the globe became eager to either provide the perfect framework conditions for innovation (sometimes also called innovation eco-systems) or even govern the national innovation systems with the aim of increasing the national, regional or local competitiveness. This is essentially the root of category 3 in our innovation indicators scheme (see Fig. 3).

Box 3: Reporting and Monitoring Systems
We have outlined the historical development of innovation systems and innovation process analyses as well as the evolution of indicators to measure and analyse them. From the beginning, one of the core aims, but also the core challenges was the international comparability of the data and the analyses. For this purpose, several activities of regular innovation monitoring were undertaken by the US National Science Foundation as early as 1973 (e.g. NSB 2020) or by the French Observatoire des Sciences et des Technologies since 1995 (OST 2000). Following the NSF, the OECD published in 1984 "Science and Technology Indicators", which in 1988 was replaced by "Main Science and Technology Indicators" (Godin 2003, p. 680). "Industrialised countries followed the NSF definition when they adopted the OECD Frascati Manual in 1963. The manual was designed to help countries in their measurement efforts, offering methodological conventions that allowed international comparisons". Godin (2006, p. 648).

Important activities for creating, discussing and testing new innovation indicators were efforts to establish regular innovation monitoring, which came up in several countries. For example, in Germany the annual "Bericht zur technologischen Leistungsfähigkeit Deutschlands (TLF)" (Report on the technological competitiveness of Germany) was initiated on behalf of the German Federal Ministry for Education and Research (Bundesministerium für Bildung und Forschung, BMBF). It started with Grupp and Legler (1987), a cooperation between Fraunhofer ISI and the Institute of Economic Research of Lower Saxony (NIW), and ended in 2007 (Egeln et al. 2007). Over the years, the number of participating institutes and of the indicators on different topics increased, for instance in the last report indicators on productivity, foreign trade, patents, R&D, scientific publications, technology transfer, skilled labour, women in science, technology and research, environmental technology, etc. were analysed and discussed. Since 2008, the activities of TLF have been pursued in parts of the work of the Commission of Experts for Research and Innovation (EFI) to the German government.

New data trends are already in full swing, especially the emancipation of science from commercial data providers. Public research itself started to produce data and non-profit organisations emerged with other than purely profit-oriented business models. The move from proprietary to open infrastructures continues and now covers a broader area of data sources—most visible in bibliometrics, but also in terms of other data like geospatial, mobility or company information.

Already in the late 1990s and early 2000s this development found a seedbed in public data providers like statistical offices or publicly funded service providers that made their previously non-disclosed micro-data accessible to researchers. New developments in anonymisation of individual data made data protection possible even when sharing micro-data. For example, R&D data in Europe became (partly) accessible as scientific use files became available or—more comprehensive—by on-site visits to Eurostat's data centre. Many national data providers established similar data centres and possibilities of data access as well. Company registers and broad data access—for example, in Sweden—allowed even more sophisticated and detailed analyses that were able to address completely new research questions (see, for example, Jung and Ejermo 2014).

All these developments were possible because of the step change in technological progress of computing power and software packages. Local servers instead of super-computers in the computing centres of—mostly only—selected universities were able to handle the relational databases in a satisfactory manner. Even desktops and laptops gained the computing power to analyse the extracted data or acted as access points to the central servers of the innovation and economics institutes that were working on these topics.

In consequence, the number of institutes using innovation data especially universities worldwide and the number of authors analysing or even developing new innovation indicators grew very quickly since the middle of the first decade of the 2000s. This is visible in the number of publications using the keywords patent, publication or bibliometrics in their title (Fig. 2)

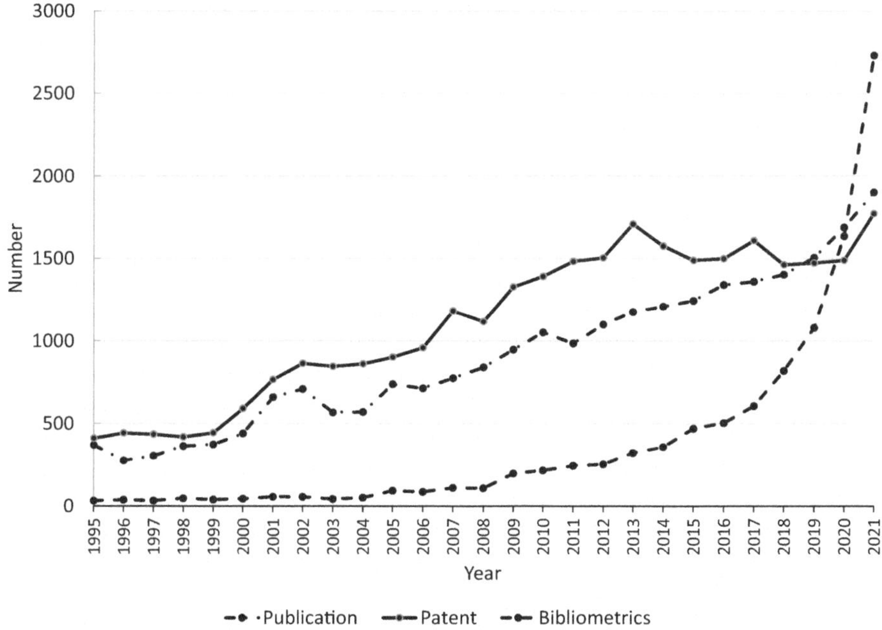

Fig. 2 Number of publications on patents, publications and bibliometrics. Source: Elsevier - Scopus; own representation

6 Summarising Conclusions: Historical Developments and Challenges of the Near Future

In this chapter, we outline and interpret the evolution and development of innovation indicators since the early 1970s. We identify four driving factors of innovation indicators provision and use, namely (1) new theories/concepts, (2) lower technical thresholds for data analyses and availability of new data, (3) increasing policy demands and (4) technological and economic developments, mainly the shift towards R&D-intensive sectors and technologies. Our discussion shows that at different stages of the indicator development different factors are the driving forces.

While in Schumpeter's time the focus was on the individual entrepreneur and his/her impact on technological progress and innovations, the post-Second World War era was characterised by a change in the innovation processes and their organisation, based on a division of labour and responsibilities. This led to a need for innovation indicators that are internationally comparable and generally applicable. In addition, the first innovation-oriented political programmes and the first innovation indicators appeared about 1990.

In this first phase of a broadening take-up of indicators in the 1960s and 1970s, various scientific entities at different universities and research institutes emerged—among them Fraunhofer ISI in 1972—that took innovations and innovation processes, as well as science and innovation policy as their subject of analysis. Individuals in these organisations pushed the conceptual and methodological foundations further ahead (driving factor 1 of our model). New disciplines, namely innovation economics, STI policy analysis and STI indicators were born. Ever since, innovation indicators and innovation policy (factor 3) have been closely intertwined and have led to evidence-based policy-making (see, for example, Dosso et al. 2018).

> Indicators (are) the instrument of an ambitious and open S&T policymaking: Indicators feed analysis and argumentation by opening the black box of the scientific community and of the political decision (Arvantis et al. 1986).

The early innovation indicators were primarily linked to the OECD and outstanding monitoring and standardisation efforts, the establishment of the Science Citation Index (now Web of Science), the engagement of national bodies such as the US National Science Foundation and more recently international bodies such as the ILO or World Bank. Innovation indicators were mainly R&D-centred with a strong focus on the manufacturing industry and R&D processes in companies as well as the science systems. The origin and target of these analyses were the monitoring and performance measure of science systems that then shifted towards the competitiveness of nations. Essentially, next to case studies and survey data, the majority of indicators were of macro-economic nature, addressing national or technological levels. The innovation system's perspective widened the focus on various actors and their interplay, which also led to the introduction of additional indicators, among them indicators on transfer and collaboration (factors 1 and 4). Data availability and better options for data treatment and analysis (factor 2) gave the indicators

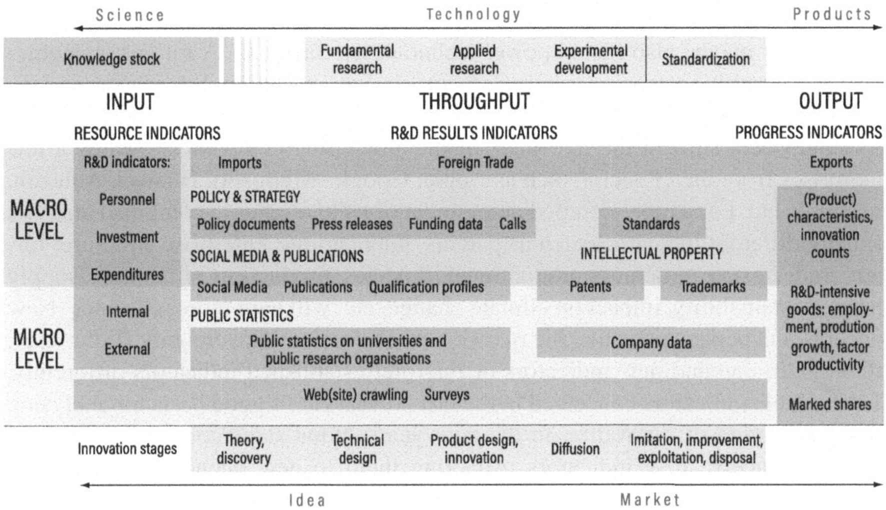

Fig. 3 Schematic representation of indicators in the innovation process. Source: own representation based on Grupp (1998) and Frietsch and Jung (2009)

development another push since about the year 2000. More recently, information and computer science methods have entered the innovation indicators scene and widened the scope even further. These latter developments supported a shift away from the macro-level to more micro-level and process-oriented indicators and analyses. Surveys are no longer the only option for micro-analyses, but company databases and their matchings with other data sources offer—in most cases even a more large-scale—option for analysing companies, research organisations, projects or even individuals. The use of unstructured (text) data is about to push the possibilities even further. More differentiated information, qualitative aspects and completely new insights might be introduced into STI analyses, based on large language models.

A schematic representation of (a selected set of) innovation indicators is depicted in Fig. 3. It tries to grasp the admittedly simplifying logic of the linear model (Bush 1945; Godin 2003) of innovation when it orders the indicators from left to right along the dimension of input, throughput and output. More conceptually, the scheme sketches the continuums from science to technology (see top) or from idea to the market (see bottom). A differentiation of the indicators that rather address the macro-level and those that rather address the micro-level is also represented. The figure shows the larger and further growing landscape of innovation indicators. Traditional indicators like R&D expenditures, patents, publications, trade or production data are supplemented by trademarks, standards, company data or unstructured data sources like web-crawling.

Among the major new trends in innovation, which will have an impact on the conception of indicators, is a further increasing relevance of science-based technologies such as batteries, fuel cells, nanotechnology, quantum technologies, materials, or maybe also fusion power technologies (factor 4). A further phenomenon is the steadily increasing number of new services and (digital) business models. Among the current challenges of innovation research as well as innovation indicators might be a shift in the relevance of manufacturing companies towards a few enterprises from the IT sector such as Apple, Google, Microsoft, Huawei, Amazon, Facebook, etc. For a more detailed assessment of technologies, a combined analysis of many different indicators such as patents, publications, enterprise structure, foreign trade, price structure, international linkages production structures, supply chains, sustainability, impact on climate change, etc. will gain more relevance. New concepts and perspectives like the question of Technology Sovereignty (Edler et al. 2020, 2023) demand new indicators or interpret established indicators differently. The available indicators conceived in the last 50 years are a good basis for analysing the development of innovation in the next years. What stays constant is a steady effort to conceive new indicators. Adapting them to new phenomena is a great challenge.

However, we should not forget that indicators are a means to measure and assess constructs which are otherwise not directly measurable. They should not become a means in itself. Recently there has been some fundamental critique on the current status of indicators. While Goodhart's law is a fundamental critique of any indicator becoming a means in itself, some more specific criticisms (e.g. Larivière and Sugimoto 2019; Moed 2018) have been raised with respect to S&T indicators. For example, Barré (2019, p. 44) sees a "landslide of instrumentalised S&T indicators" since the mid-90s to the present day. In this context, he mentions the link between new public management and bibliometrics and linked to that publication statistics as tools for competition, the increased funding of scientific research by enterprises and the emergence of an industry of science information. Barré complains that "the indicator is integrated in the social, professional and cultural norms and has become the undisputable reality of the object or phenomenon. It is forgotten that an indicator is only a proxy, but not the object". In this context, he speaks about "culturally produced ignorance". However, many researchers struggle to correct these misleading activities, e.g., the San Francisco Declaration on Research Assessment-DORA (ASCB 2012), the Leiden Manifesto (Hicks et al. 2015) or the review of the use of metrics in the UK research assessment (Wilsdon 2015).

It seems that many options, but also challenges stay ahead of indicator-based science and innovation research. This makes this field so attractive to many young researchers, who will push the frontiers even further. Artificial intelligence based on large language models already opens a new avenue of research that might be able to simplify and differentiate classification tasks that were a bottleneck for many decades. Neural networks will help to detect relations of topics, organisations or persons, helping to better understand the relations, effects, causalities and impacts of certain factors in the science and innovation systems. New questions will arise and new answers will be given. The new possibilities, even after more than 50 years of indicator development, still let it appear a rather young and dynamic field.

References

American Society for Cell Biology (ASCB) (2012) San Francisco declaration on research Assessment (DORA). University of Nebraska, Lincoln, NE

Arundel A, Smith K, Patel P, Sirilli G (1998) The future of innovation measurement in Europe: Concepts, problems and practical direction. IDEA paper series, 3. NIFU, Oslo

Arvantis R, Callon M, Latour B (1986) Evaluations des Politiques de la Recherche et de la Technique. La Documentation Française, Paris

Barré R (2019) Les indicateurs sont morts, vive les indicateurs! Towards a political economy of S&T indicators. Res Eval 28:2–6

Bauschmann M, Ahnert C (2016) Vergleich von Web of Science und Scopus im Hinblick auf den Informationsbedarf an der TU Chemnitz. TU Chemnitz, Chemnitz

Bergek A, Jacobsson S, Carlsson B, Lindmark S, Rickne A (2008) Analyzing the functional dynamics of technological innovation systems; A scheme of analysis. Res Policy 37:407–429

Blind K (2004) The economics of standards: theory, evidence, policy. Edward Elgar, Cheltenham

Blind K, Fenton A (2022) Standard-relevant publications: evidence, processes and influencing factors. Scientometrics 127:577–602. https://doi.org/10.1007/s11192-021-04210-8

Bush V (1945) Science—the endless frontier; a report to the President, Washington: Office of Scientific Research and Development, reissued by the National Science Foundation in celebration of the agency's 70th anniversary and the 75th anniversary of Science—the Endless Frontier. NSF, Alexandria

Caulcutt R (1992) Statistics in research and development. Chapman & Hall, London

Chesbrough HW (2003) Open innovation: the new imperative for creating and profiting from technology. Harvard Business School Press, Boston

Cohen WM, Levinthal DA (1990) Absorptive capacity; a new perspective on learning and innovation. Adm Sci Q 35:128–152

Coombs R (2003) The changing character of "service innovation" and the emergence of "knowledge-intensive business services". In: Dankbaar B (ed) Innovation management in the knowledge economy. Imperial College Press, London, pp 83–96

Crépon B, Duguet E, Mairesse J (1998) Research, innovation and productivity; an econometric analysis at the firm level. Econ Innov New Technol 7:115–158

de Rassenfosse G, Dernis H, Guellec D, Picci L, van Potterie P (2013) The worldwide count of priority patents; A new indicator of inventive activity. Res Policy 42:720–737

Dornbusch F, Schmoch U, Schulze N, Bethke N (2013) Identification of university-based patents; a new large-scale approach. Res Eval 22:52–63

Dosso M, Martin B, Moncada-Paternò-Castello P (2018) Towards evidence-based industrial research and innovation policy. Sci Public Policy 45:143–150

Edler J, Fagerberg J (2017) Innovation policy: what, why & how. Oxf Rev Econ Policy 33:2–23

Edler J, Blind K, Frietsch R, Kimpeler S, Kroll H, Lerch C, Reiss T, Roth F, Schubert T, Schuler J, Walz R (2020) Technology sovereignty. From demand to concept; perspectives—Policy brief no. 02 / 2020. Fraunhofer ISI, Karlsruhe

Edler J, Blind K, Kroll H, Schubert T (2023) Technology sovereignty as an emerging frame for innovation policy; defining rationales, ends and means. Res Policy 52:104765

Edquist C (ed) (1997) Systems of innovation. Technologies, institutions and organizations. Pinter, London

Egeln J, Gehrke B, Legler H, Licht G, Rammer C, Schmoch U (2007) Bericht zur technologischen Leistungsfähigkeit Deutschlands 2007. BMBF, Berlin

Etzkowitz H, Leydesdorff L (1995) The triple helix—university-industry-government relations: a laboratory for knowledge-based economic development. EASST Rev 14:14–19

Eurostat (2020) Community innovation survey CIS. Eurostat, Luxembourg

Fagerberg J (1988) International competitiveness. Econ J 98:355–374

Fagerberg J, Fosaas M, Bell M, Martin BR (2011) Christopher Freeman: social science entrepreneur. Res Policy 40:897–916. https://doi.org/10.1016/j.respol.2011.06.011

Faust K, Schedl H (1984) Internationale Patentdaten: Ihre Nutzung für die Analyse technologischer Entwicklungen. In: Oppenländer KH (ed) Patentwesen, technischer Fortschritt und Wettbewerb. BMFT, ifo Institut, Berlin, pp 151–172

Frame JD, Narin F, Carpenter MP (1977) The distribution of world science. Soc Stud Sci 7:501–516

Freeman C (1974) The economics of industrial innovation. Cassel, London

Freeman C (1987) Technology policy and economic performance; lessons from Japan. Pinter, London

Freeman C (2004) Technological infrastructure and international competitiveness. Ind Corp Chang 13:541–567

Freeman C, Soete L (1997) The economics of industrial innovation. Pinter, London

Frietsch R (2011) Bildung und Innovation. Fraunhofer IRB, Stuttgart

Frietsch R, Jung T (2009) Transnational patents—structures, trends and recent developments; Studien zum deutschen Innovationssystem Nr. 7-2009. Expertenkommission Forschung und Innovation (EFI), Berlin

Frietsch R, Schmoch U (2010) Transnational patents and international markets. Scientometrics 82:185–200

Frietsch R, Neuhäusler P, Jung T, van Looy B (2014) Patent indicators for macroeconomic growth—the value of patents estimated by export volume. Technovation 34:546–558

Frietsch R, Kladroba A, Markianidou P, Neuhäusler P, Peter V, Ravet J, Rothengatter O, Schneider J (2017) Final report on the collection of patents and business indicators by economic sector; societal grand challenges and key enabling technologies collection and analysis of private R&D investment and patent data in different sectors, thematic areas and societal challenges. Publications Office of the European Union, Luxembourg

Frietsch R, Kroll H, Jonkers K (2019) China's development of an innovation-driven economy—An intermediate assessment. Chinese Academy of Sciences, Beijing

Frietsch R, Darold D, Karaulova M, Gruber S, Neuhäusler P, Rammer C, Doherr T, Dörr J, Gottschalk S (2021) Spin-offs from public research organisations in Germany; A comprehensive analysis based on bibliometric, patent, website and company register data. Fraunhofer ISI; ZEW, Karlsruhe

Gallouj F (1997) Towards a neo-Schumpeterian theory of innovation in services? Sci Public Policy 24:405–420

Gallouj F (2002) Innovation in services and the attendant old and new myths. J Socio-Econ 31:137–154

Garfield E (1955) Citation indexes for science. Science 122:108–111

Geels FW (2002) Technological transitions as evolutionary reconfiguration processes: a multi-level perspective and a case study. Res Policy 31:1257–1274

Gerstenberger W (1992) Zur Wettbewerbsposition der deutschen Industrie im High-Tech-Bereich. ifo Schnelldienst 45:14–23

Gibbons M, Johnston R (1974) The role of science in technological innovations. Res Policy 3:220–242

Gielow G, Kuntze U, Mannsbart W, Meyer-Krahmer F (1982) Möglichkeiten einer Verbesserung des Repräsentationsgrades der Statistik über Forschung und Entwicklung in der Wirtschaft in der Bundesrepublik Deutschland. Fraunhofer ISI, Karlsruhe

Gilpin R (1975) Technology, economic growth and international competitiveness. Washington, DC, U.S. Government Printing Office

Glänzel W, Moed HF, Schmoch U, Thelwall M (eds) (2019) Springer handbook of science and technology indicators. Springer, Cham

Godin B (2003) The emergence of S&T indicators; Why did governments supplement statistics with indicators? Res Policy 32:679–691

Godin B (2006) The linear model of innovation the historical construction of an analytical framework. Sci Technol Hum Values 31:639–667

Gotsch M, Hipp C (2012) Measurement of innovation activities in the knowledge-intensive services industry; A trademark approach. Serv Ind J 32:2167–2184

Grenzmann C, Marquart R, Wudtke J (1991) Forschung und Entwicklung in der Wirtschaft 1989. Stifterverband für die deutsche Wirtschaft, Essen

Griliches Z (1979) Issues in assessing the contribution of research and development to productivity growth. Bell J Econ 10:92–116

Griliches Z (1981) Market value, R&D and patents. Econ Lett 7:187

Grupp H (1992) Competitive trade advantage of EC nations as a function of their science and technology production? In: van Raan AF (ed) Science and technology in a policy context. DSWO Press, Leiden, pp 327–355

Grupp H (1998) Foundations of the economics of innovation; Theory, measurement and practice. Elgar, Cheltenham, UK

Grupp H, Legler H (1987) Spitzentechnik, Gebrauchstechnik, Innovationspotential und Preise; Trends, Positionen und Spezialisierung der westdeutschen Wirtschaft im internationalen Wettbewerb. Verl. TÜV Rheinland, Köln

Grupp H, Schmoch U (1999) Patent statistics in the age of globalisation; new legal procedures, new analytical methods, new economic interpretation. Res Policy 28:377–396

Grupp H, Münt G, Schmoch U (1996) Assessing different types of patent data for describing high-technology export performance. In: OECD (ed) Innovation, patents and technological strategies: Proceedings of a workshop held at OECD headquarters in Paris, 8–9 December 1994. OECD, Paris, pp 271–287

Grupp H, Legler H, Jungmittag A, Schmoch U (2000) Hochtechnologie 2000. Neudefinition der Hochtechnologie für die Berichterstattung zur technologischen Leistungsfähigkeit Deutschlands

Grupp H, Dominguez Lacasa I, Friedrich-Nishio M, Jungmittag A (2005) Innovation and growth in Germany over the past 150 years. In: Cantner U, Dinopoulos E, Lanzillotti RF (eds) Entrepreneurship, the new economy and public policy: Schumpeterian perspectives. Springer, Berlin, pp 267–287

Hatzichronoglou T (1997) Revision of the high-technology sector and product classification. OECD science, technology and industry working papers. OECD, Paris

Hauknes J (1998) Services in innovation—innovation in services; SI4S Final report

Hekkert MP, Suurs RAA, Negro SO, Kuhlmann S, Smits REHM (2007) Functions of innovation systems; A new approach for analysing technological change. Technol Forecast Soc Change 74:413–432

Henriques L, Larédo P (2013) Policy-making in science policy: The "OECD model" unveiled. Res Policy 42:801–816

Hicks D, Wouters P, Waltman L, de Rijcke S, Rafols I (2015) Bibliometrics: The Leiden Manifesto for research metrics. Nature 520:429–431

Jung T, Ejermo O (2014) Demographic patterns and trends in patenting; gender, age, and education of inventors. Technol Forecast Soc Change 86:110–124

Kleinknecht A, Reijnen J, Smits W (1993) Collecting literature-based innovation output indicators. The experience in the Netherlands. In: Kleinknecht ABD (ed) New concepts in innovation output measurement. Palgrave MacMillan, London, pp 42–84

Kleinknecht A, van Montfort K, Brouwer E (2002) The non-trivial choice between innovation indicators. Econ Innov New Technol 11:109–121

Kommission der Europäischen Gemeinschaft (1982) Die Wettbewerbsfähigkeit der Industrie der Europäische Gemeinschaft, Luxembourg

Krieger B, Pellens M, Blind K, Gruber S, Schubert T (2021) Are firms withdrawing from basic research? An analysis of firm-level publication behaviour in Germany. Scientometrics 126:9677–9698. https://doi.org/10.1007/s11192-021-04147-y

Kuntze U, Müller P, Rempp H, Westermann G, Pfeiffer R (1975) Patentstatistik zur Erfassung technologischer Entwicklungstendenzen auf der Basis des im Deutschen Patentamt vorhandenen Datenmaterials und technologischen Wissens. Fraunhofer ISI, ifo Institut, Karlsruhe, München

Larivière V, Sugimoto CR (2019) The journal impact factor: a brief history, critique, and discussion of adverse effects. In: Glänzel W, Moed HF, Schmoch U, Thelwall M (eds) Springer handbook of science and technology indicators. Springer, Cham, pp 3–24

Legler H (1982a) Internationale Wettbewerbsfähigkeit der westdeutschen Chemischen Industrie; Hochschulschrift. Duncker und Humblot, Berlin

Legler H (1982b) Zur Position der Bundesrepublik Deutschland im internationalen Wettbewerb. NIW, Hannover

Legler H, Frietsch R (2007) Neuabgrenzung der Wissenswirtschaft—forschungsintensive Industrien und wissensintensive Dienstleistungen (NIW/ISI-Listen 2006); Studie zum deutschen Innovationssystem, No. 22-2007. NIW und Fraunhofer ISI, Hannover

Legler H, Grupp H, Gehrke B, Schasse U (1992) Innovationspotential und Hochtechnologie; Technologische Position Deutschlands im internationalen Wettbewerb. Physica-Verlag, Heidelberg

Leydesdorff L, Etzkowitz H (1998) The Triple Helix as a model for innovation studies. Sci Public Policy 25:195–203. https://doi.org/10.1093/spp/25.3.195

Leydesdorff L, Wagner CS (2008) International collaboration in science and the formation of a core group. J Inform 2:317–325. https://doi.org/10.1016/j.joi.2008.07.003

Lissoni F, Llerena P, McKelvey M, Sanditov B (2008) Academic patenting in Europe; new evidence from the KEINS database. Res Eval 17:87–102

Lundvall B-A (1988) Innovation as an interactive process; From user-producer interaction to the national system of innovation. In: Dosi G, Freeman C, Nelson R, Silverberg G, Soete L (eds) Technical change and economic theory. Pinter, London, pp 349–369

Lundvall B-A (ed) (1992) National innovation systems. Towards a theory of innovation and interactive learning. Pinter, London

Maclaurin WR (1954) Technological progress in some American industries. Q J Econ 44:178–189

Majer H (1978) Industrieforschung in der Bundesrepublik Deutschland; Eine theoretische und empirische Analyse. Mohr, Tübingen

Malerba F, Orsenigo L (1995) Schumpeterian patterns of innovation. Camb J Econ 19:47–65

Mansfield E (1968) Industrial research and technological innovation; an econometric analysis. Norton, New York

Mansfield E (1991) Academic research and industrial innovation. Res Policy 20:1–12

Maraut S, Martínez C (2014) Identifying author–inventors from Spain; Methods and a first insight into results. Scientometrics 101:445–476. https://doi.org/10.1007/s11192-014-1409-1

Marengo L, Dosi G (2003) Division of labor, organizational coordination and market mechanism in collective problem-solving. Scuola Superiore Sant'Anna, Laboratory of Economics and Management (LEM), Pisa

Martinez C (2011) Patent families: When do different definitions really matter? Scientometrics 86:39–63

Mendonça S, Pereira TS, Godinho MM (2004) Trademarks as an indicator of innovation and industrial change. Scuola Superiore Sant'Anna, Laboratory of Economics and Management

Mendonça S, Schmoch U, Neuhäusler P (2019) Interplay of patents and trademarks as tools in economic competition. In: Glänzel W, Moed HF, Schmoch U, Thelwall M (eds) Springer handbook of science and technology indicators. Springer, Cham, pp 1023–1035

Merrifield B (1979) Stimulating technological innovation: nurturing the inventor. Res Manag 22:12–14. https://doi.org/10.1080/00345334.1979.11756568

Meyer-Krahmer F, Schmoch U (1998) Science-based technologies: university–industry interactions in four fields. Res Policy 27:835–851. https://doi.org/10.1016/S0048-7333(98)00094-8

Miles I (2004) Innovation in services. In: Fagerberg J, Mowery DC, Nelson RR (eds) The Oxford handbook of innovation. Oxford University Press, Oxford, pp 433–458

Moed HF (2017) From Francis Narin's science-technology linkages to double boom cycles in technology. In: Moed HF (ed) Applied evaluative informetrics. Springer, Cham, pp 209–227

Moed HF (2018) Assessment and support of emerging research groups. FEMS Microbiol Lett 365. https://doi.org/10.1093/femsle/fny189

Narin F, Noma E, Perry R (1987) Patents as indicators of corporate technological strength. Res Policy 16:143–155
Narin F, Hamilton KS, Olivastro D (1997) The increasing linkage between U.S. technology and public science. Res Policy 26:317–330. https://doi.org/10.1016/S0048-7333(97)00013-9
National Science Board (NSB) (2020) Science and engineering indicators, Washington
Nelson RR (ed) (1993) National innovation systems. Oxford University Press, Oxford
Nelson R, Winter S (1977) In search of a useful theory of technical progress. Res Policy 6
Nelson RR, Winter SG (1982) An evolutionary theory of economic change. The Belknap Press of Harvard University Press, Cambridge, MA
Neuhäusler P, Frietsch R, Mund C, Eckl V (2016) Identifying the technology profiles of R&D performing firms—A matching of R&D and patent data. Int J Innov Technol Manag 14
Neuhäusler P, Feidenheimer A, Frietsch R, Kroll H (2021a) Generating a classification for EUIPO trademark filings—A string matching approach; Fraunhofer ISI discussion papers innovation systems and policy analysis nr. 69, Karlsruhe
Neuhäusler P, Rothengatter O, Feidenheimer A (2021b) Patent applications—structures, trends and recent developments. Expertenkommission Forschung und Innovation (EFI), Berlin
Noyons ECM, van Raan AFJ, Grupp H, Schmoch U (1994) Exploring the science and technology interface: inventor-author relations in laser medicine research. Res Policy 23:443–457
OECD (1985) Trade in high technology products; An initial contribution to the statistical analysis of trade patterns in high technology products, Paris
OECD (1992) Oslo manual. The measurement of scientific and technological activities; proposed guidelines for collecting and interpreting technological innovation data. OECD, Paris
OECD (ed) (1996) Innovation, patents and technological strategies. In: Proceedings of a workshop held at OECD headquarters in Paris, 8–9 December 1994. OECD, Paris
OECD (2003) OECD science, technology and industry scoreboard. OECD, Paris
OECD (2015) World corporate top R&D investors: innovation and IP bundles. OECD, Paris
OECD (ed) (2018) Oslo manual 2018; guidelines for collecting, reporting and using data on innovation, 4th edn. OECD, Paris
OECD, Eurostat (2005) Oslo manual. Proposed guidelines for collecting and interpreting innovation data. OECD/Eurostat, Paris
OST (2000) Indicateurs 2000, Paris
Oztemel E, Gursev S (2018) Literature review of Industry 4.0 and related technologies. J Intell Manuf 1–56. https://doi.org/10.1007/s10845-018-1433-8
Patel P, Pavitt K (1991) Large firms in the production of the world's technology: an important case of "non-globalisation". J Int Bus Stud 22:1–21
Pavitt K (1985) Patent statistics as indicators of innovative activities: Possibilities and problems. Scientometrics 7:77–99
Perkmann M, Tartari V, McKelvey M, Autio E, Broström A, D'Este P, Fini R, Geuna A, Grimaldi R, Hughes A, Kitson M, Krabel S, Llerena P, Lissoni F, Salter A, Sobrero M (2013) Academic engagement and commercialisation; a review of the literature on university-industry relations. Res Policy 42:423–442. https://doi.org/10.2139/ssrn.2088253
Price WJ, Bass LW (1969) Scientific research and the innovative process. Science 164:802–806. https://doi.org/10.1126/science.164.3881.802
Rammer C, Doherr T, Krieger B, Marks H, Niggemann H, Peters B, Schubert T, Trunschke M, von der Burg J (2021) Innovationen in der deutschen Wirtschaft. Indikatorenbericht zur Innovationserhebung 2020. ZEW, Mannheim
Rappa MA, Debackere K (1992) Technological communities and the diffusion of knowledge. R&D Manag 22:209–220
Reiss T (2001) Drug discovery of the future: the implications of the human genome project. Trends Biotechnol 19:496–499. https://doi.org/10.1016/s0167-7799(01)01811-x
Romer PM (1990) Endogenous technological change. J Polit Econ 98:S71–S102
Rosenberg N (1976) Perspectives on technology. Cambridge University Press, Cambridge

Rosenberg N (2010) Why do firms do basic research (with their own money)? In: Studies on science and the innovation process, pp 225–234
Schmoch U (1997) Indicators and the relations between science and technology. Scientometrics 38:103–116
Schmoch U (2003) Service marks as novel innovation indicator. Res Eval 12:149–156
Schmoch U (2007) Double-boom cycles and the comeback of science-push and market-pull. Res Policy 36:1000–1015
Schmoch U (2008) Concept of a technology classification for country comparisons. Final report to the World Intellectual Property Organization (WIPO). Fraunhofer ISI, Karlsruhe
Schmoch U, Gauch S (2009) Service marks as indicators for innovation in knowledge-based services. Res Eval 18:323–335
Schmoch U, Gehrke B (2022) China's technological performance as reflected in patents. Scientometrics 127:299–317
Schmoch U, Khan M (2019) Methodological challenges for creating accurate patent indicators. In: Glänzel W, Moed HF, Schmoch U, Thelwall M (eds) Springer handbook of science and technology indicators. Springer, Cham, pp 907–927
Schmoch U, Thielmann A (2012) Cyclical long-term development of complex technologies—premature expectations in nanotechnology? Res Eval 21:126–135
Schmoch U, Grupp H, Mannsbart W, Schwitalla B (1988) Technikprognosen mit Patentindikatoren. TÜV Rheinland, Köln
Schmoch U, Grupp H, Reiss T, Strauss E (1991) Monitoring science-based biotechnology by means of patent indicators. Res Eval 1:61–68. https://doi.org/10.1093/rev/1.2.61
Schmoch U, Licht G, Reinhard M (2000) Wissens- und Technologietransfer in Deutschland. Fraunhofer IRB Verlag, Stuttgart
Schmoch U, Laville F, Patel P, Frietsch R (2003) Linking technology areas to industrial sectors; Final report to the European Commission, DG Research. Fraunhofer ISI, OST, SPRU, Karlsruhe, Paris, Brighton
Schmookler J (1966) Innovation and economic growth. Harvard University Press, Cambridge, MA
Schumpeter J (1942) Capitalism, socialism and democracy. Harper & Brothers, New York
Schumpeter JA (1997) Theorie der wirtschaftlichen Entwicklung—Eine Untersuchung über Unternehmergewinn, Kapital, Zins und den Konjunkturzyklus. Duncker & Humblot, Berlin
Sigurdson J (ed) (1990) Measuring the dynamics of technological change. Pinter, London
Solow RM (1956) A contribution to the theory of economic growth. Q J Econ 70:65–94
Stroetmann KA (1977) Innovation, economic change and technology policies. In: Proceeding of a seminar on technological innovation. Birkhäuser, Basel
Sundbo J (1995) Three paradigms in innovation theory. Sci Public Policy 22:399–410
Sundbo J (1997) Management of innovation in services. Serv Ind J 17:432–455
Sundbo J (1998) The theory of innovation—entrepreneurs, technology and strategy. Edward Elgar, Cheltenham
Sundbo J (2001) The strategic management of innovation; a sociological and economic theory. Edward Elgar, Northampton, MA
Sundbo J, Gallouj F (1998) Innovation in services. SI4S Project synthesis paper
Teece DJ (1986) Profiting from technological innovation: Implications for integration, collaboration, licensing and public policy. Res Policy 15(6):285–305
Teece DJ (1998) Capturing value from knowledge assets: the new economy, markets for know-how, and intangible assets. Calif Manag Rev 40(3):55–79
Teece DJ (2007) Explicating dynamic capabilities; The nature and microfoundations of (sustainable) enterprise performance. Strateg Manag J 28:1319–1350
Thelwall M (2019) Online indicators for non-standard academic outputs. In: Glänzel W, Moed HF, Schmoch U, Thelwall M (eds) Springer handbook of science and technology indicators. Springer, Cham, pp 835–856
Thursby J, Thursby M (2006) Research and development. Where is the new science in corporate R&D? Science (New York, NY) 314:1547–1548. https://doi.org/10.1126/science.1134257

Tidd J, Hull FM (2003) Service innovation: organizational responses to technological opportunities & market imperatives. Imperial College Press, London
Tijssen RJW (2006) Universities and industrially relevant science: Towards measurement models and indicators of entrepreneurial orientation. Res Policy 35:1569–1585
UNCTAD (1992) World investment report 1992; transnational corporations as engines of growth. UNCTAD, Geneva, Switzerland
Utterback JM, Abernathy WJ (1975) A dynamic model of process and product innovation. Omega 3:639–655
van der Panne G (2007) Issues in measuring innovation. Scientometrics 71:495–507. https://doi.org/10.1007/s11192-007-1691-2
van Raan AFJ (ed) (1989) Science and technology indicators: Their use in science policy and their role in science studies. In: Selected proceedings of the first international workshop on science and technology indicators. DSWO Press, University of Leiden, Leiden
van Raan AF (2005) Measuring science. In: Moed HF, Glänzel W, Schmoch U (eds) Handbook of quantitative science and technology research: the use of publication and patent statistics in studies of S&T systems. Springer, Dordrecht, pp 19–50
Verbeek A, Debackere K, Luwel M, Zimmermann E (2002) Measuring progress and evolution in science and technology—I: The multiple uses of bibliometric indicators. Int J Manag Rev 4:179–211. https://doi.org/10.1111/1468-2370.00083
von Hippel E (1988) The sources of innovation. Oxford University Press, New York
von Hippel E, Krogh G (2013) Open innovation and the private-collective model for innovation incentives. In: Dreyfuss R, Strandburg K (eds) The law and theory of trade secrecy: A handbook of contemporary research. Edward Elgar, Cheltenham, pp 201–221
Warnke P, Koschatzky K, Dönitz E, Zenker A, Stahlecker T, Som O, Cuhls K, Güth S (2016) Opening up the innovation system framework towards new actors and institutions. Fraunhofer ISI, Karlsruhe
Weber M (1972) Wirtschaft und Gesellschaft. Mohr, Tübingen
Weingart P (1975) Das Verhältnis von Wissenschaft und Technik im Wandel ihrer Institutionen. In: Stehr N, König R (eds) Wissenschaftssoziologie: Studien und Materialien. VS Verlag für Sozialwissenschaften, Wiesbaden, pp 393–418
Wilsdon J (2015) We need a measured approach to metrics. Nature 523:129. https://doi.org/10.1038/523129a
WIPO (2019) The 2019 world intellectual property report—the geography of innovation: local hotspots. Global Networks, WIPO, Geneva

Rainer Frietsch has business training in wholesale and import/export trade. Studied Social Science at the University of Mannheim. Ph.D. in Economics at the Karlsruhe Institute of Technology (KIT). Since October 2000 employed at the Fraunhofer Institute for Systems and Innovation Research. From 2008 till June 2019, he was the Deputy Head of the Competence Centre Policy—Industry—Innovation and Coordinator of the Business Unit Innovation Indicators. Since July 2019 he is the Head of the Competence Centre Innovation and Knowledge Economy. Since 2008 he is the Visiting Professor at the Institutes of Science and Development of the Chinese Academy of Sciences in Beijing.

Thomas Reiß is a senior scientist at the Fraunhofer Institute for Systems and Innovations Research ISI. He studied Biology and Chemistry at Freiburg University, Germany, and holds a doctoral degree in Biology. Following a post-doctoral fellowship at the DOE Plant Research Laboratory, Michigan State University (USA) he joined Fraunhofer ISI in 1987. From 1996 to 2004, he was the head of the department Innovations in Biotechnology, from 2005 to 2020 head of the Competence Centre Emerging Technologies. From August 2006 to March 2007 Thomas Reiß was the Deputy Head of Fraunhofer ISI.

Ulrich Schmoch holds a doctoral degree in Social and Political Sciences (University of Hannover), a diploma in Mechanical Engineering (University of Hannover) with modules at the University of Grenoble, France, and a state doctorate in Sociology of Science and Technology (Karlsruhe Institute of Technology KIT). He works at Fraunhofer ISI since 1986. From 1991 to 2002 he was the Deputy Head and from 2002 to 2004 Head of the department Technology Assessment and Innovation Strategies. Between February 2012 and September 2014 he served as stand-in professor in Science Management at the German University of Speyer, Germany.

Foresight: Fifty Years to Think Your Futures

Kerstin Cuhls, Ewa Dönitz, Lorenz Erdmann, Bruno Gransche, Simone Kimpeler, Elna Schirrmeister, and Philine Warnke

Abstract Foresight started with the notion of (Science and) Technology Foresight as a part of Technology Assessment (TA) but is now an independent scientific area. For a long time, Foresight mainly aimed at detecting determined futures but has developed into different directions under different headings. Most processes make use of a combination of methods to explore and develop different possible, probable or desirable futures. Foresight is more and more embedded in different institutions. There, it is carried out together with clients, and serves their specific purposes in the preparation of decision-making or science, technology and innovation policies. It offers the space to bring the different actors in the respective innovation system together. Foresight concepts are more and more accepted, and the results are distributed and used—in companies, ministries, associations, NGOs or the European Commission. Our contribution describes how Foresight has changed during the last 50 years and explains some of the aspects researchers have addressed. We conclude by highlighting two crosscutting emerging dimensions of change in Foresight, i.e. the engagement with transformative, mission-oriented agendas and the meaningful integration of machine-based approaches. Foresighters have to be aware that not only the results of their projects change but also the methods and the actors who work with them.

K. Cuhls (✉) · E. Dönitz · L. Erdmann · S. Kimpeler · E. Schirrmeister · P. Warnke
Fraunhofer Institute for Systems and Innovation Research ISI, Karlsruhe, Germany
e-mail: Kerstin.Cuhls@isi.fraunhofer.de; Ewa.Doenitz@isi.fraunhofer.de; lorenz.erdmann@isi.fraunhofer.de; Simone.Kimpeler@isi.fraunhofer.de; elna.schirrmeister@isi.fraunhofer.de; philine.warnke@isi.fraunhofer.de

B. Gransche
Karlsruhe Institute of Technology KIT, Karlsruhe, Germany
e-mail: bruno.gransche@kit.edu

J. Edler, R. Walz (eds.), *Systems and Innovation Research in Transition*, Sustainability and Innovation, https://doi.org/10.1007/978-3-031-66100-6_4

1 Introduction

Preparing for the future, dealing with futures' uncertainty and the question of what may be in store for us is what drives people to engage in looking forward. Like Futures Research, Foresight does not aim to predict a determined single future, but supports people in exploring alternative future pathways (Amara 1974; Kreibich 2006; Voros 2017, 2019): possible, probable and desirable futures open up different perspectives and approaches. With each perspective, different objectives in science, technology and innovation (STI) are addressed. "Foresight" in a broader sense started in operations research of the 1950s, especially in military support (RAND Corporation, Santa Monica, see e.g. Helmer 1975, 1983; Helmer and Rescher 1959. And with expert support (Dayé 2020) to consider the future of science and technology as a driver of the economy and societal development. During the 1960s, many researchers in the world assumed that it could be possible to develop world simulation models even though the first models at that time were only able to process a limited number of variables (Forrester 1961; Meadows et al. 1972). A milestone in this way of futures thinking was the report "Limits of Growth" written by the Club of Rome, which raised the awareness of resource limits. This coincided with new institutional foundations, for example, in Germany, the Fraunhofer Institute for System Technology and Innovation (ISI) or the Science and Policy Research Unit (SPRU) in the United Kingdom. Although working on the long-term view, the researchers in the early 1970s were already aware that predictive outlooks are assumptions about the future, mere means or "working material", not facts, and the major task of a Foresight researcher is to deal with uncertainties about future issues and transfer new ideas and technology into real life (Krupp 1972; ISI 1973).

While the existing simulation models were updated (see, e.g., in Meadows et al. 2009) and new models were added, the toolbox of searching for potential futures, assessing assumptions about or consequences of these different futures as well as developing new options has been growing tremendously over time and was discussed internationally, most visibly in the AGARD project for NATO in 1977 (Hetman 1977). Under the umbrella of "technology assessment" (see Heyen et al. 2024 in this anthology), many Foresight studies were performed during the 1970s and early 80s to assess the future impacts of certain technological developments on the economy, society and policy-making (see, e.g., Jochem 1973, 1975; Jochem and Wiesner 1977; Krupp 1984 and many others). The 1980s rather saw a decrease in science and technology-based foresight, whereas participatory future workshops (Zukunftswerkstätten) were still actively used in civil society and spatial planning (Jungk and Müllert 1987).

At that time, Foresight and Futures Research of all kinds were often perceived as predictive (Which future may come true?), as part of planning processes (strict planning in the sense of the "planning decade"), but also as policy and political processes (Flechtheim 1966; Jungk 1986; Toffler 1990; Steinmüller 2012, 2013, 2014a, b, Godet 1986; Radkau 2017; Seefried 2015). Even the notion of "social technology" was used (Helmer 1966), but sociologists were not very active in Foresight, "the analysis of the future has been ... neglected in sociological theory and research" (Mische 2009). First overview studies in early Foresight were scenarios (Kahn and Wiener 1977) or Delphi surveys (Helmer 1967, 1983). After a single broad study in the USA (Dalkey and Helmer 1963; Helmer 1983; Helmer-Hirschberg and Gordon 1964), Japan was the first country to establish regular Delphi surveys and later fully-fledged Foresight processes in science, technology, innovation and society on a national level (Cuhls 1998; Kuwahara et al. 2008; NISTEP 2019). In other countries, Delphi surveys fell into oblivion after harsh criticism by Sackman (1975), a criticism that rather aimed at the practical use of surveys, less at the method itself.

From the content point of view, science, technology and innovation were often at the forefront of Foresight-like activities—human-induced and described as drivers for "progress". Policy advice based on the findings and to support innovation policy beyond pure science and technology policy was and is still intertwined with Foresight. Triggered by Irvine and Martin's report on comparing Foresight in several countries in 1984, a broad way of conducting science and technology outlook studies with Key Technologies approaches to learn for the present attained momentum at the beginning of the 1990s and gained traction when the first national Foresight activities in Japan were taken over to be repeated at first in Germany, then in the UK, France and South Korea. This marks the beginning of national Foresight activities all over the world (Georghiou et al. 2008; Grupp 1999) and the enhancement of the methodological toolbox, encompassing analytical, participatory and anticipatory tools (Cuhls et al. 2002; Cuhls 2008; Glenn and Gordon 2009; Slaughter 2005; UNIDO 2002). First regional, national and international networks emerged. International collaborations started with the exchange of knowledge and were extended to framework contracts and nowadays cooperative online projects.

Several aspects and dimensions of Foresight have changed drastically over time, for example the development from more deterministic views to open and diverse future perspectives, or the tools and the way researchers collaborate and with whom, from single projects in teams to participative approaches, even integrating the clients or the general public. Foresight started struggling with the scarcity of data and is now integrating AI approaches in a flood of "big data". Foresight is more and more (also) working in virtual settings. With hindsight, we point at some of these developments during the last 50 years.

We structure the sections according to key aspects of change and proceed chronologically within the subsections. The key aspect in the first section explains how Foresight started without a theory by researchers who had the will to practically apply first methods. The terminology they used was still scattered and—at least in Germany—it took until 2020 to be officially acknowledged as a scientific area "futures research" at all. The second section describes the way towards a broad variety of perspectives and different conceptions of futures with open processes that also demand different methodological concepts. Section 2.3 highlights the evolution of participatory approaches within Foresight. At the same time, new ways of making use of data and information in the digital age developed (Sect. 2.4), leading to a certain degree of automation in Foresight. The underlying struggle between quantitative and qualitative approaches is illustrated in Sect. 2.5. Foresight has played different roles in innovation systems research and policy—in Sect. 2.6, we describe this development. Networks in Foresight supported all the developments (Sect. 2.7) and changed in nature over time. In the last section, we offer a summary and give a brief outlook on potential further developments of the field.

2 50 Years of Exploring Futures and Dealing with Uncertainty

Foresight—starting with the notion of (Science and) Technology Foresight as a part of technology assessment—has broadened and developed in different directions under different headings. Foresight is dealing with different, complex futures and some uncertainties in imagining and assessing the different issues or "things to come". The following sections describe some of the changes over time having selected different aspects and following them over historical times.

2.1 Towards a Scientific Discipline with Accepted Terminology

The terminology for Foresight, Futures Research or Futures Studies has always been in flux and developed in different communities all over the globe (cf. Gransche 2015). In France, "la prospective" emerged as a similar concept (Godet 1986, 2000). Since 1992, the term "Foresight" has been used more often in the sense Martin (1995) defined it: "(technology) foresight is the process involved in systematically attempting to look into the longer-term future of science, technology, the economy and society with the aim of identifying the areas of strategic research and the emerging of generic technologies likely to yield the greatest economic and social benefits" or Coates (1985, p. 30) who formulated more broadly "Foresight is the overall process of creating an understanding and appreciation of information generated by looking ahead. Foresight includes qualitative and quantitative means for monitoring clues and indicators of evolving trends and developments and is best and most useful when directly linked to the analysis of policy implications. Foresight prepares us to meet the needs and opportunities of the future. Foresight in government cannot define policy, but it can help condition policies to be more appropriate, more flexible, and more robust in their implementation, as times and circumstances change. Foresight is, therefore, closely tied to planning. It is not planning—merely a step in planning". Thus, Foresight is "the systematic debate of complex futures" (Cuhls 2013), contains much more than scenario building (see examples in Jochem et al. 2024 in this anthology) and includes full processes (EFFLA 2013).

The need to be clearer in the definition, to explain why Foresight needs an epistemological explanation or theory behind it, and the need to communicate this to others (disciplines, communities, institutions) became obvious when researchers of the field met at conferences worldwide and had to explain what they are doing. It was also observed during the 1990s that it was difficult to be accepted by other fields or by scientific disciplines. At this time, first projects, processes and communities started to reflect on this epistemological area. Whether or not Foresight should be a discipline that is taught at universities was not yet discussed and is even now, in 2024, up for debate.[1] But what became clear was that Foresight needs an epistemic frame to be understood and to be taught—otherwise, it would be difficult to apply for projects or to recruit personnel for further projects. During the 1990s, Foresight also became institutionalised in academic journals with rigorous peer reviews, dedicated conferences, specialised networks of professionals (see below) and acknowledged in public research organisations and universities but was still belittled as "unscientific" or an "art" (de Jouvenel 1967).

Within a European network (ASTPP-TSER, see Kuhlmann et al. 1999) and a special issue about "Foresight" in the Journal of Forecasting (Cuhls and Salo 2003), more agreement was achieved on terminology, especially concerning Foresight, Planning and Strategic or Anticipatory Intelligence. Foresight and Forecasting were competing notions at that time, and although there was an agreement to differentiate in framing both notions in the EU context (Cuhls 2003; Kuhlmann et al. 1999), in many regions of the world, Forecasting and Foresight terminology remained the same. In the first decades of this century, "Futures Studies" has been more and more used in other regions of the world (especially in Australia, the term is congruent with Foresight, see, e.g., Slaughter 2005, etc.) by researchers organised in the World Futures Studies Federation (WFSF). "Futures Literacy" (Miller 2018) and "Anticipation Studies" (Poli 2017, 2019) represent another part of the more recent communities in rather qualitative future-oriented studies framing futures thinking and working on the present by "using the future" (Miller 2018). For 30 years, the UNESCO has been active in Foresight and since 2012 has gained influence again (https://en.unesco.org/futuresliteracy/about). Since then, UNESCO chairs have been nominated in all regions of the world and a High-level Committee on Programmes (HLCP) Strategic Foresight Network was established in 2020 to coordinate global activities (https://unsceb.org/foresight-network). The discussions on Futurology or Futures Research as "a real discipline" in the scientific sense or as a scientific subject (Seefried 2015; Steinmüller 2012, 2013, 2014a) are thus still ongoing and the research community is highly interdisciplinary. They rarely use the term "anticipation", which experiences a renaissance in the context of RRI and mission-oriented innovation policy (OPSI 2020; Poli 2017).

With time, the number of communities and scientific literature in Foresight has increased. The first journals, in which Foresight results were published, came from

[1] For example, the Bundesakademie für Sicherheitspolitik (BAKS) started new online events called "Foresight Breakfast" with the question if Foresight needs to be a discipline (May 10, 2022, 8:30).

operations research, modelling and quantitative approaches (*Long-Range Planning* since 1968, *European Journal of Operational Research* since 1977 or the *International Journal of Forecasting* which started in 1985), or Business Administration journals (e.g. *Business Horizons* since 1957; *Journal of Business Research* since 1973). Later, a more strategic focus was in the forefront of publications (like the *Strategic Management Journal* since 1980 or *Technology Analysis & Strategic Management* since 1989). Foresight articles were also included in policy-oriented journals (like in *Research Policy*, which started in 1971) and more general societal, economic, political and technology discussions (*Technological Forecasting & Social Change* since 1969, *Futures* since 1968 and *Foresight* since 1999, or the *Russian Foresight Journal*), often with a partly quantitative "touch" were established. Others have a more technological or engineering direction like *Technovation* since 1981, or *IEEE Transactions on Engineering Management* since 1968.

To establish a community and to further strengthen and institutionalise Foresight towards a "scientific discipline", the *European Journal of Futures Research* was founded in 2013 at the same time as the *German Zeitschrift für Zukunftsforschung*. The latest journal is *Futures & Foresight Science*, which started in 2019. All three journals address methods, theoretical approaches and scientific projects on futures.

Since 2019, Futures Research is an official "Small Scientific Subject" acknowledged by the German Federal Ministry for Research and Education (BMBF) even though chairs at universities labelled "Foresight" are still rare. There is no UNESCO Chair of Futures Literacy or similar in Germany yet, although there are more and more departments at universities offering courses or integrating Foresight knowledge—as well as many faculties working in Foresight-like thematic areas without calling it Foresight (e.g. in environmental studies, climate change modelling, innovation research or philosophy).

In sum, the field of Foresight has seen a growing institutionalisation in different research organisations and into a scientific area acknowledged by universities. Foresight and futures research are now highly differentiated and have developed from more forecasting-like activities in models, assessments of future topics or clear trend lines to methods working with open and varied future perspectives and combinations thereof.

2.2 From Detecting "the" Future to Exploring Multiple Open Futures

Looking back on looking forward, two major approaches to Foresight can be distinguished (van Asselt et al. 2012: 24–25): The *predictive approach* to Foresight (forecasting) dominated from the 1950s to the 1970s—and is still prevalent today (Kreibich 2007c), in particular in the realm of business economics and macro-economics, and the *explorative approach* to Foresight that surveys multiple futures (scenarios) which emerged in the 1960s (e.g. Godet 1986). Foresight researchers worked towards a view that sees the future as essentially open, and Foresight processes as a way of reflecting on possible futures to derive insights for the present. The first modern futures science and futures research approaches emerged during the 1930s–1940s in the USA. The pioneers of "Futurology", children of their time, assumed a linear trajectory of human development and tried to predict that trajectory accurately. Not surprisingly, the focus was on the perceived main driver of change of modern societies: scientific and technological progress, its expectations and assessments. The 1950s were the time, when the RAND Corporation—founded in 1948—developed scenarios (Kahn 1977; Kahn and Wiener 1977), Delphi surveys (Dalkey 1968; Dalkey and Helmer 1963; Helmer 1983), models and simulations.

The original "forecasting approaches" (Cetron 1970) broadened into general societal questions, and emerging science and technology were identified (Helmer and Rescher 1959; Jantsch 1967; de Jouvenel 1967). In the 1960s, the early years of Foresight and Futures planning in West Germany, the Batelle-Institute was leading in the development of new methods, combining quantitative and qualitative approaches. In East Germany, Foresight as such did not exist (Steinmüller 2014a) but deterministic future planning was an integral part of state policy. All over the world, different prognostic (Beinhauer and Schmacke 1970; Picht 1967) and forecasting approaches (Martino 1983) were tested. It took until the 1960s, when futures conceptions were developed more openly, identifying impacts and risks of the technological-economic dynamics were brought into play mainly by citizen movements (Kreibich 2007a, b: 177–181).

Although not explicitly framed as a scenario study, "The Limits of Growth" explored different futures through various prospective simulations (Meadows et al. 1972). It is considered to be one of the earliest scenario studies in policy-oriented Foresight (Kosow and Gaßner 2008; van Asselt et al. 2012). It is no coincidence that at the time, when global environmental and climate concerns were first systematically assessed and widely modelled for contributing to the creation of new policies and attitudinal changes. Research institutes with similar missions like the Fraunhofer ISI (in Germany) and SPRU (in the UK) were founded. Following the example of other international institutions, innovation researchers already worked with world models and "modelling the future" approaches, mainly by extrapolating variables from the present (Krupp 1972). The ISI founding proposal (Krupp 1972) mentions the collection, inventory and assessment of research issues, connected with technological foresight and technology transfer, as a principal task. Accordingly, complex systems were to be viewed with system technology methods, in particular dynamic simulation, to fill a gap for future modelling in applied research at that time (ISI 1973; Krupp 1972), which required interdisciplinary competences.

The earliest projects in the 1970s were still US-dominated, but first German contributions to a world model with decision layers existed, and technology assessment with the support of dynamic simulation started (Bossel and Hughes 1973). At the same time, the relevance of technological forecasts for industrial long-term planning was reflected (Fischer 1976). Not surprisingly, there was no uniform reasoning about the future at that time. While the exploration of the potential of deterministic simulations was attempted, the limitations of dynamic systems modelling to represent complex systems have been apparent right from the beginning. The contingency of perceptions and of future developments was already considered. However, at that time, researchers searched for *likely* scenarios in-between "cornerstone scenarios" and the deliberation of viewpoints rather than fully-fledged participatory scenario processes (e.g. on a morphological basis or at least thinking in alternative pathways into the future) embracing the uncertainties in assessing "the future".

The following three examples illustrate how uncertainty was dealt with in the early 1970s: In a study on waste, three alternative future scenarios were built based on an analysis of statistics and trends—however without describing in detail how[2] and to what end the scenarios were built (Jochem 1973). The implications were discussed and assessed against a set of criteria. The author concluded that the most likely scenario would be somewhere in-between. From today's point of view, these scenarios are the single author's personal plausible assumptions on how particular single trends or developments could unfold in the future, but at that time, it was already a huge achievement to figure out futures at all. In 1976, "chemical scenarios" were published, that already used a terminology similar to environmental factors ("Umfeldbereiche") and included different experts in the scenario creation (Batelle Institut e.V. 1976).

In a second project, a value-controlled decision-making process on energy policy was modelled and simulated (Bossel and Hughes 1973) with the flow of decision-making represented as a causal logical sequence (e.g. "Is Dissatisfaction Less Than 100?"—YES/NO, p 73). In addition, three scenarios were built to account for the complexity of the energy system: worst case, best case and most likely case. These scenarios had the nature of plausible sets of parameter values, and it was simulated how these assumptions propagated to the results. According to the authors, who were aware of the pioneering character of their work, "A great deal remains to be done in the creation of a fully acceptable value-based decision-making structure" (Bossel and Hughes 1973). This kind of Foresight work was not very actively followed in the next 10 years and the communities split into those who worked on a sectoral basis (mainly in the field of energy) with new ways of modelling, and those who were searching for future issues in a broader sense.

In the 1980s, new indicators for measuring progress in science and technology, competitiveness studies and innovation for a new ecosociety (Krupp 1989) were paramount. Indicators described the presence, the here and now. They could only be extrapolated into one single future. Country comparisons and outlooks for the different regions of the world (Bierhals 1980; Grupp et al. 1987)—always with a view to staying competitive ("Standortpolitik")—were started to guarantee growth and jobs. Under the heading of "early detection of technologies" (Technologiefrüherkennung) the development levels of industrial countries were

[2] Today, one would ask if they are constructed as desk research or with expert participation, combining different pathways systematically or just assuming one most likely pathway etc.

measured by technometrics, patent analysis, bibliometric studies and literature reviews (e.g. Grupp et al. 1987; Grupp 1997; Schmoch 1988, 1990, see also Frietsch et al. 2024 in this anthology). The community of futures researchers was split into those coming from academia, making use of indicators and data, and others who emphasised the diverse interests of actors in different futures and the empowerment of civil society in shaping futures (like Flechtheim or Jungk, see, e.g., Flechtheim 1990; Jungk 1983, 1986; Seefried 2015; Steinmüller 2014a).

During the later years of the 1980s with the experience of nuclear accidents (esp. Harrisburg, Chernobyl), the discussions shifted towards problems and negative effects of technology. But it remained obvious that also the opportunities of science and technology needed attention. Researchers started to learn from other countries, especially Japan, which was en vogue at that time—partly propagated as a threat, a competitor on global markets, and partly admired, as Japanese scientists were successful in learning from others and getting more out of the lessons. Thus, the idea was to learn from the USA and Japan by performing Foresight—regarded as a positive view towards futures, to induce complementary thinking to the more negative aspects often dealt with in TA and to learn their way of shaping futures with innovation.

Therefore, in the 1990s, Delphi expert surveys[3] were added to the German and later to other European countries' Foresight repertoire (again), this time on a larger level and with communication purposes. Studies on the national level to gain an overview of future science and technology developments and their potential became well-known, especially in industry. Corporate and organisational Foresight also spread (Gordon et al. 2020) with the aim of competitiveness in a global world and the identification of strengths and weaknesses of the innovation system to support "basic technologies" or foster those science areas in one's own country. As in Delphi surveys, the time horizon of realisation is one major estimation to ask for, the method was criticised as being predictive and/or too specific. This led to the attempt of integrating a "megatrend" assessment to analyse the participants' views on large societal developments as well (Cuhls et al. 1998, 2002). The factor analysis on the megatrends made it also possible to raise the question if there were hidden biases or values behind the judgements on science and technology in general (Blind et al. 2001).

But political questions changed and so did the methods in Foresight—from more forecasting and focusing on the supply side of science and technology to problem- and demand-driven approaches. The German national activities of the 2000s started with more society-oriented (demand-driven) Foresight attempts (e.g. FUTUR of the BMBF, see Cuhls and Georghiou 2004), integrating different participants, broadening the scope of issues in question and even re-uniting the technology assessment of the German TAB with "Future Reports" (ITAS and Fraunhofer ISI 2002). In 2007,

[3] Expert surveys with future statements assessed in several so-called rounds, in which later rounds feedback was given on the assessments so that the same experts could assess once more with the psychological anchor of other results. These surveys started a large communication with many experts from different backgrounds.

BMBF started a first Foresight "programme" to broaden its range of interdisciplinary research themes beginning with the technology fields of the German High-Tech Strategy (Cuhls et al. 2009a, b). In 2012, the second, more demand-driven approach followed (Warnke and Schirrmeister 2016; Zweck et al. 2015a, b). The third German Foresight programme is combined with a "Horizon Scanning", as the search part of Foresight is called meanwhile, supported by a Committee (Zukunftskreis) identifying thematic issues relevant for diving deeper.[4]

One of the major methods in Foresight were and are scenarios. As there are many different scenario methods and "schools" (see for example Bradfield et al. 2005; Fink et al. 2001; Fink and Siebe 2016; Kosow and Gaßner 2008) in Foresight, most of the scenario methods use different inputs to combine future paths into coherent and plausible images of the future. Morphological approaches with different key factors (or drivers) and their potential developments into the future (named also: assumptions, options, projections) involving just data or experts or literature or other sources illustrate how the different pieces of future paths can be combined to new coherent pictures. Scenario work and thinking in different future paths, in fully-fledged processes or as scenario sprints, are the backbone of many Foresight processes nowadays in Germany and all over the world.

On the European Union level, Foresight in RTDI developed along similar lines as in Germany, but started later. Systems dynamics and other models were used for the long-term view of the interconnections in and between different systems in single projects. The same is true for qualitative scenario work, which started in a division called K2, a division that was given up later. The European Forum of Forward-Looking Activities brought Foresight back into the Commission processes in 2012 and the following years. Foresight gained attention by being used in the preparation of Horizon 2020, the Framework programme on RTDI of the European Commission. There were already many qualitative approaches available to identify new topics for science, technology and innovation (policy). Open, facilitated workshops for systematic Foresight were used in the preparation of Horizon Europe, the 9th Framework Programme. All of these processes used assumptions about different futures under uncertainty—to identify futures, to prepare for, or to make them possible, but less to predict them. The European Commission itself got more and more involved in the Foresight processes via workshops, interviews or internal consultations, expert consultations (European Commission 2017b), and the European Commission's Foresight Correspondent's Network, which was later established as the Horizon Europe Network (Cuhls et al. 2021; European Commission 2017a; Kimpeler et al. 2022) to support in-house processes. With a new Commissioner responsible for Foresight, the European Commission broadened its strategic Foresight work and meanwhile also publishes regular reports (European Commission 2020, 2021, 2022).

[4] https://www.bmbf.de/bmbf/de/forschung/zukunftstrends/foresight/mit-foresight-in-die-zukunft-schauen.html

To recapitulate, Foresight started with the expectations of detecting "the" determined "future" but is now—and partly always has been—a debate about futures under uncertainty, with remaining unknowns, not predictive, but about understanding and handling uncertainties, preparing decisions under uncertainty as well as working with different futures (often in form of different kinds of "scenarios"), less with "worst case" and "best case" but with different plausible scenarios "somewhere in between or beyond". Starting with expert involvement, the stakeholder groups that take part in the processes have diversified and expanded over time.

2.3 From Narrow to Broad Perspectives: Evolving Demand for Participation

The step-by-step integration of Foresight activities into on-going policy-making and strategy development routines over the past 50 years has developed towards opening up Foresight for different types of actors (Cuhls 2003). Before, it was mainly based on desk research of intelligence units and selected expert assumptions, used for policy or business consultation and planning in the 1970s. Scenarios and futures work were often regarded as part of technology assessment (Verein Deutscher Ingenieure 1991, see Heyen et al. 2024 in this anthology) in the 1970s and 80s. The opportunities and limits of technology assessment (TA) were reflected early on (Jochem 1975) with the impossibility of an "objective or neutral" TA (e.g. causal relationships are only partially known), methodological difficulties (e.g. subjective impact perceptions) and technocratic versus democratic management of technology (e.g. negligence of poorly organised groups). This calls for a participatory process approach. It is argued that (1) the integration of stakeholders benefits the identification of unintended and unforeseen side effects, (2) group-specific views of possible developments differ and (3) controversial steps in TA could be resolved through participation (Jochem 1975).

Accordingly, participatory Foresight processes are future-oriented activities that encourage the integrated, focused engagement of interdisciplinary experts, stakeholders and citizens at multiple points in the Foresight research process and recognise interactively created artefacts as an important mode of developing and communicating "bottom-up" imaginaries of the future and their inherent diverse aspirations. The participatory turn is similar to that in technology assessment (see Heyen et al. 2024 in this anthology). Different from TA studies, Foresight processes are creating these images. The plausible images of the future and the paths to them should be described in easy-to-understand narratives and in a way that is comprehensive, inspiring, evoking interest, provocative and nice to read. The science behind it involves the addressees in the creation of the images and narratives and combines it with data and information pieces that already exist or are assumed to be possibly based on current findings. Information could be gained from experts and stakeholders—and that is why they are often included in scenario processes or surveys (e.g. Delphi surveys).

Despite the identified shortcoming of a Foresight process that uses scenarios based on a few expert opinions only, i.e. to illustrate impacts of certain technologies, policies or actions at a systems level (Jochem 1975), participatory approaches were not at the centre of methodological improvements in the 1970s and 80s. During the formation years of the Innovation Systems approach in the mid-1980s (Irvine and Martin 1984), the quantitative analysis of the technological performance of national or regional economies was a main interest in "prospective analysis" (Grupp et al. 1987; Irvine and Martin 1984; Schmoch 1988). Consequently, the heuristic innovation system model attributes a key role in technological progress to actors from science and industry, and policy actors set the framework conditions. Other groups of actors are only considered as stakeholders, consumers or technology users (Kuhlmann and Arnold 2001).

One merit of the systems approach was that the need for interdisciplinary knowledge and for stakeholder involvement in Foresight was slowly increasing. It reflected the understanding that different actors and their roles are key for the development, diffusion and use of innovation (Dosi et al. 1988; Edquist 1997). Therefore, expert surveys like large Delphi surveys, for example, the Delphi 1993, 1995 and Delphi '98, spread not only in Germany (BMFT 1993; Bundesministerium für Forschung und Technologie 1993; Cuhls et al. 1996, 1998), but also in Hungary, France and the UK (Georghiou et al. 2008)). Alternatively, the so-called Key Technology Lists with national priorities were created, e.g. in Germany, the Czech Republic or France (Grupp 1994; Klusacek 2002; Ministère de l'Économie 2006; Wagner and Popper 2003), including international comparisons (Cuhls et al. 2002). Delphi surveys were already performed in the 1960s in the USA as large communicative and participative surveys. Since 1970 every 5 years surveys have been carried out in Japan without knowing the notion of the "innovation system", but they have been regarded as a tool which has taken into account the opinions of different (if possible all) actor groups in the innovation system.

Interviews with stakeholder representatives like in the INTERDIS Project in the 1990s were conducted to gain deeper insights and involve a broader audience—as the participants in surveys and interviews were also the multipliers of the processes (Schmoch et al. 1996). With time, the notion "expert" broadened, and more and more citizen involvement was asked for by the European Commission or the national ministries. In the 1990s, participatory Foresight gained momentum, led to some participatory frameworks (Inayatullah 2000; Rosa et al. 2018) and was even discussed in high-level circles in the European Commission (Brussels talks on opening up Foresight processes: Cuhls 2002). Since the beginning of this century, societies have increasingly faced complex, interrelated, grand challenges with high uncertainty. The realisation spread that more actors than knowledge providers need to be involved in foresight (Cuhls 2000, 2003; Martin and Johnston 1999) and that led to a new German Foresight approach: The first FUTUR activity tried to involve stakeholders "from bottom-up" in a workshop-based process combining "open space" and "focus group" activities with heterogeneous actors. The FUTUR conference on participation in other countries' Foresight processes (Cuhls and Jaspers 2004) demonstrated the huge interest in experiments with participative Foresight and citizen involvement in countries like Sweden or the Netherlands. From then on, more and more actors with a broad and general, interdisciplinary view were involved in Foresight processes to discuss and find solutions for the broad societal challenges, mainly via conceptualised workshops or surveys.

With the emerging concept of "Responsible Research and Innovation (RRI)" (see Bührer et al. 2024 in this anthology) and its guiding principles of inclusiveness, anticipation, reflexivity and responsiveness, more actors in the system, namely science and society, were supposed to work intensively together to develop responses to the grand societal challenges and to identify new challenges. The idea of harnessing the wisdom of the crowd, in the sense of "many people know more", even if they are not specialists (Surowiecki 2005) like in the large Delphi surveys (Belton et al. 2021), was gaining momentum, not only in open innovation activities in industry (Baldwin and von Hippel 2011) but also in innovation systems research, in particular for energy and sustainability transitions. As a consequence, the understanding of actor constellations and their specific roles in innovation systems was revised (Warnke et al. 2016) and led to the concept of *open organisational Foresight* (Wiener 2018). In Germany, the BMBF Foresight cycles I and II mobilised different actors, new types of experts and citizens for emerging trends, and initiated dialogues between them (Warnke and Schirrmeister 2016), for example in "strategic dialogues" (BMBF Foresight Process I) and different co-creative workshop formats to support mission-oriented (innovation) policy.

The development illustrates that it is not only crucial to use a wide range of sources to search for signals of continuity or change in Foresight, but also to include a variety of perspectives and knowledge from different backgrounds, key actors and affected stakeholder groups in developing visions, analysing alternative futures and developing the necessary actions. Opening up Foresight to more participants and to involve stakeholder representatives and citizens as experts of everyday life has two objectives: first, to provide a more valid knowledge base for strategic decision-making by harnessing the cognitive diversity and varying perspectives of different actors; and second, to broaden the engagement of different actors in agenda-setting, prioritisation of action needs, development of ideas for solutions, and in the implementation of measures (Rosa et al. 2021; Nikolova 2014). In addition to the benefits of a more valid information base, despite uncertainties, there is also a societal function of participatory Foresight. It can support pluralistic information and communication about emerging societal challenges, "which builds political coherence and trust, increases commitment to joint action and reduces resistance to change (...), and highlights tensions in society in a constructive way" (Committee for the Future 2020). Specifically the societal function of participatory Foresight nowadays receives increased attention, in particular in the context of resilience in times of crises (Kononenko 2021).

The key to participatory Foresight that builds upon citizen engagement is the strengthening of people's capacity to recognise and embrace uncertainty while collectively shaping a preferable vision of the future. Engaging with citizens in

Foresight requires specific methods suitable for the different phases of a Foresight process, for the development of joint visions, for dialogues to share perspectives and priorities and for scenario generation and analysis to identify challenges or develop ideas for actions (Rosa et al. 2021). Visioning methods can support diverse groups of actors, including citizens, to develop a shared vision of their preferred future as a community, often shaped by normative principles, e.g., well-being or sustainability (Rosa et al. 2018). Futures dialogue methods enable discussions with a strict future-orientation on complex issues like bioeconomy or sustainable consumption (Kimpeler et al. 2022; Zweck et al. 2015b). Co-created storylines for consistent scenarios about possible futures in everyday life of the people are more comprehensible than expert reports. They serve as narrations in the communication about possible futures (Kimpeler et al. 2021). Examples like the storyboards on the application of Artificial Intelligence in new environments (see, e.g., https://www.uba-ki-storyboard.de) or the bioeconomy scenario stories "How do we want to live in the future?" can evoke discussions to negotiate desirable futures, be it in workshops with young people, in museum exhibitions, or other arenas (Kimpeler et al. 2021). These different forms of engagement can be considered at multiple scales of governance, from local communities or regions, to national, European or global levels (Rosa et al. 2021).

In addition, a gradual expansion of expert assumptions and advice towards more co-generation of knowledge about possible futures and collaboration in the development of ideas for action to tackle the challenges can be observed. This has taken a long time to become acknowledged and is aligned with the engagement of more people in the co-creation of the urgently needed societal transitions to meet the UN Sustainable Development Goals. One example is the participation process for the co-creation of ideas for action in the context of mission-oriented innovation policies for the German High-Tech Strategy (Trénel et al. 2020).

For this kind of co-thinking and co-creation of futures, a thorough stakeholder analysis that identifies not only dominant but also affected and "silent" system stakeholders (Haegeman et al. 2012; Schmidt et al. 2020) is crucial. It should be the first step of any participatory Foresight process. A particular emphasis is on recognising potential future stakeholders who may emerge along with the change of the system (dormant or latent stakeholders, Clausen et al. 2020), stakeholders with a particular role in transformations (Lyon et al. 2020) and long-term stakeholders such as future generations. Also, as we are dealing with complex emergent and transforming systems, a continuous critical reflection of system boundaries and subsequent revision of stakeholder mapping is crucial (Achterkamp and Vos 2007).

Due to the urgency of change in society as a whole, there is a demand for participatory approaches in Foresight that go beyond describing trends or exploring the space of possibilities and opportunities. Approaches in the 2020s try to support the various actors and experts in critically reflecting on their expectations for the future, and learning from each other to be able to shape transformations together. Participatory Foresight is asked for to include citizens in critical thinking about futures and the co-creation of goals, priorities and actions for mission-oriented policies. This is essential for bridging citizen needs and policy requirements to finally increase the reflexivity of innovation systems.

2.4 *Towards Conscious Integration of Machine-Learning Based Approaches*

In the 1980s and 90s, literature and patent data were the most important data publicly available to analyse innovation activity (Blind et al. 1999, 2001; Cuhls 1998; Cuhls et al. 2002; Grupp 1997; Schmoch 1990). Technology Foresight was closely interlinked with these activities and extensively used scientometrics methods along with expert interviews and surveys. In addition to the Delphi surveys for the BMBF or the EU (Bundesministerium für Forschung und Technologie 1993; Cuhls et al. 1996, 1998; Dreher et al. 2005; Grupp 1995), the production innovation survey, later on the European Manufacturing Surveys (EMS), can be mentioned as extensive primary surveys connected to Foresight (see Lerch and Jäger 2024 in this anthology).

Since the 1990s and even more so the beginning of the 2000s, researchers have explored the mining of more diverse but structured and clearly defined databases for environmental scanning and Horizon Scanning, for example in Cuhls et al. (1995). Bibliometric analyses offer the possibility to broaden the perspective and to include scientific developments outside the patent realm such as social sciences. At the same time, the use of new data sources became possible with much less effort than the collection of primary data, allowing the definition of keywords for further searches even in different languages (Cuhls et al. 1995, 2009a, b). To enlarge the diversity of sources even further, Foresight researchers combine qualitative and quantitative sources and pay closer attention to the biases at play. As an example, in the second BMBF Foresight process, the Fraunhofer ISI team developed a sophisticated approach for working with fringe sources to identify seeds of change across fields of human needs (Warnke and Schirrmeister 2016).

At the same time, the very notion of objectively "observing" signals of change is challenged. Especially in Finland, a lively discourse emerged to advance capacities for identifying and assessing "weak signals" (Hiltunen 2010; Ilmola and Kuusi 2006 based on the early work of Igor Ansoff (1975)). Others investigated the role of rare and unexpected high-impact events under labels such as "wildcards" or "black swans" (Taleb 2007).

Another group of Futurists and Foresighters emphasised challenging of anticipatory assumptions in order to recognise "change in the conditions of change" (Miller 2007). Foresight researchers increasingly adopt a constructivist approach towards futures thinking (Rossel 2012) and subsequently focus on the cognitive framings determining the way futures are perceived (Schirrmeister et al. 2019, 2020) and build on seminal work in cognitive science (Gigerenzer and Gaissmaier 2011;

Tversky and Kahneman 1974). Human perception can be distorted by such biases at the individual, group or organisational level.

In addition, data artefacts that root in signal thresholds or decision rules coded into the algorithms can be misleading when judged by humans. The deliberate explication of such biases and distortions has become a focus of Foresight researchers within the last 10 years (Day and Schoemaker 2004).

The availability of new, unstructured data in different sources and the possibility of using this data via machine-learning algorithms have developed very dynamically in the last 5 years. Algorithms to analyse patterns in unstructured data sources such as natural language processing (NLP), topic-modelling such as Latent Dirichlet Allocation (LDA) and deep learning methods based on artificial neural networks are increasingly available (Daas and van der Doef 2020; LeCun et al. 2015; Muhlroth and Grottke 2022; Porter 2019). These new possibilities are incorporated mainly at the beginning of the Foresight process. One example is the "hybrid AI-expert foresight framework" (Geurts et al. 2021), which serves to support the questioning of anticipatory assumptions by feeding a wider range of (emerging) aspects into users' conceptualisations of the future. In current and recently completed projects, researchers extensively analyse unstructured data (especially news sites) without a predefined search realm and use the results as a starting point and reflection opportunity for future dialogues with experts. However, in most cases, the outcomes require substantial human sense-making activities. In addition, the development of quality criteria for AI-based approaches to deliver futures insights is still under way.

In the future, the development and choice of suitable algorithms will be a key challenge for generating meaningful insights for Foresight processes in machine-based approaches. The composition of sources for the search with sufficient diversity and quality is similarly important. It will be key to generate results in a format that lends itself to sense-making and assessment by humans in participatory futures dialogues. Examples are projects like the Radical Innovation Breakthrough Inquirer (Warnke et al. 2018) or Fraunhofer Foresight (Ganz and Schirrmeister 2019) where future technologies and other potential innovations were identified semi-automatically by algorithms and then assessed in dialogues by human beings. In these "sense-making processes" humans excel with their abilities to identify relationships across concepts and generating novel ideas. The cognitive biases of human perception can partly be counter-acted by the machine-based input. But AI algorithms and the underlying databases are also subject to biases on several levels (Friedman and Nissenbaum 1996), and sole reliance on algorithms in Foresight might even lead to self-fulfilling data prophecies (Gransche 2016).

To sum up, there is no such thing as an unbiased assessment of "signals of change", yet it is possible to soften the shortcomings of both the human and machine side by combining the two approaches reflexively, and thereby broaden our perspectives when constructing possible futures. In short, we are inclined towards cautious optimism when it comes to data-driven anticipatory methods if complemented by human sense-making.

2.5 Towards New Combinations of Quantitative and Qualitative Methods

Quantitative forecasting methods are helpful within Foresight, when themes are identified and investigated in more detail (Cuhls 2003). They can be relatively precise in the short term, but cannot adequately capture qualitative factors such as political, social, ecological and technological future developments. Moreover, it is often challenging to take the structural discontinuities into account, since mathematical-statistical models assume that patterns observed in the past will continue to be valid in the future (Helm and Satzinger 1999). An important feature of qualitative Foresight and forecasting methods is the collection of non-quantifiable expert knowledge about science, technology or innovation issues, including subjective opinions, in expert-based approaches. The focus is less on "explaining" a certain issue (causally) and more on "understanding" its meaning (Dilthey 1968). However, purely qualitative statements about the future are often difficult to communicate as a basis for decision-making.

Over time, the collaboration between more quantitatively oriented research and more qualitatively oriented Foresight has differed substantially. In the 1970s and 80s, both approaches existed alongside and first attempts were made to bring issues about science and technology, changes in technology directions or emerging technologies into the frameworks of economic—often input-output—models. Later in the 1980s and 90s, qualitative approaches and quantitative simulations were often used independently of each other, because both have their advantages depending on the specific research question (Alcamo 2008; Lamnek and Krell 1993). There was an increasing distinction between the two research directions (even in one single institute), differing scientific communities (Foresight versus forecasting) and controversial discussions about their respective performance and possible applications (Kardorff v. 1995). However, it soon became obvious that there are many overlaps and shared concepts between the methodological approaches (Moschner and Anschütz 2010; Stummer et al. 2021), especially when single methods are combined in a larger process.

Thus, the combination of the two approaches with qualitative Foresight as the open view into the future and quantitative data from the past and present simulated into the future by different models was increasingly discussed as reasonable. It was tested in different projects because problems can be described more comprehensively and the disadvantages of each type of approach can be overcome (Smolenaars et al. 2021). For the identification of new topics (from problem fields to thematic research areas), quantitative approaches are mainly used in bibliometrics (mapping) or later in semi-automated Horizon Scanning approaches—as phases of broader Foresight activities (see previous section).

Different types of combinations for qualitative and quantitative approaches are used nowadays: Firstly, narratives containing qualitative statements are created with the involvement of experts and then transformed into quantitative parameters or model variables or input values for simulation models (Mallampalli et al. 2016;

Shaaban et al. 2022; Voglhuber-Slavinsky et al. 2022; Erdmann and Priebe 2022). And secondly, quantitative results from modelling are placed in context by developing narratives to explain the model and the results of the modelling (Rogge et al. 2020).

An example for a combination is the project Time Rebound, Time Prosperity and Sustainable Consumption (ReZeitKon), where a combined scenario narrative and systems dynamics modelling approach was used. Scenarios were framed through guideline-based interviews challenging peoples' mindsets and by identifying influencing factors from Foresight or futures studies. The initial causal loop diagrams of the system under study served to develop scenario storylines, from which implications on future parameter values were estimated in focus groups and led to the refinement of the model structure. In addition, a representative survey was used to calibrate the system dynamics model with empirical data (Erdmann and Priebe 2022). Another example is the creation of scenarios for the maritime industry's pathway to a greener future, which are mainly available as stories. They also contain quantitative statements, and they are subsequently transferred into model parameters of the MATISSE-SHIP simulation model (MAN Energy Solutions 2020). The scenarios were discussed in a workshop with international experts, verified with stakeholders and updated. In a further round, the experts' feedback for the final results was incorporated.

The future will see more combinations of scenario stories or single assumptions about futures with quantitative models or backed up by surveys, because they have the advantage of linking subjective judgement and rational analysis, which is meaningful only to a certain degree. Mathematical models are able to contribute to the interpretation of qualitative statements and allow quantitative conclusions about possible social, ecological or political effects and conditions in the future. In addition, the usually static representation at the selected point (of time) in the future gains analytical depth in dynamics through the combination with simulation models. In the combination of qualitative scenarios and quantitative modelling, there are many new opportunities, which require researchers to collaborate intensively.

2.6 *Foresight and Innovation Systems: From Wiring Up to Rewiring*

From the mid-1970s onwards, in the context of "evolutionary economics", systems approaches have gained considerable importance in innovation research along with a new recognition of the complexity and non-linearity of innovation processes (c.f. Dosi 1982; Dosi et al. 1988; Nelson and Winter 1977). This resulted in the central notion of the "national innovation system" that was supposed to comprise "... all important economic, social, political, organisational, institutional, and other factors that influence the development, diffusion, and use of innovation" (Edquist 2005). It implies that in order to foster competitiveness, national governments should strengthen not only the technological infrastructures but also the interlinkages between system elements, in particular between research and industry actors.

This concept has been further expanded to include regional, sectoral and even technological innovation systems. The image developed by Kuhlmann and Arnold (2001) in the context of the evaluation of the Norwegian Research Council has become one of the most cited elements in innovation research, and different teams have developed countless variants for specific policy arenas and tested them in projects. The emergence of the innovation system framework is also a defining element of the rise of Foresight in the innovation policy discourse. Foresight became one of the "systemic" instruments (Smits et al. 2010; Smits and Kuhlmann 2004) with the core function of "wiring up" innovation systems (Martin and Johnston 1999) to better evolve in phase with changing framework conditions, aligning actors behind shared priorities and creating "distributed anticipatory intelligence".

This understanding remained dominant for almost two decades and fuelled several large Foresight exercises in countries around the globe. The framework began to evolve again along with changes both in the understanding of innovation processes and in the innovation policy framework. The crucial role of actors beyond research and industry became increasingly obvious and the "triple helix" of university, industry and government (Etzkowitz and Leydesdorff 2000) broadened to include societal actors (Knappe et al. 2019) and even ecosystems actors in a "quintuple helix" approach (Galvao et al. 2019). In parallel, in the 1990s, the notion of "social shaping" of technology gradually diffused from its origin in "Science and Technology Studies (STS)" into the innovation research community. The result was that closer attention was paid to societal actors and downstream phases of innovation trajectories (Warnke and Heimeriks 2008). Subsequently, concepts like user innovation, collaborative innovation, social innovation as well as social and relational capital became important also in the innovation systems framework. An internal research project involving researchers from all competence centres at Fraunhofer ISI about opening up the innovation system framework resulted in a new graphical representation of the innovation system (Warnke et al. 2016). At the same time, in the innovation policy arena, the "Lund declaration" in 2009 marked the beginning of the orientation towards addressing "grand challenges", while internal Fraunhofer Foresight processes were also carried out (Cuhls 2012). This shift later resulted in the "mission orientation" (Mazzucato 2021) of the European and EU member states' research and innovation policy (cf. Lindner et al. 2024). This again has major implications for the "systemic instruments" and especially for Foresight (Daimer et al. 2012).

Since 2010, it has no longer been enough to "wire up" the existing structures but essential to form new ones, that are better suited to accomplish "missions" in the sense of "desirable futures". This "rewiring" includes also actively challenging or breaking up existing non-sustainable innovation trajectories (Kivimaa and Kern 2016). The increasingly broadening notion of innovation systems enables innovation policy-oriented Foresight processes to expand their long-established practice of expert and stakeholder involvement.

While this aspect of opening up is well addressed in Foresight theory and practice (Nikolova 2014), the "mission orientation" is still posing challenges. Foresight is increasingly requested to generate "transformative pathways" or "normative scenarios" for defining targets or achieving predefined goals. Even though first approaches have been developed (Brunori et al. 2020; Erdmann et al. 2013; Schirrmeister and Warnke 2013; van den Ende et al. 2021; Warnke and Schirrmeister 2018) some tensions between open and oriented scenario building remain. Predefined normative orientations do not go hand in hand with key notions of the openness of the future. A bridging concept may be the notion of "resilience" which has long been an important pillar of Foresight thinking. Resilience entails both the ability of systems to react to unexpected futures (coping/ adaptive capacity) and the capacity to transform and develop new narratives that reach beyond dominant frameworks (Brunori et al. 2020; Lorenz 2013; Roth et al. 2021). A focus on such a transformative capacity in reflexive resilient innovation systems may become the next interface between Foresight and innovation system thinking. It can lead to the incorporation of Foresight into governance structures or an institutionalisation of Foresight (Warnke et al. 2021).

2.7 *From Loose Networks to International Collaboration and Networking*

During the 1980s and 90s, an international innovation policy network existed and futures research discussions already took place with an exchange of knowledge on methods and approaches organised in irregular workshops and conferences (e.g. in Kyôto in 1992). The regular monitoring of science and technology as an official network started with the European Science and Technology Observatory (ESTO), the first project of the European Commission's Joint Research Centre's Institute for Prospective Technological Studies based in Seville, Spain. It was set up to attempt to "create a platform of experts engaged in monitoring and analysing scientific and technological developments and their relation and interaction with society". ESTO developed into two new networks: the ERAWATCH Network, a web-based service that presented information on national research policies, actors, organisations and programmes, and the ETEPS Network (European Techno Economic Policy Support Network), a network which supports Foresight of European organisations. ETEPS managed projects in all 27 EU Member States, covering policy subjects such as agriculture, consumer protection, energy, environment, enterprise, health, information society, innovation, research and transport and their respective futures. Through these networks, the European Commission financed several projects to foster Foresight and observations and laid the ground for further developments.

From 2001 on, under the heading of "Future-oriented Technology Analysis (FTA)" a series of conferences in Seville and Brussels were organised by JRC, and many papers concerning Foresight and FTA were published by authors from very different backgrounds. The last of these conferences took place in Brussels in 2018. They were the major meeting place for Foresight experts from all over the world.

The unification of Foresight in science and technology but also towards making use of "strategic intelligence" on a European level gained momentum with the project TSER-ASTPP as a European Network that defined terminology and raised questions about making use of actor networks and their strategic intelligence (Kuhlmann et al. 1999). The ForLearn Foresight Learning Platform (https://knowledge4policy.ec.europa.eu/foresight/topic/forlearn-online-foresight-guide_en) has been built up since 2005 to foster a shared understanding of Foresight, teach newcomers and facilitate mutual learning among practitioners. The platform was followed by the European Foresight Monitoring Network and funded by the European Commission. It formed the basis for a platform to collect Foresight studies and processes from all over the world. The network described Foresight projects in "briefs" to give a short overview of different approaches and their implementations and developed a sandbox for retrieval. The European Foresight Platform was its successor (http://www.foresight-platform.eu/community/forlearn/what-is-foresight/). The European network FOREN for mutual Foresight learning in the Regions (Gavigan et al. 2001) served to bring Foresight to the regions.

In Germany, the Federal Ministry of Education and Research (BMBF) was one of the drivers of Foresight starting in 1992 with larger activities to connect Foresight colleagues internationally. Several of the single projects in the different ministries could be labelled with "Foresight", but there was no co-ordination or coherent understanding of what Foresight could mean for ministries and agencies. The external impetus came from the Stiftung Neue Verantwortung (Foundation New Responsibility, SNV) that tried to bring Foresight into different government agencies and ministries. This also initiated new conversations about futures and Strategic Foresight first on a rather informal level at the German Bundesakademie für Sicherheitspolitik (BAKS), later more formalised in the Federal Chancellery. More and more ministries or agencies now institutionalise Foresight in their departments (often as "Strategic Foresight", see also Warnke et al. 2021). The SNV also started the first teaching programmes that were taken over by the BAKS, which developed towards a node connecting the ministerial network. The seminars on Foresight started as a trial—but in 2022, they are still on-going and always fully booked.

A German network for Futures Research has been already established since 2007 as a club ("Verein"; see https://netzwerk-zukunftsforschung.de/) with the aim of uniting the understanding of Futures Research in Germany, giving the researchers a joint "Leitbild" (kind of vision) and mutual exchange in working groups. The Foresight Europe Network (https://feneu.org/) tries to connect the Foresight community for a "better Europe". The Millennium project (https://www.millennium-project.org/) can be located between a project and a network: It connects "futurists" around the world to improve global foresight, has so-called "nodes" in many countries and carries out global projects. National private activities like the D2030 (https://www.d2030.de) contribute to the picture.

Internationally, the OECD, very active in data gathering, quantitative analyses and reporting information to the member states started a Government Foresight Network (GFN) with a broadened perspective in Foresight. To bring together worldwide approaches in anticipatory monitoring for policy-makers and governments under the headline of "anticipatory innovation governance", the observatory OPSI intends to give new impetus to the strategic orientation of policy processes and is driving Strategic Foresight further within the respective organisations (OPSI 2020). The international collaboration and networking of Foresighters worked on joint book projects, International Advisory Boards of journals, the collaboration with the nodes of the Millennium Project and was present in institutions' Advisory Boards. Also expert groups of the European Commission like the European Forum of Forward-Looking Activities (EFFLA), the Strategic Foresight Group (SFRI) or the Research, innovation and science expert group (RISE) cooperate for independent scientific advice in Foresight. Major associations, in which the researchers organise themselves globally are the World Future Society (WFS), founded in 1966, and the World Futures Studies Federation (WFSF) which has existed since 1973.

The new possibilities of video meetings together with joint working tools like boards, fast surveys or games offer many new possibilities of international exchange, networking, collaboration on projects, training Foresight or fast information collection as well as project acquisition. Even though many researchers are looking forward to meeting physically again, there will be huge opportunities in offline, online and hybrid workshops of all kinds for connecting and conducting future Foresight. Researchers will go on working globally and in their respective networks. OECD and EU have the intention to broaden their networks further.

3 Reflection and Outlook

We have outlined key dimensions in the development of Foresight theory and practice within the last 50 years. This included institutional maturing such as the development towards an established academic field and the forming of a differentiated community with dedicated networks, organisations, conferences and journals. The evolution of innovation systems theory and related policy approaches also impacted Foresight. Its role evolved from informing to wiring up and transforming innovation systems. At the same time, the underlying epistemological basis of Foresight shifted from a predictive forecasting mode to an emphasis on better dealing with uncertainty and complexity by broadening present perceptions of the future through challenging dominant cognitive framings. In line with this, Foresight processes became increasingly participatory and included a wider range of diverse voices and perspectives. In parallel, the increasing availability of data, computing capacities and algorithms for their machine-based analysis led to a plethora of new approaches for identifying emerging changes under the umbrella term of "horizon scanning". This added up to an already on-going integration of quantitative modelling approaches into qualitative Foresight scenarios.

Across these dimensions of change, a few emerging issues can be highlighted:

As indicated in Sect. 2.6, Foresight will be increasingly requested to position itself vis-à-vis transformative mission-oriented agendas. One of the major contributions will certainly be the well-honed set of participatory deliberative Foresight approaches outlined in Sect. 2.3. Already now Foresight methods like visioning, backcasting or transformative scenario development increasingly underpin the deliberation of goals and transition pathways. Established Foresight approaches are adopted in many transformative policy arenas such as those fostering sustainability. Balancing this normative orientation with Foresight's insistence on the openness of the future and the inherent uncertainty of complex systems (Sect. 2.2) will most likely remain a highly dynamic and contested field of evolution in the coming years. Reflexivity and resilience may function as bridging concepts.

The second "game changer" under way is certainly the uptake of machine-based approaches and especially the analysis of unstructured data through natural language processing (NLP). As discussed in Sect. 2.4 this is opening up a whole range of new possibilities and brings in various perspectives. At the same time, the development of adequate sense-making activities is posing major challenges. There is a risk that the impressive possibilities create a false sense of certainty about future pathways. This could bring about a reversal of the focus on reflexive processes for challenging today's anticipatory assumptions and may lead to subsequent re-emergence of purely predictive and deterministic approaches.

Finally, as indicated especially in Sect. 2.2, Foresight is increasingly embedded into wider policy and strategy processes. In the European Commission and many countries around the world, Foresight units are being set up on high political level and also in corporate strategy departments. Driven by the increasing number of unexpected developments the need for futures literacy is becoming ever more obvious. This may open up new inroads to actual implementation of Foresight insights and strengthening of anticipatory culture in organisations. At the same time, it pushes Foresight to reflect more on the institutional framework required to fulfil its function of opening up arenas for reflections and deliberations that may well go beyond well-accepted paradigms and not lend themselves to immediate implementation. It poses a challenge to "institutionally embedded Foresighters" to carefully balance institutional power and scientific autonomy as well as the risks and opportunities of driving change from within versus reflecting on change unhindered from inside constraints.

Summing up one could say that all three key developments in Foresight bring major opportunities for strengthening outreach and impact of Foresight but also carry a tendency to draw Foresight back into the deterministic paradigms of the "planning decade". Foresight activities penetrate deeper into persisting and new power structures (again). Reaping these opportunities while avoiding the risks will most certainly be a major challenge for Foresight in the coming decade.

Addressing this challenge will require Foresight to further extend its co-operations and to continuously ask underlying epistemological questions. Existing networks and terminologies (Sects. 2.1 and 2.7) will most certainly evolve. A growing alignment with transformative R&I policy as well as transition-oriented networks, in particular with a sustainability focus, is already visible. On its way is also increased co-operation with researchers from public management and organisational innovation. The recent prominence of the notion of "anticipatory innovation governance" terminology in the OECD context could be seen as an indicator. Less developed yet but of at least the same urgency is to strengthen ties with research communities in artificial intelligence, big data analysis and natural language processing. This would need to include critical reflection on these approaches within the social sciences and humanities. Foresight was a fragile field but we, the authors, are convinced that it will continue to evolve and even expand—maybe under different headings and most certainly linked to an increasing number of diverse research fields.

References

Achterkamp MC, Vos JFJ (2007) Critically identifying stakeholders. Syst Res 24:3–14. https://doi.org/10.1002/sres.760

Alcamo J (2008) Chapter six: The SAS approach: combining qualitative and quantitative knowledge in environmental scenarios. Dev Integr Environ Assess 2:123–150

Amara RC (1974) The futures field: functions, forms and critical issues. Futures 6(4):289–301

Ansoff HI (1975) Managing strategic surprise by response to weak signals. Calif Manag Rev 18:21–33. https://doi.org/10.2307/41164635

Baldwin C, von Hippel E (2011) Modeling a paradigm shift: from producer innovation to user and open collaborative innovation. Organ Sci 22:1399–1417

Batelle Institut e.V. (1976) Szenarien Chemische Technik. With assistance of Eberhard Jochem, Harald Legler, Siegfried Lange. Batelle Institut e.V., Frankfurt am Main

Beinhauer H, Schmacke E (1970) Fahrplan in die Zukunft: Digest internationaler Prognosen. Droste, Düsseldorf

Belton I, Cuhls K, Wright G (2021) A critical evaluation of 42, large-scale, science and technology foresight Delphi surveys. Foresight Future Sci 4(2):e2118

Bierhals R (1980) Technischer Fortschritt vor Akzeptanzproblemen—am Beispiel der automatischen Datenverarbeitung, Bürotechnik, pp 1239–1242

Blind K, Cuhls K, Grupp H (1999) Current foresight activities in Central Europe. Technol Forecast Soc Change (Special Issue on National Foresight Projects) 60:15–37

Blind K, Cuhls K, Grupp H (2001) Personal attitudes in the assessment of the future of science and technology: A factor analysis approach. Technol Forecast Soc Change 68(2):131–149. https://doi.org/10.1016/S0040-1625(00)00083-4

BMFT (1993) Deutscher Delphi-Bericht zur Entwicklung von Wissenschaft und Technik. Bundesministerium für Forschung und Technologie (BMFT). With assistance of Sibylle Breiner, Kersin Cuhls, G. Jaeckel, Peter Georgieff, Knut Koschatzky, Thomas Reiß et al. Bonn. Available online at http://publica.fraunhofer.de/documents/PX-45525.html. Checked on 2 Sept 2021

Bossel HH, Hughes BB (1973) Simulation of value-controlled decision-making: Approach and prototype. Institut für Systemtechnik und Innovationsforschung ISI

Bradfield R, Wright G, Burt G, Cairns G, van der Heijden K (2005) The origins and evolution of scenario techniques in long range business planning. Futures 37:795–812. https://doi.org/10.1016/j.futures.2005.01.003

Brunori G, Hudson RL, Baldí A, Bisoffi S, Cuhls K, Kohl J, Treyer S, Ahrné L, Aschemann Witzel J, de Clerck F, Duncan J, Otte Hansen H, Ruiz B, Siebelec G (2020) Resilience and transformation: Report of the 5th SCAR Foresight Exercise on natural resources and food systems: transitions towards a safe and just operating space, Brussels

Bührer S, Edler J, Kuhlmann S, Seus S (2024) Evaluating public research and innovation policies: A short history of co-evolution. In: Edler J, Walz R (eds) Systems and innovation research in transition: Research questions and trends in historical perspective. Springer, Heidelberg

Bundesministerium für Forschung und Technologie (1993) Deutscher Delphi-Bericht zur Entwicklung von Wissenschaft und Technik. With assistance of Sibylle Breiner, Kersin Cuhls, G. Jaeckel, Peter Georgieff, Knut Koschatzky, Thomas Reiß et al. Bonn. Available online at http://publica.fraunhofer.de/documents/PX-45525.html. Checked on 2 Sept 2021

Cetron MJ (1970) A method for integrating goals and technological forecasts into planning. Technol Forecast Soc Chang 2:23–51. https://doi.org/10.1016/0040-1625(70)90005-3

Clausen LPW, Hansen OFH, Oturai NB, Syberg K, Hansen SF (2020) Stakeholder analysis with regard to a recent European restriction proposal on microplastics. PLoS One 15:e0235062. https://doi.org/10.1371/journal.pone.0235062

Coates JF (1985) Foresight in federal government policymaking. In: Futures research quarterly summer, pp 29–53

Committee for the Future (2020) Strategic foresight - charting the course towards a more reslilient Europe: strategic foresight report from the commission 2020, Finland

Cuhls K (1998) Technikvorausschau in Japan. Ein Rückblick auf 30 Jahre Delphi-Expertenbefragungen. Physica, Heidelberg

Cuhls K (2000) Opening up foresight processes. Économies et Sociétés, Série Dynamique technologique et organisation 5:21–40

Cuhls K (2002) Participative foresight—How to involve stakeholders in the modelling process. In: European Commission (ed) Future directions of innovation policy in Europe, Proceedings of the innovation policy. Workshop held in Brussels on 11th July 2002 by the "Innovation Policy Unit" of the European Commission (Directorate-General Enterprise), Brussels

Cuhls K (2003) From forecasting to foresight processes—new participative foresight activities in Germany. J Forecast 22(2–3):93–111. https://doi.org/10.1002/for.848

Cuhls K (2008) Methoden der Technikvorausschau—eine internationale Übersicht. Fraunhofer IRB Verlag, Stuttgart

Cuhls K (2013) Foresight in Germany: implications for policy making. In: Meissner D, Gokhberg L, Sokolov A (eds) Science, technology and innovation policy for the future. Potentials and limits of foresight studies. Springer, Heidelberg, pp 199–218. Available online at https://doi.org/10.1007/978-3-642-31827-6

Cuhls K, Georghiou L (2004) Evaluating a participative foresight process: 'Futur—the German research dialogue'. Res Eval 13(3):143–153

Cuhls K, Jaspers M (eds) (2004) Participatory priority setting for research and innovation policy. IRB, Stuttgart

Cuhls K, Salo A (2003) Preface, technology foresight—past and future. New participative foresight activities in Germany. Cuhls, Kerstin und Salo, Ahti (guest editors). J Forecast 22(2–3):79–83. https://doi.org/10.1002/for.848

Cuhls K, Grupp H, Hinze S, Reiß T, Schmoch U (1995) Hauptlinien der zukünftigen Technikentwicklung in der Meeresforschung und Meeresüberwachung. Beitrag zum Projekt "High-Tech-Transfer in die Meeresforschungs- und Meeresüberwachungstechnik". Karlsruhe

Cuhls K, Breiner S, Grupp H (1996) Delphi-Bericht 1995 zur Entwicklung von Wissenschaft und Technik—Mini-Delphi. Bundesministerium für Bildung und Forschung (BMBF). Bonn. Checked on 27 Aug 2020

Cuhls K, Blind K, Grupp H, Bradke H, Carsten D, Harmsen DM et al (1998) Delphi '98—Studie. Befragung zur globalen Entwicklung von Wissenschaft und Technik. Zu-sammenfassung der Ergebnisse. Fraunhofer ISI, Karlsruhe. Checked on 27 Aug 2020

Cuhls K, Blind K, Grupp H (2002) Innovations for our future. Physica-Verlag HD, Heidelberg. Checked on 8 June 2017

Cuhls K, Ganz W, Warnke P, Allmendinger K, Ardilio A, Bienzeisler B et al (2009a) Foresight process on behalf of the German Federal Ministry for Education and Research: new future fields. Cuhls K, Ganz W, Warnke P (eds). Fraunhofer, Karlsruhe. Checked on 20 Aug 2020

Cuhls K, Ganz W, Warnke P (eds) (2009b) Foresight-Prozess—Im Auftrag des BMBF: Zukunftsfelder neuen Zuschnitts. Karlsruhe, Stuttgart. Available online at www.isi.fraunhofer.de/bmbf-foresight.php; http://publica.fraunhofer.de/starweb/pub09/servlet.starweb. Checked on 28 May 2019

Cuhls, K, Rosa A, Weber M, Giesecke S, Wasserbacher D, Könnölä T (2021) After the new normal: Scenarios for Europe in the post Covid-19 world. A Foresight on Demand Project. European Commission, Brussels

Cuhls K (2012) Zu den Unterschieden zwischen Delphi-Befragungen und „einfachen“ Zukunftsbefragungen. In: Popp R (ed) Zukunft und Wissenschaft: Wege und Irrwege der Zukunftsforschung, Bd. 2. Springer, Berlin, pp 139–159

Daas PJH, van der Doef S (2020) Detecting innovative companies via their website. SJI 36(4):1239–1251. https://doi.org/10.3233/SJI-200627

Daimer S, Hufnagl M, Warnke P (2012) Challenge-oriented policy-making and innovation systems theory: reconsidering systemic instruments. In: Fraunhofer ISI (ed) Innovation system revisited: experiences from 40 years of Fraunhofer ISI research. Fraunhofer-Verlag, Fraunhofer ISI, Stuttgart, Karlsruhe

Dalkey NC (1968) Predicting the future. RAND, Santa Monica

Dalkey N, Helmer O (1963) An experimental application of the Delphi-method to the use of experts. J Inst Manag Sci 9:458–467

Dayé C (2020) Experts, social scientists, and techniques of prognosis in Cold War America. Springer, Cham

Day GS, Schoemaker P (2004) Peripheral vision: sensing and acting on weak signals. Long Range Plan 37:117–121. https://doi.org/10.1016/j.lrp.2004.01.003

de Jouvenel B (1967) Die Kunst der Vorausschau (original: L‘Art de Conjecture). Luchterhand

Dilthey W (1968) Der Aufbau der geschichtlichen Welt in den Geisteswissenschaften, vol 7. Teubner Verlagsgesellschaft, Stuttgart

Dosi G (1982) Technological paradigms and technological trajectories. Res Policy 11:147–162. https://doi.org/10.1016/0048-7333(82)90016-6

Dosi G, Freeman C, Nelson R, Silverberg G, Soete L (eds) (1988) Technical change and economic theory. Pinter, London

Dreher C Armbruster H, Butter M, Jung-Erceg P, Schirrmeister E, Warnke P et al (2005) Manufacturing visions. Policy summary and recommendations. Deliverable D17. Contract no. NMP2-CT-2003-507139 MANVIS. Fraunhofer ISI. Karlsruhe (ManVis Report, No. 6). Checked on 20 Aug 2020

Edquist C (1997) Systems of innovation: technologies, institutions and organizations. Pinter, London

Edquist C (2005) Systems of innovation: perspectives and challenges. In: Fagerberg J, Mowery DC, Nelson RR (eds) The Oxford handbook of innovation. Oxford University Press, Oxford, pp 181–208

EFFLA (2013) Towards standards in forward looking activities for the EC. EFFLA policy brief 14. European Forum on Forward Looking Activities. European Forum on Forward Looking Activities. Brussels (14). Available online at http://ec.europa.eu/research/innovation-union/pdf/expert-groups/effla-reports/effla_pb_14_-_towards_a_foresight_standard.pdf. Accessed 10 Jan 2014

Erdmann L, Priebe M (2022) Prospektive Modellierung: Prospektive Modellierung: Der zukünftige Einfluss von zeiteffizienten Praktiken und von Zeit-Rebound-Effekten auf die CO2-Emissionen. Karlsruhe (ISSN: 2702-1947). Available online at http://www.rezeitkon.de/wordpress/wp-content/uploads/2022/08/Erdmann-Priebe-2022-Prospektive-Simulierung.pdf

Erdmann L, Schirrmeister E, Warnke P, Weber M (2013) New ways of doing research—from explorative to transformative scenarios. Deliverable D2.1: Modular Scenario Report—Annexes. Fraunhofer ISI, Karlsruhe. Checked 20 Aug 2020

Etzkowitz H, Leydesdorff L (2000) The dynamics of innovation: from national systems and “mode 2” to a triple helix of university–industry–government relations. Res Policy 29:109–123. https://doi.org/10.1016/S0048-7333(99)00055-4

European Commission (2017a) New horizons. Future scenarios for research & innovation policies in Europe, Brüssel

European Commission (2017b) Open innovation, open science, open to the world—a vision for Europe, Brussels, Belgium

European Commission (ed) (2020) 2020 strategic foresight report: charting the course towards a more resilient Europe, Brussels

European Commission (2021) 2021 Strategic foresight report. The EU's capacity and freedom to act. Brussels. Available online at https://ec.europa.eu/info/strategy/strategic-planning/strategic-foresight/2021-strategic-foresightreport_en

European Commission (2022) 2022 strategic foresight report. Twinning the green and digital transitions in the new geopolitical context, Brussels

Fink A, Siebe A (2016) Szenario Management: Von strategischem Vorausdenken zu zukunftsrobusten Entscheidungen. Campus, Frankfurt

Fink A, Schlake O, Siebe A (2001) Erfolg durch Szenario-Management. Prinzip und Werkzeuge der strategischen Vorausschau [Success through scenario management. Principles and instruments of strategic foresight]. Campus, Frankfurt

Fischer M (1976) Systemanalytische Arbeiten auf dem Gebiet der Wasserreinhaltung. Karlsruhe, Fraunhofer ISI

Flechtheim OK (1966) History and futurology. Anton Hain, Meisenheim am Glan

Flechtheim OK (1990) Ist die Zukunft noch zu retten? Mit einem Vorwort von Oscar Lafontaine. Wilhelm Heyne, München

Forrester JW (1961) Industrial Dynamics. Cambridge

Friedman B, Nissenbaum H (1996) Bias in computer systems. ACM Trans Inf Syst 14:330–347. https://doi.org/10.1145/230538.230561

Frietsch R, Reiß T, Schmoch U (2024) Development of innovation monitoring and innovation indicators in the past 50 years. In: Edler J, Walz R (eds) Systems and innovation research in transition: Research questions and trends in historical perspective. Springer, Heidelberg

Galvao A, Mascarenhas C, Marques C, Ferreira J, Ratten V (2019) Triple helix and its evolution: a systematic literature review. JSTPM 10:812–833. https://doi.org/10.1108/JSTPM-10-2018-0103

Ganz W, Schirrmeister E (2019) Foresight Fraunhofer. München. Available online at https://www.innovationsforschung.fraunhofer.de/de/presse/foresight-fraunhofer.html. Checked on 29 Oct 2019

Gavigan J, Scapolo F, Keenan M, Farhi F, Lecoq D, Capriati M, Di Bartolomeo T (2001) Foresight for regional network: a practical guide to regional foresight, Brussels

Georghiou L, Cassingena-Harper J, Keenan M, Miles I, Popper R (eds) (2008) The handbook of technology foresight. Edward Elgar, Cheltenham

Geurts A, Gutknecht R, Warnke P, Goetheer A, Schirrmeister E, Bakker B, Meissner S (2021) New perspectives for data-supported foresight: The hybrid AI-expert approach. Future Foresight Sci. https://doi.org/10.1002/ffo2.99

Gigerenzer G, Gaissmaier W (2011) Heuristic decision making. Annu Rev Psychol 62:451–482. https://doi.org/10.1146/annurev-psych-120709-145346

Glenn JC, Gordon TJ (2009) Futures research methodology—version 3.0—The Millennium Project. https://www.millennium-project.org/publications-2/futures-research-methodology-version-3-0/. Accessed 26 Feb 2022

Godet M (1986) Introduction to la Prospective. Seven key ideas and one scenario method: special issue La prospective. Futures 18:134–157

Godet M (2000) The art of scenarios and strategic planning: tools and pitfalls. Technol Forecast Soc Change 65:3–22

Gordon AV, Ramic M, Rohrbeck R, Spaniol MJ (2020) 50 Years of corporate and organizational foresight: Looking back and going forward. Technol Forecast Soc Change 154

Gransche B (2015) Vorausschauendes Denken. Philosophie und Zukunftsforschung jenseits von Statistik und Kalkül. transcript, Bielefeld

Gransche B (2016) The Oracle of big data—prophecies without prophets. Int Rev Inf Ethics IRIE 24:55–62

Grupp H (1994) Technology at the beginning of the 21st century. Teehnol Anal Strat Manag 6:379–409

Grupp H (1995) Der Delphi-Report. Stuttgart

Grupp H (1997) Messung und Erklärung des Technischen Wandels. Springer, Berlin

Grupp H (1999) Special issue on national foresight activities. Technol Forecast Soc Change (Special Issue on National Foresight Projects) 60
Grupp H, Hohmeyer O, Schmoch U (1987) Internationale Märkte und Zukunft der Industrieländer 1987. io Management
Haegeman K, Pérez SE, Marinelli E, Warnke P, Teufel B, Degelsegger A, Weiss G (2012) Communication strategy: deliverable D6.2. Report of the Forward Visions on the European Research Area (VERA) project. Sevilla, Karlsruhe. Available online at https://eravisions.archiv.zsi.at/attach/D6-2_Communication_Strategy-Final.pdf
Helm R, Satzinger M (1999) Strategische Unternehmensplanung mittels Szenario-Analysen. Wisu 7
Helmer O (1966) Social technology. Basic Books, New York
Helmer O (1967) Analysis of the future: The Delphi method. Rand, Santa Monica
Helmer O (1975) An agenda for futures research. Futures 7(1):3–14
Helmer O (1983) Looking forward. A guide to futures research. Sage, Beverly Hills
Helmer O, Rescher N (1959) On the epistemology of the inexact sciences. Manag Sci 6:47ff
Helmer-Hirschberg O, Gordon TJ (1964) Report on a long-range forecasting study. Rand, Santa Monica, CA. Checked on 26 Apr 2022
Hetman F (1977) Methods of technological forecasting. AGARD Report no. 655. Edited for the NATO Advisory Group. Paris
Heyen NB, Friedewald M, Jochem E, Reiß T, Thielmann A (2024) Assessing technological innovations: From early warning to the governance of socio-technical transformations. In: Edler J, Walz R (eds) Systems and innovation research in transition: Research questions and trends in historical perspective. Springer, Heidelberg
Hiltunen E (2010) Weak signals in organizational futures learning. Acta Universitatis Oeconomicae Helsingiensis. A, vol 365. Helsinki School of Economics, Helsinki
Ilmola L, Kuusi O (2006) Filters of weak signals hinder foresight: Monitoring weak signals efficiently in corporate decision-making. Futures 38:908–924. https://doi.org/10.1016/j.futures.2005.12.019
Inayatullah S (2000) Alternative futures: methodology, society, macrohistory and the long-term future. Tamkang University Press, Taipei
Irvine J, Martin BR (1984) Foresight in science, picking the winners. Dover, London
ISI (1973) Gegenwärtiger Stand des ISI. Ziele, Tätigkeiten, Finanzierung. Institut für Systemtechnik und Innovationsforschung ISI
ITAS and Fraunhofer ISI (2002) Angebot zum Betrieb des Büros für Technikfolgen-Abschätzung beim Deutschen Bundestag. Checked on 28 Feb 2022
Jantsch E (1967) Technological forecasting in perspective. OECD, Paris
Jochem E (1973) Abfall- und Abwasserwirtschaft. Studie über neue Technologien für den Geschäftsbereich GIA der BBC Mannheim. Institut für Systemtechnik und Innovationsforschung ISI
Jochem E (1975) Technology assessment and participation. In: OECD (ed) Methodological guidelines for social assessment of technology. OECD, Paris, pp 139–145
Jochem E, Bradke H, Dütschke E, Klobasa M, Wietschel M, Plötz P, Fleiter T (2024) Energy demand and modelling of energy systems: five decades from little knowledge to differentiated know-how. In: Edler J, Walz R (eds) Systems and innovation research in transition: Research questions and trends in historical perspective. Springer, Heidelberg
Jochem E, Wiesner J (1977) Beurteilung von F&E-Projekten im Bereich Umweltfreundliche Technik. Beiträge zur Umweltgestaltung 11:76–82
Jungk R (1983) Menschenbeben. Der Aufstand gegen das Unerträgliche. C. Bertelsmann, München
Jungk R (1986) Und Wasser bricht den Stein: Streitbare Beiträge zu drängenden Fragen der Zeit. Taschenbuchausgabe München 1988, Freiburg
Jungk R, Müllert N (1987) Future workshops: How to create desirable futures. Institute for Social Inventions, London
Kahn H (1977) Vor uns die guten Jahre (original: The next 200 years): Ein realistisches Modell unserer Zukunft. Molden

Kahn H, Wiener AJ (1977) Ihr werdet es erleben. Voraussagen der Wissenschaft bis zum Jahre 2000. rororo, Reinbek bei Hamburg

Kardorff v E (1995) Qualitative Sozialforschung - Versuch einer Standortbestimmung: Verlags Union, Weinheim. In: Flick, v. Kardorff, Keupp, v. Rosenstiel, Wolff (eds) Handbuch Qualitative Sozialforschung.: Flick / v. Kardorff / Keupp / v. Rosenstiel / Wolff (Hrsg.): Handbuch Qualitative Sozialforschung. 2. Auflage 1995, Psychologie Verlags Union, Weinheim., 2nd edn. Psychologie Verlags Union, Weinheim

Kimpeler S, Höfling C, Hüsing B, Lux A, Parrisius M, Pohsner U et al (2021) Diskurse über Transformation anregen. Die Nutzung von Zukunftsszenarien für partizipative Dialogformate in Museen und Schulen am Beispiel Bioökonomie. BioKompass, Karlsruhe

Kimpeler S, Cuhls K, Freudenberg C, Guiffrè G Galvini G, Ricci A et al (2022) Impact of the COVID-19 pandemic on European Consumer Behaviour—Foresight study. Foresight on demand. European Commission, Brussels

Kivimaa P, Kern F (2016) Creative destruction or mere niche support? Innovation policy mixes for sustainability transitions. Res Policy 45:205–217. https://doi.org/10.1016/j.respol.2015.09.008

Klusacek K (2002) Technology foresight in the Czech Republic. In: UNIDO (ed) International practice in technology foresight, Vienna, pp 191–207

Knappe H, Holfelder A-K, Löw Beer D, Nanz P (2019) The politics of making and unmaking (sustainable) futures: introduction to the special feature. Sustain Sci 14:891–898. https://doi.org/10.1007/s11625-019-00704-w

Kononenko V (2021) Participatory foresight: Preventing an impact gap in the EU's approach to sustainability and resilience. Scientific Foresight Unit (STOA). European Parliamentary Research Service, Brussels

Kosow H, Gaßner R (2008) Methods of future and scenario analysis. DIE, Bonn. Available online at http://www.die-gdi.de/uploads/media/Studies_39.2008.pdf. Checked on 18 Dec 2016

Kreibich R (2006) Zukunftsforschung. ArbeitsBericht Nr. 23/2006. Berlin. Available online at https://www.izt.de/fileadmin/downloads/pdf/IZT_AB23.pdf. Checked on 2 Oct 2019

Kreibich R (2007a) Planen für das Jahr 2025—Können die Zukunftsforschung und die Agenda 21 Orientierungshilfe leisten? In: Monheim H, Zöpel C (eds) Räume für Zukunft. Zur Innovationsfähigkeit von Stadtentwicklungs- und Verkehrspolitik, Essen

Kreibich R (2007b) Über Transparenz und Verantwortung in einer immer komplexer werdenden Welt. Ufu themen und informationen 2/2008 + 1/2009

Kreibich R (2007c) Wissenschaftsverständnis und Methodik der Zukunftsforschung. Zeitschrift für Semiotik 29:177–198

Krupp H (1972) Die Funktion der Fraunhofer-Gesellschaft im Innovationssystem der Bundesrepublik Deutschland: Studie für das Bundesministerium für Bildung und Wissenschaft, Karlsruhe

Krupp H (1984) Werden wir's erleben? Mitteilungen aus der Arbeitsmarkt- und Berufsforschung (MittAB) 17:1–18

Krupp H (1989) Re-Instrumentalisierung der anwendungsgerichteten Forschung in einer ressourcenbewußteren Gesellschaft. Vortrag, Karlsruhe

Kuhlmann S, Arnold E (2001) RCN in the Norwegian research and innovation system. Background report no. 21 in the evaluation of the Research Council of Norway. Fraunhofer ISI; Technopolis, Karlsruhe, Brighton

Kuhlmann S et al (1999) Improving distributed intelligence in complex innovation systems. Final report of the Advanced Science & Technology Policy Planning Network (ASTPP). ISI, Karlsruhe

Kuwahara T, Cuhls K, Georghiou L (2008) Foresight in Japan. In: Georghiou L, Cassingena-Harper J, Keenan M, Miles I, Popper R (eds) The handbook of technology foresight. Edward Elgar, Cheltenham, pp 170–184

Lamnek S, Krell C (1993) Qualitative Sozialforschung. Mit Online-Material, 2nd edn. Psychologie Verlags Union, Weinheim

LeCun Y, Bengio Y, Hinton G (2015) Deep learning. Nature 521:436–444. https://doi.org/10.1038/nature14539

Lerch C, Jäger A (2024) Industry in a changing era: Production paradigms during the last 50 years. In: Edler J, Walz R (eds) Systems and innovation research in transition: Research questions and trends in historical perspective. Springer, Heidelberg

Lindner R, Edler J, Daimer S (2024) Understanding paradigm change in science, technology and innovation policy: Between science push and policy pull. In: Edler J, Walz R (eds) Systems and innovation research in transition: Research questions and trends in historical perspective. Springer, Heidelberg

Lorenz DF (2013) The diversity of resilience: contributions from a social science perspective. Nat Hazards 67:7–24. https://doi.org/10.1007/s11069-010-9654-y

Lyon C, Cordell D, Jacobs B, Martin-Ortega J, Marshall R, Camargo-Valero MA, Sherry E (2020) Five pillars for stakeholder analyses in sustainability transformations: The global case of phosphorus. Environ Sci Policy 107:80–89. https://doi.org/10.1016/j.envsci.2020.02.019

Mallampalli VR, Mavrommati G, Thompson J, Duveneck M, Meyer S, Ligmann-Zielinska A, Druschke CG, Hychka K, Kenney MA, Kok K, Borsuk ME (2016) Methods for translating narrative scenarios into quantitative assessments of land use change. Environ Model Softw 82:7–20. https://doi.org/10.1016/j.envsoft.2016.04.011

MAN Energy Solutions 2020 (2020) AHOY2050: exploring scenarios for the maritime industry's pathway to a greener future: Augsburg 2020

Martin BR (1995) Foresight in science and technology. Tech Anal Strat Manag 7:139–168

Martin BR, Johnston R (1999) Technology foresight for wiring up the national innovation system: experiences in Britain, Australia, and New Zealand. J Technol Forecast Soc Change (Special Issue) 60:37–54

Martino JP (1983) Technological forecasting for decision making, 2nd edn. North Holland, New York

Meadows DH, Meadows DL, Randers J (1972) The limits to growth. A report for the club of Rome's project on the predicament of mankind. Universe Books, New York

Meadows DH, Meadows DL, Randers J (2009) The limits to growth: the 30-year update. Earthscan, London

Miller R (2007) Futures literacy: A hybrid strategic scenario method. Futures 39:341–362. https://doi.org/10.1016/j.futures.2006.12.001

Miller R (ed) (2018) Transforming the future: Anticipation in the 21st century. Routledge, Taylor & Francis Group, London

Ministère de l'Économie (2006) Technologies clés 2010. Paris. Available online at https://www.vie-publique.fr/var/storage/rapports-publics/064000771.pdf

Mische A (2009) Projects and possibilities: researching futures in action. Sociol Forum 24:694–704. https://doi.org/10.1111/j.1573-7861.2009.01127.x

Moschner B, Anschütz A (2010) Kombination und Integration von qualitativen und quantitativen Forschungsmethoden in einem interdisziplinären Forschungsprojekt: n I. Diethelm, C. Dörge, C. Hildebrandt, & C. Schulte (Eds.), Didaktik der Informatik. Möglichkeiten empirischer Forschungsmethoden und Perspektiven der Fachdidaktik (pp. 11–20). Gesellschaft für Informatik e.V. In: Dörge, Diethelm, I., Hildebrandt C, Schulte (Eds.) C (eds) Didaktik der Informatik. Möglichkeiten empirischer Forschungsmethoden und Perspektiven der Fachdidaktik, pp 11–20

Muhlroth C, Grottke M (2022) Artificial intelligence in innovation: how to spot emerging trends and technologies. IEEE Trans Eng Manag 69:493–510. https://doi.org/10.1109/TEM.2020.2989214

Nelson RR, Winter SG (1977) In search of useful theory of innovation. Res Policy 6:36–76. https://doi.org/10.1016/0048-7333(77)90029-4

Nikolova B (2014) The rise and promise of participatory foresight. Eur J Future Res 2. https://doi.org/10.1007/s40309-013-0033-2

NISTEP (2019) 第11 回科学技術予測調査 S&T Foresight 2019 総合報告書 (The 11th science and technology foresight 2019, full volume). National Institute of Science and Technology Policy. NISTEP, Tokyo. Available online at https://nistep.repo.nii.ac.jp/?action=pages_view_

main&active_action=repository_view_main_item_detail&item_id=6657&item_no=1&page_id=13&block_id=21. Checked on 26 May 2020
OPSI (ed) (2020) Anticipatory innovation governance. what is, how it works, and why we need more that ever before. Observatory of Public Sector Innovation, OECD OPSI, Paris. Checked on 28 Feb 2022
Picht G (1967) Prognose, Utopie, Planung: Die Situation des Menschen in der Zukunft der technischen Welt, Stuttgart
Poli R (2017) Introduction to anticipation. Springer, Dordrecht
Poli R (ed) (2019) Handbook of anticipation: theoretical and applied aspects of the use of future in decision making. Springer, Cham
Porter AL (2019) Data analytics for better informed technology & engineering management. IEEE Eng Manag Rev 47:29–32. https://doi.org/10.1109/EMR.2019.2928265
Radkau J (2017) Geschichte der Zukunft. Prognosen, Visionen, Irrungen in Deutschland von 1945 bis heute, München
Rogge KS, Pfluger B, Geels FW (2020) Transformative policy mixes in socio-technical scenarios: The case of the low-carbon transition of the German electricity system (2010–2050). Technol Forecast Soc Chang 151:119259. https://doi.org/10.1016/j.techfore.2018.04.002
Rosa A, Gudowsky N, Warnke P (2018) But do they deliver?: Participatory agenda setting on the test bed. Eur J Future Res 6:285. https://doi.org/10.1186/s40309-018-0143-y
Rosa AB, Kimpeler S, Schirrmeister E, Warnke P (2021) Participatory foresight and reflexive innovation: setting policy goals and developing strategies in a bottom-up, mission-oriented, sustainable way. Eur J Future Res 9. https://doi.org/10.1186/s40309-021-00171-6
Rossel P (2012) Early detection, warnings, weak signals and seeds of change: A turbulent domain of futures studies. Futures 44:229–239. https://doi.org/10.1016/j.futures.2011.10.005
Roth F, Warnke P Nissen P, Edler J (2021) Insights into systemic resilience from innovation research. Fraunhofer ISI (Perspectives, 03 / 2021). Available online at https://www.isi.fraunhofer.de/content/dam/isi/dokumente/policy-briefs/policy_brief_resilience_en_web.pdf
Sackman H (1975) Delphi critique. Expert opinion, forecasting, and group process. Rand, Toronto
Schirrmeister E, Warnke P (2013) Envisioning structural transformation—lessons from a foresight project on the future of innovation. Technol Forecast Soc Chang 80:453–466. https://doi.org/10.1016/j.techfore.2012.10.008
Schirrmeister E, Meißner L, Gutknecht R, Göhring A-L (2019) Analyse organisationsspezifischer Biases im Kontext von Innovationsmanagement und Foresight. In: Gausemeier J, Bauer W, Dumitrescu R (eds) Vorausschau und Technologieplanung. Heinz Nixdorf Institut Univ, Paderborn, pp 361–380
Schirrmeister E, Göhring A-L, Warnke P (2020) Psychological biases and heuristics in the context of foresight and scenario processes. Future Foresight Sci 2. https://doi.org/10.1002/ffo2.31
Schmidt L, Falk T, Siegmund-Schultze M, Spangenberg JH (2020) The objectives of stakeholder involvement in transdisciplinary research. A conceptual framework for a reflective and reflexive practise. Ecol Econ 176:106751. https://doi.org/10.1016/j.ecolecon.2020.106751
Schmoch U (1988) Technikprognosen mit Patentindikatoren. Zur Einschätzung zukünftiger industrieller Entwicklungen bei Industrierobotern, Lasern, Solargeneratoren und immobilisierten Enzymen. Verl. TÜV Rheinland, Köln
Schmoch U (1990) Wettbewerbsvorsprung durch Patentinformationen. Handbuch für die Recherchepraxis. Schriftenreihe Zukunft der Technik. Verlag TšV Rheinland, Koln
Schmoch U, Breiner S, Cuhls K, Hinze S, Münt G (1996) The organisation of interdisciplinarity—research structures in the areas of medical lasers and neural networks. In: Reger G, Schmoch U (eds) Organisation of science and technology at the watershed. The academic and industrial perspective. Physica, Heidelberg
Seefried E (2015) Zukünfte. Aufstieg und Krise der Zukunftsforschung 1945-1980, Oldenburg
Shaaban M, Voglhuber-Slavinsky A, Dönitz E, Macpherson J, Paul C, Mouratiadou I, Helming K, Piorr A (2022) Understanding the future and evolution of agri-food systems: a combination of qualitative scenarios with agent-based modelling. Futures 149:103141

Slaughter RA (2005) Knowledge base of futures studies CD-ROM professional edition. http://www.foresightinternational.com.au/
Smits R, Kuhlmann S (2004) The rise of systemic instruments in innovation policy. Int J Foresight Innov Policy 1:4–32
Smits RE, Kuhlmann S, Shapira P (eds) (2010) The theory and practice of innovation policy: an international research handbook. Edward Elgar, Cheltenham
Smolenaars WJ, Lutz AF, Biemans H, Dhaubanjar S, Immerzeel WW, Ludwig F (2021) From narratives to numbers: spatial downscaling and quantification of future water, food & energy security requirements in the Indus basin. Futures 133. https://doi.org/10.1016/j.futures.2021.102831
Steinmüller K (2012) Zukunftsforschung in Deutschland—Versuch eines historischen Abrisses (Teil 1). Zeitschrift für Zukunftsforschung 1:6–19
Steinmüller K (2013) Zukunftsforschung in Deutschland: Versuch eines historischen Abrisses (Teil 2). Zeitschrift für Zukunftsforschung 2:5–21
Steinmüller K (2014a) Aufstieg und Niedergang der Prognostik. Zeitschrift für Zukunftsforschung 2
Steinmüller K (2014b) Zukunftsforschung in Deutschland—Versuch eines historischen Abrisses (Teil 3). Zeitschrift für Zukunftsforschung 3:5–24
Stummer C, Lüpke L, Günther M (2021) Beaming market simulation to the future by combining agent-based modeling with scenario analysis. J Bus Econ 91:1469–1497. https://doi.org/10.1007/s11573-021-01046-9
Surowiecki J (2005) The wisdom of crowds. Anchor Books, New York
Taleb NN (2007) The black swan: the impact of the highly improbable. Allen Lane, London
Toffler A (1990) Powershift. Bantam Books, New York
Trénel M, Kimpeler S, Fitschen K, Priebe M, Röß A, Shahgholi S (2020) Kernbotschaften aus dem Beteiligungsprozess zur Weiterentwicklung der Hightech-Strategie 2025: Bericht in die Runde der Staatssekretäre am 28. Oktober 2020, Berlin. 2 March 2022
Tversky A, Kahneman D (1974) Judgment under uncertainty: heuristics and biases. Science 185:1124–1131. https://doi.org/10.1126/science.185.4157.1124
UNIDO (ed) (2002) International practice in technology foresight. Economy environment employment, Vienna
van Asselt M, van't Klooster SA, van Notten PW, Smits LA (2012) Foresight in action. Routledge, London
van den Ende M, Wardekker A, Mees H, Hegger D, Vervoort J (2021) Towards a climate-resilient future together: A toolbox with participatory foresight methods
Verein Deutscher Ingenieure (1991) VDI-Report 15, Technikbewertung—Begriffe und Grundlagen, Erläuterungen und Hinweise zur VDI-Richtlinie 3780. VDI, Düsseldorf
Voglhuber-Slavinsky A, Zicari A, Smetana S, Moller B, Dönitz E, Vranken L, Zdravkovic M, Aganovic K, Bahrs E (2022) Setting life cycle assessment (LCA) in a future-oriented context: the combination of qualitative scenarios and LCA in the agri-food sector. Eur J Future Res 10. https://doi.org/10.1186/s40309-022-00203-9
Voros J (2017) The voroscope. The futures cone, use and history https://thevoroscope.com/2017/02/24/the-futures-cone-use-and-history/. Accessed 12 Apr 2022
Voros J (2019) Big history and anticipation: using big history as a framework for global foresight June 2017. In: Poli R (ed) Handbook of anticipation: Theoretical and applied aspects of the use of future in decision making. Springer, Cham
Wagner CS, Popper SW (2003) Identifying critical technologies in the United States: a review of the Federal effort. J Forecast 113–128
Warnke P, Heimeriks G (2008) Technology foresight as innovation policy instrument: learning from science and technology studies. In: Cagnin C (ed) Future-oriented technology analysis: strategic intelligence for an innovative economy. Springer, Berlin, pp 71–87
Warnke P, Schirrmeister E (2016) Small seeds for grand challenges—Exploring disregarded seeds of change in a foresight process for RTI policy. Futures 77:1–10. https://doi.org/10.1016/j.futures.2015.12.001
Warnke P, Schirrmeister E (2018) Transition-scenarios towards socially sustainable global value chains: Insights from the SONA WSK foresight. Karlsruhe. Working paper sustainability

and innovation, 10/2018. Available online at http://publica.fraunhofer.de/dokumente/N-506838.html
Warnke P, Koschatzky K, Dönitz E, Zenker A, Stahlecker T, Som O et al (2016) Opening up the innovation system framework towards new actors and institutions. Karlsruhe. Fraunhofer ISI discussion papers innovation systems and policy analysis, 49
Warnke P, Cuhls K, Schmoch U, Daniel L, Andreescu L et al (2018) 100 radical innovation breakthroughs for the future (RIBRI). Brussels. Available online at https://ec.europa.eu/info/sites/info/files/research_and_innovation/knowledge_publications_tools_and_data/documents/ec_rtd_radical-innovation-breakthrough_052019.pdf
Warnke P, Priebe M, Veit S (2021) Studie zur Institutionalisierung von Strategischer Vo-rausschau als Prozess und Methode in der deutschen Bundesregierung. Karlsruhe. Available online at https://www.bundesregierung.de/breg-de/aktuelles/studie-strategische-vorausschau-2059782
Wiener M (2018) Open foresight: The influence of organizational context. Creat Innov Manag 27:56–68. https://doi.org/10.1111/caim.12238
Zweck A, Holtmannspötter D, Braun M, Cuhls K, Hirt M, Kimpeler S (2015a) Forschungs- und Technologieperspektiven 2030. Düsseldorf. Checked on 19 Jan 2022
Zweck A; Holtmannspötter D, Braun M, Hirt M, Kimpeler S (2015b) Geschichten aus der Zukunft 2030. Ergebnisband 3 zur Suchphase von BMBF-Foresight Zyklus II. BMBF, Düsseldorf. Available online at https://www.bmbf.de/files/VDI_Band_102_C1.pdf. Checked on 23 May 2020

Kerstin Cuhls has been a scientific project manager at the Fraunhofer Institute for Systems and Innovation Research (ISI) since 1992 and is Professor for Japanology at the University of Heidelberg, Centre for East Asian and Transcultural Studies (CATS). She combines East Asian studies with Futures concepts, Foresight and time research. In national, regional and international studies for different contracting authorities, she built up a huge repertoire of Foresight methods and their combinations, that is applied in projects. Kerstin is teaching in Foresight seminars at the Freie Universität Berlin (Master Futures Research) and the Bundesakademie für Sicherheitspolitik (BAKS).

Ewa Dönitz has been working as a scientific project manager at the Competence Centre Foresight of the Fraunhofer ISI since 2009. Since June 2013, she has been heading the Business Unit of Foresight for strategy development. The focus of her work is on the application of future research in foresight-based strategy processes, in particular the development of scenario-based roadmaps, visions and trend analysis in the context of industrial and public as well as regional foresight projects to support decision-making under uncertainty. She is also a lecturer in future research methods at the Karlsruhe Institute of Technology KIT.

Lorenz Erdmann leads the Business Unit Foresight for Organisations from Research and Civil Society at Fraunhofer ISI's Competence Centre Foresight. He graduated at the Technical University Berlin in Environmental Sciences and Technology. Then he worked at the Institute for Futures Studies and Technology Assessment IZT, before joining Fraunhofer ISI's Competence Centre Foresight in 2011. Among his major research topics are the development and application of foresight methods (e.g. Horizon Scanning, Scenario Technique, Roadmapping) and their integration into inter- and transdisciplinary research processes (e.g. combination with quantitative modelling and participatory action research).

Bruno Gransche conducts research and lectures as a philosopher and foresight expert in the fields of philosophy of technology/ethics and anticipatory thinking with focus on emerging human-technology relations, AI & digitalisation, imaginations and images of technology, social appropriateness, modal shaping of futures and transformative philosophy. Gransche is associate professor at the Institute of Technology Futures ITZ at the University of Karlsruhe KIT (since 2020); study and PhD in Heidelberg, habilitation in Karlsruhe. He is Fellow at the Fraunhofer Institute for Systems and Innovation Research ISI in Karlsruhe, where he worked in the Competence Centre Foresight until 2016.

Simone Kimpeler is the head of Competence Centre Foresight at Fraunhofer ISI, background in communication research, sociology and economic policy (University Münster) and doctorate on discourse analysis. She has been working in the field of innovation and future research at Fraunhofer ISI since 2000. Her research is focused on strategic foresight for policy-making, including participatory foresight and foresight for transitions

Elna Schirrmeister studied industrial engineering and worked in a chemical company after graduating before joining Fraunhofer in 1999. Her research focused on the further development of methods such as bibliometrics, scenario development, visioning, Delphi surveys and future dialogues to reflect cognitive biases and address complex future issues. She has led various foresight processes to support the strategic development of the Fraunhofer-Gesellschaft and has been in charge of the Corporate Strategy department at Fraunhofer headquarters since 2024.

Philine Warnke is deputy head of the Competence Centre Foresight at Fraunhofer ISI. She holds a degree in mechanical engineering (Dipl. Ing.) and a PhD in philosophy of technology. Since 2002 she worked as a senior researcher and group leader at Fraunhofer ISI, the "Institute for Prospective Studies" of the European Commission (JRC-IPTS), the Austrian Institute of Technology AIT and the German Federal Chancellery. Philine's research focuses on theory and practice of participatory futures dialogues (Foresight).

Evaluating Public Research and Innovation Policies: A Short History of Co-evolution

Susanne Bührer, Jakob Edler, Stefan Kuhlmann, and Sarah Seus

Abstract In this article, we present the mutually development of R&I policies and R&I policy evaluation with a specific focus on Germany and the EU. We show that there has been considerable development in the field of R&I policy evaluation over the past decades, in terms of underlying theories, methodological approaches and the emergence of an R&I evaluation community. The field of R&I policy evaluation has naturally interacted and grown in parallel with the changes in R&I policy and practice as well as scientific advances. In line with the actual claims of R&I policy to support transformation, R&I policy evaluation has engaged in a debate around more responsible and inclusive evaluation procedures. The discussion about "responsible metrics" and societal stakeholder engagement in evaluation studies are examples of it. Our overview of evaluation practice has shown that the R&I policy evaluation community need continuous, in-built critical reflection on the (incessantly changing) role of evaluation in R&I policy.

S. Bührer (✉) · S. Seus
Fraunhofer Institute for Systems and Innovation Research ISI, Karlsruhe, Germany
e-mail: Susanne.Buehrer-Topcu@isi.fraunhofer.de; sarah.seus@isi.fraunhofer.de

J. Edler
Fraunhofer Institute for Systems and Innovation Research ISI, Karlsruhe, Germany

Manchester Institute of Innovation Research (MIoIR), University of Manchester, Manchester, United Kingdom
e-mail: jakob.edler@isi.fraunhofer.de

S. Kuhlmann
University of Twente, Enschede, The Netherlands
e-mail: s.kuhlmann@utwente.nl

J. Edler, R. Walz (eds.), *Systems and Innovation Research in Transition*, Sustainability and Innovation, https://doi.org/10.1007/978-3-031-66100-6_5

1 Introduction

The objective of this chapter is to discuss the interplay between the major lines of Research and Innovation (R&I) policy[1] and the role of explicit evaluation exercises that aim to capture the function and influence of such policies. Obviously, evaluation concepts and instruments correspond with the respective "fashions" and waves in R&I policy, such as cluster policies or competence centres at the end of the 1990s, the new mission orientation of recent years but also, with regard to institutional funding, the introduction of New Public Management approaches in public research organisations. Also partly considered in our chapter are the goals, instruments and approaches to evaluating research institutions. In this chapter, we focus on the German and the European level with regard to evaluation practice, however, with regard to evaluation theory we also include the Anglo-American literature (Guba and Lincoln 1989; Mayne 2001; Patton 1997; Scriven 1991; Weiss 1997).

The chapter is structured along a chronology of decades. For each decade, we describe core features and whether or how they differ from previous decades. The individual sections are subdivided according to the following aspects: (a) political ambitions and developments; (b) main aims and topics of evaluation; (c) applied methods and concepts and (d) emergence and development of an R&I evaluation community.

1.1 What Are We Talking About? Some Definitions

The German Evaluation Society (DeGEval) understands evaluations as "the systematic examination of the quality or utility of an object of evaluation" (DeGEval—Gesellschaft für Evaluation e.V. 2016; cited from Kohlweg 2019, 5). According to Kohlweg (2019), an evaluation is characterised by "a comprehensible systematic procedure based on empirically obtained data, which contrasts it with everyday evaluation processes, by a transparent, criterion-driven assessment that is carried out against the background of a specific context of use (investigation of utility) or across the board (investigation of quality), which contrasts it with pure research studies, and by its applicability to different objects. The most important objects of evaluation include projects, policy measures and other interventions (programmes), organisations, products, and evaluations themselves (meta-evaluation)" (ibid.).

Evaluations can be distinguished according to the timing (ex ante, interim, ex post), the purpose (summative, formative), the object of evaluation (see above,

[1] If not otherwise stated, we use "research and innovation policy" as a generic term to cover the various policy activities aiming to foster science, technology, research and innovation activities (Boekholt 2010). The terminology and foci of those policies have changed over time, and we will refer to those changes as needed. We focus on policies by ministries of research, science, innovation or economic affairs, i.e. we do not cover policies in other ministries that may support or influence the generation and diffusion of innovation.

projects, programmes, organisations, but also whole innovation systems) and the content-related (impact) dimensions (scientific, economic, societal, political, systemic impacts). Evaluation and impact analysis can hardly be separated, since questions about intended and unintended effects are typically part of every (programme) evaluation. At the same time, other communities, concepts and methodological approaches have emerged in the field of impact analysis, and these will be taken into account in the following wherever it is useful.

2 Beginnings: From the 1960s and the Evaluation of Research to the 1980s with the First Comprehensive Approaches to Evaluating R&I Policies

2.1 Political Ambitions and Developments

Starting in the 1960s, in Western Europe, North America and beyond, societal, economic and political actors increasingly perceived needs to reform and strengthen post-World War II structures of economies, to advance social and welfare policies, to expand and modernise education systems and to make political-administrative systems more effective. Across the "varieties of capitalism"—from liberal to coordinated market economies—such reforms assumed quite different shapes (Hall and Soskice 2001). Altogether though, "modernisation" became a key term, in industry, politics and academia, often closely related to the term "democratisation" (e.g. Lipset 1959). Governmental public policies were expected to function as key resources in modernisation processes, understood as "problem solving". To effectively solve problems, governments and their agencies would have to engage with "political planning" (Friedmann 1987; a term only partly overlapping with the French political concept of "planification", see, e.g., Crozier 1965, or with the socialist-soviet concept of economic planning, see, e.g., Mandel 1986).

In the 1960s to 1980s, most authors understood "political planning" as rational end-means and decision-making action by political actors or political systems (see Deutsch 1973; Etzioni 1968). The conception is essentially based on an economic understanding of rationality, according to which "policy problems are seen as instrumental in nature, and policy makers are seen as rational to the extent that they do the best they can to satisfy the combined welfare functions of those affected by their policies" (Schön and Rein 1994, 11). This understanding has also been developed and partially practiced in Germany since the late 1960s as a targeted state modernisation strategy (e.g. Naschold 1969; Krauch 1970). Political planning presupposed a special role for the state: the central political authorities are ascribed the ability to conceptually shape the social environment (Martinsen 1992, 53). On the basis of a thorough reform of the political-administrative system (in terms of competencies, decision-making processes, structural and operational organisation), equipped with intelligent information about current and foreseeable future social, economic and technological developments, "political planning" should enable effective state control and problem solving. The necessary generation and processing of information should take on a scientific character: "... effective policy-making will have to be future-oriented and will increasingly depend upon medium- and long-range forecasting and policy-planning capabilities" (Mayntz and Scharpf 1975, 5).

This is the political and intellectual context where governments made increasing efforts to support education, science, research and technological innovation no longer mainly through institutional funding of related organisations (higher education; research) but additionally also through policy programmes with defined aims, timing and formats.

- *Higher education and research*: Since the early 1960s higher education (and related research capacity) has seen a worldwide expansion: " … growth patterns are similar in all types of countries, [they] are especially high in countries more linked to world society, and sharply accelerate in virtually all countries after 1960" (Schofer and Meyer 2005, 898). In the same vein, in Germany the diagnosis of an educational crisis ("Bildungskatastrophe", Picht 1964) marked the beginning of an educational expansion and an era of reform in the education (and related research) system, continuing to this day.
- *Science and technological innovation in industry*: After World War II, science policy was based on the belief that publicly sponsored scientific research would result in beneficial social and economic outcomes (see the influential report Science the Endless Frontier, Bush 1945). In the late 1970s and throughout the 1980s, in Western European countries, certainly in Western Germany, the competitiveness of national industries in international markets was perceived as weak, not at least in view of the rising economic power of Japan, so governmental funding policies were launched to "push" technological innovation in research organisations and industry. From the mid-1960s, empirical studies criticised such supply-push policy concepts for underestimating the role of market forces in the innovation process; as a consequence, demand-pull models of innovation were supposed to guide policy designs. As from the late 1970s, an increasing number of innovation studies scholars have argued that innovation should be understood as a two-sided process involving complex interactions between supply and demand (e.g. Freeman and Soete 1997; Boekholt 2010; Molas-Gallart and Davies 2006). In the Federal Republic of Germany, the innovative capacities of small and medium-sized enterprises (SMEs) were considered a crucial backbone with respect to competitiveness and employment. Consequently, governments supported the research, development and innovation capacities in SMEs.

The more publicly funded support measures were taken by governments and their agencies, the more the request grew—by the public, parliaments, media—to legitimise the investments by properly examining their efficiency and effectiveness.

2.2 *Main Aims and Topics of Evaluation*

Since the 1960s, in Germany as in other industrialised countries, heterogeneous lines of evaluation concepts and methods in the field of R&I policy have been prevalent. They can be represented as a shell model (Kuhlmann 1998b, 86pp, 2009, 285p): A first shell relates to individual research achievements: the "core" is formed by peer-review procedures and later additional procedures for measuring the research performance of individual researchers and groups (bibliometrics, etc.) as internal scientific instruments for deciding on the allocation of funding in research institutions (Daniel 1989). The second shell consists of the evaluation of political support programmes that, in contrast to institutional support, pursue politically decided and specifically designed intentions. A third shell relates to the performance of research institutions or larger scientific areas, as they have been carried out since the 1990s, e.g. by the Science Council (Wissenschaftsrat) (see also below).

In particular the second shell, the evaluation of research and innovation programmes gained traction in the 1970s and 1980s. "A case in point is the long and fruitless discussion among economists about the relative impact of direct and indirect government measures for promoting R&D on the national economy and technical and industrial development. This debate is mainly carried out under aspects of principles of political control and economic order, with little empirical backing and verification" (Meyer-Krahmer 1981, 358). Since the mid-1970s, however, research has been carried out by Fraunhofer ISI (Bräunling and Harmsen 1975) that has tried to overcome this "abstract either/or type argument and showed that the reality of government funding programmes must be viewed in a much more differentiated way" (Meyer-Krahmer 1981, 358).

2.3 *Applied Methods and Concepts*

The reform's hopes conveyed by political planning efforts, on the one hand, and increasing disappointment about unintended effects of planning and programming, on the other hand, (Pressman and Wildaysky 1984) triggered a growing interest in understanding the actual *implementation* processes of programmes in bureaucracies vis-à-vis societal stakeholders. In Germany, a research project group "Implementation of political programmes" around Renate Mayntz (Mayntz 1979) studied basic aspects of policy implementation, both with internal administrative enforcement problems and with the difficulties of carrying out political programmes with the help of intermediary organisations. It became clear that effective policy implementation needs to understand, also from a methodological point of view, issues of norm formation, the effect of different types of programmes (regulatory policies, incentive programmes), the calculus of action at the local level in the implementation process, the coordination problems in the implementation process, etc. (Mayntz 1980).

Following a similar reasoning, colleagues at Fraunhofer ISI developed novel concepts and methods for the *evaluation of impacts* of governmental research and innovation programmes, in particular those providing R&D personnel cost grants to SMEs. With the help of surveys, interviews and statistical analyses the evaluation studies addressed the (1) acceptance of the funding by SMEs and the relevance to their internal decision-making; (2) quantitative and qualitative effects on R&D personnel; (3) impact on innovative activities of firms; (4) impact on the economy as a whole, such as employment, stimulation of innovation and diffusion processes and (5) efficient implementation and application of the measure. At the same time, the studies considered consequences for the governmental funding strategy, in the short and long term (Meyer-Krahmer 1981, 359).

Since the mid-1970s, in parallel with implementation and impact studies, the interest in *research and technology indicators* has grown sharply worldwide. Numerous studies have suggested indicators and used them to describe the level of performance in an inter-individual, inter-institutional and international comparison. While in early years the focus was mainly on input indicators for research and technology, in the 1980s the interest in result and impact indicators increased. Daniel (1989) provides an overview of the status and prospects of this field of science and innovation research, including an extensive bibliography of bibliometric, scientometric (e.g. Moed et al. 1985; Weingart et al. 1988) and technometric literature (e.g. Grupp and Hohmeyer 1986), the latter exploring the technological competitiveness of national economies, as an input in research and innovation policymaking and programme design.

2.4 Examples of Typical Research and Innovation Policy Evaluation Studies

Throughout the three decades, first slowly then accelerating, a growing number of research and innovation (policy) evaluation studies were conducted. Daniel (1989) gives an overview of science-related evaluation efforts, Becher and Kuhlmann (1995) and Holland and Kuhlmann (1995) present numerous cases of technology and innovation policy evaluations in Germany. In the following, two typical examples are presented.

Fraunhofer ISI became a forerunner in Germany and Europe in the evaluative study of the actual impacts of SME research and innovation programmes. Drawing on conceptual work by Bräunling and Harmsen (1975) and Meyer-Krahmer (1981), in the early 1980s, Meyer-Krahmer et al. (1983) on behalf of the federal government conducted a seminal series of evaluation studies of the largest West-German government programmes offering grants to cover the costs of R&D personnel in SMEs. Subjects of investigation were "the quantitative and qualitative changes in R&D activities caused by the programme in the firms … and the contribution the programme can make to strengthen innovative and competitive capabilities"

(Meyer-Krahmer et al. 1983, 153). Also the problem assumptions underlying the programme were reviewed critically, in particular to better understand "whether the barriers to innovation and the weak spots assumed did in fact exist in the firms examined" (Meyer-Krahmer et al. 1983, 153). This was important for the political legitimation of public funding; at the same time genuinely empirical insights into the innovation processes in SMEs were generated—a crucial contribution to the emerging academic field of "innovation studies" (see, e.g., Fagerberg et al. 2013). The administrative management, "free rider" effects, and factors restricting impacts were investigated along with such questions as to whether firms located in peripheral regions were reached by the programme to the same extent as firms in conurbation areas. Furthermore, the reasons for non-participating firms were analysed.

In 1983, the German Federal Ministry of Research and Technology (BMFT) started a pilot project to promote technology-oriented start-ups (TOU). In order to support this, in particular to expand the information basis for the technology-political decision-making in this new funding field, a scientific monitoring project was launched in 1984 and conducted for several years by Fraunhofer ISI (Kulicke 1993; see also Holland and Kuhlmann 1995, 175–187). Research work was carried out with three functions: as a service to the pilot project's "actor network" in particular the sponsor BMFT; as an accompanying evaluation and as an ex-ante analysis for potential subsequent measures. During the monitoring, the evaluation team had to coordinate and process a variety of tasks, partly conflicting between implementation and evaluation. Over time, the intensity of evaluation activities increased. Targeted analyses of the equity capital supply were initiated early on, based on findings from the accompanying research. On this basis, operational ex-ante investigations were carried out with a view to designing a subsequent pilot project "Participation Capital for Young Technology Companies" (BJTU). Overall, the accompanying research and evaluation of the TOU pilot project was a comparatively time-consuming and complex process, the experimental character which included intensive investigations of the funding clientele over an extensive period of time and active participation in the funding measure, as well as the difficulty of maintaining a distanced observer status within this process. This tension is probably unavoidable in the actor-oriented development of complex new funding models, as a kind of mediator between governmental agencies, beneficiaries and industry.

2.5 *Evaluation Community*

The 1980s saw an upsurge of professionalisation and community building in the field of evaluation experiences and competencies. Most prominently, in 1986 the *American Evaluation Association (AEA)* was established, a merger of the former Evaluation Research Society and the Evaluation Network (Kingsbury 1986).

In the field of science and research evaluation the Institute for Science and Technology Research *(Institut für Wissenschafts- und Technikforschung)* at Bielefeld University has been a network hub for evaluative science studies since the mid-1970s.

In the course of the 1980s other institutes, like the Centre for Science and Technology Studies (*CWTS*) at Leiden University, helped to boost this growing field of competence and expertise.

In the field of innovation policy studies, the *Six Countries Programme (6CP*[2]*)*, established in 1974 became for many years a network hub, which provided the opportunity for professional exchange about related policy evaluations. "Within the broad area of innovation policy, the Six Countries Programme is an example for a non-issue-specific Forum with emphasis on an open-minded exchange of information, reflection and the identification of new issues/assessment of new developments" (Edler et al. 2006), including the question "how innovation policies have to be designed to create effective out-comes of such public policy".

With financial seed support from the European Union, the University of Twente in the Netherlands established an annual *R&D Evaluation Course* in the late 1980s, to foster European and international capability and capacity-building in the field of professional research evaluation (with other European evaluation institutes as partners); the course ran until 2022.

3 The 1990s: Complex Multi-actor/Multi-agency Programmes and Systems-Oriented Evaluations with Formative Claims

3.1 Political Ambitions and Developments

In retrospect, the decade of the 1990s was characterised by a high willingness to develop and implement innovative R&I measures and a generally critical reflection on the strengths and weaknesses of national innovation systems. From a system's perspective, the question of cooperation between the different subsystems has played a particularly important role. Accordingly, R&I policy had increasingly launched programmes since the mid-1990s that were intended to bring about structural change in science and industry through the formation of innovation networks. Partnerships of a wide variety of actors (research institutions, universities, industrial enterprises, banks, technology centres, further education institutions, etc.) aimed to effectively re-shape local, regional or sectoral innovation systems with a whole bundle of coordinated measures. Such multi-actor and/or multi-measure programmes covered either regions and/or specific missions and/or selected technologies and sectors. Unlike simple cooperation projects, such as those that have been funded since the mid-1980s in national programmes of collaborative research, complex network programmes intended to achieve a self-sustaining dynamic of the selected clusters with sustainable economic effects in global competition through start-up financing. Competence centres and cluster initiatives represent outstanding

[2] The core group of countries were Germany, Austria, Finland, the Netherlands, Sweden and the UK.

examples of these new funding instruments. One of the very challenging key funding prerequisite in those programmes was a network-like cooperation of the actors involved in the cluster that goes beyond concrete individual projects (OECD Proceedings 1999; Koschatzky 1996, 2000).

Further, since the 1990s a reorientation in the relationship between the state and publicly funded research organisations has taken place in numerous (Western) European countries, which can be summarised by the catchword New Public Management (NPM). In essence, this meant that the state reduced its (operational) control over public research institutions and granted them more autonomy in their decisions, but in return expected performance to increase (Ferlie et al. 1996). In the course of these changes in governance between scientific institutions and the state, various mechanisms were introduced, like global budgeting, performance-oriented resource allocation and goal agreements. The introduction of NPM also led to an increasing relevance of competition and performance-oriented funding, either as part of institutional funding or through increasingly important competitive third-party funds (Geuna 2001; Hicks 2012; Jongbloed and Lepori 2015; Jongbloed and Vossensteyn 2001).

3.2 *Main Aims and Topics of Evaluation*

In the 1990s, evaluation research and practice gained considerable momentum, driven by innovative policy approaches, on the one hand, and the increased need to legitimise public funding, especially of research and research institutions, on the other. The latter applied both in Germany and at the European level.

In Germany in the early 1990s a first meta-analysis of extant R&I policy evaluation studies showed that the number of such exercises was rising while the methodological quality of some studies and their use in policy practice remained limited (Kuhlmann et al. 1995). At the same time, however, now more often systematic so-called accompanying or monitoring evaluations were launched. They were meant to serve as a learning instrument for the continuous improvement of innovative funding approaches like EXIST, a federal programme to promote start-ups based on scientific research (Kulicke et al. 2000, 2002), Competence Centres for Nanotechnology (Bührer et al. 2000, 2001) or the Interdisciplinary Centres for Clinical Research at University Hospitals (IZKF) (Braun et al. 1997; Bührer et al. 2004).

As mentioned above, the performance of research institutions became an important topic in R&I policy and thus evaluation also turned increasingly to individual science institutions. One of the most important examples is the UK approach, today the Research Excellence Framework (REF), formerly the Research Assessment Exercise (RAE), which started as early as the 1980s, closely linked to the development of sophisticated bibliometric and other quantitative indicators (see also below). In Germany, performance of research organisations has been studied, on the one hand, with regard to the missions of research institutions (Kuhlmann and Holland

1995) and, on the other, with regard to the interaction of different research organisations within the national innovation system in the context of the so-called system evaluations (Evaluierungskommission 1998; Internationale Kommission 1999; Wissenschaftsrat 2000, 2001). Furthermore, in 1990, shortly after the German reunification, the Science Council (Wissenschaftsrat) was tasked by the government to develop recommendations for the *reorganisation of more than 130 East German non-university research institutions* (institutes of the former GDR Academy of Sciences). The aim of the recommendations was "to create as soon as possible new structures to facilitate internationally competitive basic and applied research". (Krull 1992). The Science Council had formed an evaluation committee and discipline-oriented working groups. Their recommendations for the reorganisation of East German non-university research institutions had far-reaching consequences for the development of the innovation capacity in East Germany in the 1990s (Holland and Kuhlmann 1995).

At the European level, the 1990s showed a significant dynamic not only in R&I policy, but also in evaluations and impact analysis that covered not only the European Framework Programmes (see, for example, Georghiou and Meyer-Krahmer 1992; Bach and Lambert 1992; Stern 1993; Georghiou 1995a, b; Fayl et al. 1998, 1999; Airaghi et al. 1999; Guy and Polt 1999) but also initiatives such as COST (Cunningham et al. 1997; Cunningham and Nedeva 1999) and EUREKA (Airaghi et al. 1995; Bobe et al. 1999; Dale and Barker 1994; Dekker et al. 1991; Ormala 1993). Furthermore, various impact studies were conducted on the benefits of participating in the European Framework Programmes (for example, Larédo and Callon 1990; Georghiou et al. 1993; Reger and Kuhlman 1995; Reger et al. 1998; Luukkonen and Niskanen 1998). A special feature of the evaluation system developed within the framework of EUREKA, a European cooperation scheme formally outside the constitutional framework of the European Community, was that it systematically surveyed long-term impacts by asking the beneficiaries to answer a questionnaire not only shortly after the completion of their project, but also 3 and 5 years after the end of the project. These market impact reports were implemented in recognition of the fact that (economic) impacts often manifest themselves only a long time after the funding (Georghiou 1999).

The European Commission also supported a couple of research projects that aimed to improve the European evaluation systems. For example, the TSER-Network "Advanced Sciences for Technology Policy Planning" (ASTPP, 1997–1999), that was led by Fraunhofer ISI.

In the academic context, this decade was characterised by intensive exchange and comparative work on the respective national evaluation cultures, which in turn found their way into the further development of the respective national R&I policies (see, for example, Da Silva and Henriques 1995; Hills and Dale 1995; Kastrinos and Katsoulacos 1995; Kuhlmann 1995; Larédo and Mustar 1995; Rip and van der Meulen 1995; Sanz-Menéndez 1995; Silvani and Sirilli 1995; Steiner and Sturn 1995; van Steen and Eijffinger 1998). This work has led to the transparency of national patterns in R&I evaluation on the one hand, but also to stimulating mutual learning and the emergence of an evaluation community on the other.

3.3 Applied Methods and Concepts

Overall, evaluation research in the 1990s benefited from a scientification of evaluation practice across policy fields. American scholars were considered to be leading in this respect (Guba and Lincoln 1989; Patton 1997; Scriven 1991). In the German-speaking countries, however, considerable efforts were also made in this decade to establish a scientific basis for evaluation practice (e.g. Stockmann 2000; Widmer 1996; Wollmann 1998).

One of the new concepts that shaped this decade was the understanding of evaluation as a learning medium for the actors involved, which focused on the formative aspects of evaluation (Kuhlmann 1998a, b). The key concept of the new, expanded understanding of evaluation is "negotiation" in actor arenas.[3] The result of evaluations designed accordingly is no longer "a set of conclusions, recommendations, or value judgements, but rather an agenda for negotiation of those claims, concerns, and issues that have not been resolved in the hermeneutic dialectic exchanges" (Guba and Lincoln 1989, 13). Decisions are rather made as an on-going process in which competing actors interactively reach a consensus, or not, and evaluation results are one piece of information among many. Here the evaluation process, more precisely the communications of the involved actors, come to the fore and the process is deliberately designed to be "participatory" ("Participatory Evaluation") (Patton 1997; Worthen et al. 1997). Such evaluation concepts aim above all to reframe (Schön and Rein 1994) the orientations of corporatist and political-administrative actors. In the context of the research and innovation system, they can act as an "intelligent" contributor to the negotiation and as a coping strategy of the responsible political actors and the interested public. "Intelligent" policy development processes in this sense can furthermore be enriched by a combination with Technology Foresight and Technology Assessment (Kuhlmann et al. 1999, but also Martin 1996; Meyer-Krahmer and Reiss 1992). A concrete practical example of the combination of an (institutional) evaluation and foresight activities was the use of the results of a large-scale Delphi survey for the system evaluation of the Fraunhofer Society (Bührer 2001).

In line with the policy trends described at the beginning of this section (measures to improve the relationships between various actors within national innovation systems and a stronger performance orientation of research institutions), specific data collection and analytical methods were (further) developed in the 1990s, namely network analyses and STI (Science, Technology & Innovation indicators.

Various articles (Bührer 2002; Cabo and Bijmolt 1992; Noyons and van Raan 1996; Removille and Clarysse 1999) reflected the growing importance of network analysis at the time and this has become a standard approach in many evaluation studies. One German example, the evaluation of the interdisciplinary centres for clinical research at University hospitals (Bührer and Peter 1999), analysed in detail the cooperation behaviour of the funded individuals. This example illustrates well

[3] The following paragraph is mostly a translation of Bührer and Kuhlmann (1999), page 387 ff.

the scope and ambition of network analysis in evaluation. The aim of the survey was to examine the extent to which the implementation of the funding measure has (already) succeeded in initiating and establishing new, interdisciplinary collaborations. Among other things, the aim was to elicit possible consolidation zones within the professional relationship networks of the scientific staff active in the respective centres and to identify the central actors. In addition to describing the formal aspects of the individual networks (size, density/closeness, transitivity, number of redundant contacts, centrality, cliques, homogeneity measures), the aim was also to describe the "quality" of the individual relationships. The quality of a relationship includes the duration of the relationship, the origin of the relationship, the frequency of interaction, the (perceived) strength of the relationship, the role relationship and the transactional content.

As regards STI indicators, numerous overviews were published in the 1990s, building upon the ground-breaking work that was done during the 1980s, as described above (Brisolla 1996; Gabolde 1998; Grupp et al. 1995; Tomizawa and Niwa 1996; van Raan 1988; van Steen 1995; Schmoch et al. 1991; Schmoch 1999). In the field of bibliometrics, specific indicators on (co-)citations and inter- or multi-disciplinary collaboration have been further developed (van Raan 1993; Zulueta and Bordons 1999). The 1990s also saw the beginnings of comprehensive STI indicator reports at the European level (European Commission 1997). However, there is also a growing trend in the 1990s to critically reflect the use of quantitative indicators (for example, Pavitt 1998; Hicks 1991; Luukkonen 1991).

3.4 Evaluation Community

The 1990s were decisive in building an R&I evaluation community. In 1992, the *European Evaluation Society (EES)* was initiated and registered in 1996 as a non-profit association, and soon other associations followed, such as the *German Evaluation Society (DeGEval)* in 1997, with the significant participation of Stefan Kuhlmann (Fraunhofer ISI). The main "raison d'être" of the DeGEval is the promotion of evaluation research and practice by pursuing three main goals: professionalisation of evaluation, bringing together different perspectives and promoting information and exchange. The DeGEval is subdivided into individual working groups that are dedicated to discussing specific evaluation challenges in selected areas, including the R&I Policy Working Group that has existed from the very beginning.

At the same time, also specialised associations were set up, such as the *Austrian Platform Research & Technology Policy Evaluation* in 1996. This was initially an informal cooperation, deliberating on approaches and methods of evaluation and discussing evaluation practice on national and international levels. Meanwhile, the Austrian fteval platform has become a pan-European resource for evaluators and their clients, offering a variety of important services such as the systematic collection of evaluation reports, the publication of the fteval journal, which provides timely reports on relevant evaluations and, most importantly, the organisation of major international evaluation conferences (see also section on 2010s below).

A further important milestone at the European level was the constitution of the "European RTD Evaluation Network" in 1997 that aims to promote regular exchange and cooperation between evaluation actors from the EU Member States and the Commission services. This network[4] is still operating.

At least two large evaluation conferences should also be mentioned as they offered the platforms for an increasing exchange on the challenges and potential solutions of R&I evaluations: the OECD organised a discourse-shaping conference on "Policy Evaluation Practices in Innovation and Technology" in 1997 (see OECD 1997) and Fraunhofer ISI organised, on behalf of the European Commission and the German Federal Ministry of Education and Science, a European conference on "Science and Technology in the New Europe" in Berlin on 7–8th June 1999 (Bührer and Kuhlmann 1999).

4 The 2000s: Further Development of Methods and Concepts for Capturing Complexity

4.1 Political Ambitions and Developments

This decade in some ways was a transitional decade. R&I policies further differentiated (Polt et al. 2021) as the analysis of systems failures became more sophisticated and more and more schemes for improved, more tailored cooperation and networking were being developed. This system building approach was a further expression of the overall paradigm of national competitiveness through constantly improving the science and innovation systems which was still dominant at the beginning of the decade. At the European level, the Lisbon agenda and the Barcelona target (3% of GDP to be invested for research in Europe) were the most obvious expression of the ambition to mobilise science and innovation for the international competition of systems. Accordingly, the EU and its member States were increasingly ranked by their innovation activities and capabilities, most notably demonstrated by the establishment of the European Innovation Scoreboard in 2001. Though this was not a tool to evaluate policy, it signified the focus of R&I policy and its supporting evaluation across Europe. Only once the Lisbon agenda did not deliver on its claim, moves were made towards an orientation of science and innovation policy towards "Grand Challenges" (European Commission and Directorate-General for Research and Innovation 2006). Slowly, but accelerated by the fallout of the financial crisis in 2008, European and national R&I policies moved towards a policy mode that linked

[4] EUevalnet, https://research-and-innovation.ec.europa.eu/strategy/support-policy-making/shaping-eu-research-and-innovation-policy/evaluation-impact-assessment-and-monitoring/evaluation-network_en

support for research and innovation to societal goals (see the next section) and thus linked policies which were system enhancing with those that were system directing (see below).

4.2 Main Aims and Topics of Evaluation

A first major development was to continue and further develop the formative and summative evaluation of multi-actor/multi-measure systems. This included cluster evaluations, the evaluation of competence centres (Bührer et al. 2002), approaches to knowledge and technology transfer (Kulicke 2002) and the comprehensive evaluation of the Austrian programmes Kplus which at the time were advanced multi-actor programmes (Edler et al. 2004; Edler and Lo 2004; Biegelbauer 2007). Given the complexity of those multi-actor, multi-measure programmes, policy makers and evaluators co-generated principles of design, implementation and evaluation[5] (Bührer et al. 2004). Although the evaluations had a clear summative mandate, the formative dimension gained considerable weight, as evaluation became part of creating mutual understanding and expectation management at the level of individual projects, but much more so at the level of joint system building.

In this decade, the system level became increasingly important for evaluators and R&I policy analysts. More and more, systematic, robust and valid evaluation at the systems level was called for (Arnold 2004; Feller 2007). At the end of the previous decade, the evaluation of the Norwegian Research Council was, given the broad remit of the council, a comprehensive systems evaluation (Arnold et al. 2001; Kuhlmann and Arnold 2001). One decade later, the evaluation of the Austrian science and innovation system was, and still is as of today, the most comprehensive and holistic evaluation of a science and innovation system, its governance structures and instrumentation.[6] Similar developments were to be witnessed in the USA, with attempts to focus on the multi-level interaction of policy interventions and the overall systems level additionality (see the special issue by Rogers and Jordan 2010). However, those in-depth systematic system-level analyses remained the exception rather than the rule at the national level.

In line with the policy focus on the systems level, in the early 2000s, both the EU and the OECD developed and financed benchmarking approaches to assess and compare how policies influence the innovation system (Molas-Gallart and Davies 2006, pp 73–74). In this approach, rather complex relations between input and output within innovation systems are operationalised, standardised and compared between countries. The most basic, but most continuous example is the aforementioned Innovation Scoreboard. Another prominent example is the benchmarking of

[5] Edler et al. (2004) provide an overview of many of those measures and develop design and evaluation principles.

[6] For a summary of major aspects, see the Newsletter of the Austrian Platform for Research and Technology Evaluation, 34, 2010. https://repository.fteval.at/72/%20

industry-science relationships that is based on theoretical assumptions of the interaction between science and industry and a set of indicators developed to map these relationships. While those benchmarking exercises have limited analytical value, they have strongly influenced the R&I policy discourse across the OECD world (OECD 2002).

To further support system-level evaluation and steering, country benchmarking and national and regional review exercises were intended to provide learning opportunities for policy makers. At the European level, in the 2000s the so-called "policy mix" reviews performed a similar learning exercise in a number of countries "on demand", a light-touch approach based on international peer reviews. Those European country reviews did not amount to the expert led application of sophisticated methods, rather they were expert supported peer-review exercises underpinned by a number of background analytical reports most often provided by the country reviewed. Also in the 2000s, the OECD launched their "innovation policy reviews" which claim to be a "comprehensive assessment of the innovation system of individual OECD member and partner countries, focusing on the role of government. [...] Each review identifies good practices from which other countries can learn".[7] Those OECD reports, while also underpinned by self-assessment and background material of the countries under review, were—and still are—led by R&I policy evaluation experts.

Below the level of entire R&I systems, but in fact as a building block for systems evaluation, a line of work intensified that tried to evaluate the interplay of diverse interventions. A special issue of Research Evaluation highlighted and supported the claim to analyse systematically the interplay of instruments (Vonortas et al. 2007). In particular in the USA there were a few attempts to mobilise evaluation techniques for portfolio management (Bozeman and Rogers 2001) and a number of quantitative portfolio evaluations were conducted for public research organisations (Ruegg 2006, 2007; Hage et al. 2007), with the claim to mobilise portfolio evaluation to better understand system level impact (Wagner and Jordan 2009).

At the beginning of the decade, evaluators and policy makers intensified efforts to understand wider societal impacts of publicly funded research. Rather than developing further, more sophisticated quantitative analyses, more qualitative, indirect approaches were established. A body of literature emerged that focused on tracing the processes supporting longer-term change and analysing the conditions influencing and interacting with the initial research process and the use of research results. The most prominent effort in this regard was the SIAMPI project funded by the European Commission (Cozzens 2010; Spaapen et al. 2011; Spaapen and van Drooge 2011), which focused on productive interactions between science and non-scientific stakeholders as the key driving force and a necessary precondition for the effective use of research results within the science system but also in other policy sectors. The term "productive" characterises the interactions and is only seen as successful once the research results (including knowledge) have been actively used and

---

[7] See https://www.oecd.org/sti/inno/oecd-reviews-of-innovation-policy.htm

have induced a change in behaviour. This approach was subsequently applied to a range of examples (Molas-Gallart and Tang 2011).

However, many developments outlined for the previous decade continued and started to create tensions in the R&I Policy community. On the one hand, as indicated above, there were an increasing number of evaluations that tried to capture dynamics at the systems level and understand the role of policy and policy portfolios for system developments. This was also linked to an intensified effort to support policy making throughout the entire cycle with in-depth formative approaches. On the other hand, the policy-making system, with notable exceptions, continued to demand a summative assessment, a quantitative number, to justify and legitimise R&I policy spending (Molas-Gallart and Davies 2006; Edler et al. 2010). Further, within the R&I policy evaluation community, calls for more robust quantitative and experimental methods emerged, asking for randomised controlled trials also in R&I policy evaluation. Especially as the evaluation practice had gained momentum and diversified in the previous years, the diverse claims of R&I policy evaluation and the growing tension between those claims led to an increased self-reflection in the community, with key collective events (2006 in Austria) and stock-taking and analysis exercises (Edler et al. 2010; Polt 2003).

4.3 Applied Methods and Concepts

In the field of bibliometrics, many studies began to integrate the individual level by studying mobility patterns based on CV data. This was not only an innovation in terms of the database and methodology but also the level of observation (individuals instead of institutions) (Cañibano and Bozeman 2009; Lepori and Probst 2009; Sandström 2009; Zellner 2002).

The INNO-Appraisal analysis of evaluations in innovation policy conducted between 2002 and 2007 found that most evaluations focused on economic and sometimes technological impact, societal impact, in contrast, was not very common and if so was concerned with the number of jobs created. Further societal or even transformational issues were still the exception (Edler et al. 2010). The analysis further demonstrated a tension in the R&I policy evaluation practice just outlined. While there was an increasing claim that evaluation is supposed to be formative and take into account the complex system conditions, the INNO-Appraisal study found that summative evaluations using quantitative methods, thus delivering a quantitative result, a "number" for policy makers, were more likely to be used in the policy-making process. This is in line with the analysis of Molas-Gallart and Davies (2006), who also cite two examples of evaluation in the early 2000s in which policy makers as clients disregarded the formative and system sensitive approach of the evaluation and asked for a summative, ideally quantitative assessment to be used in the political process. Thus in the 2000s a tension intensified between the claim and ambition of evaluation experts rooted in the innovation systems paradigm and policy makers in need of quantified, ideally unambiguous and short-term impacts of their interventions.

4.4 *Evaluation Community*

This decade was characterised by a growing interest in the scope and effects of evaluation practices themselves, both within the evaluation community (Edler et al. 2008), in Europe and North America (Shapira and Kuhlmann 2003), and within the community of policy makers who commissioned evaluation studies (Edler et al. 2010). The diversification of evaluation claims and evaluation methods, in particular but not exclusively at the European level, led to a number of projects taking stock of evaluation practices, making those diverse approaches widely accessible to the community, assessing the appropriateness of approaches and suggesting further adjustments to meet the needs of a changing policy environment. At the beginning of the 2000s, two European funded projects analysed the evaluation practices in Europe (Polt 2003) at the European and the national level and provided access to the breadth of evaluation methods (Fahrenkrog et al. 2002, EPuB[8]) as well as setting out future developments to bring evaluation practice in line with R&I policy developments across Europe (Georghiou et al. 2002, ASIF[9]). ASIF demonstrated that the policy and evaluation rationales were now firmly grounded in the innovation systems and "failure" paradigm. It also showed that the ambition of evaluation was increasingly to determine the net effect of interventions and to capture the broader economic and in particular societal effects of R&I policy (Polt 2003, also Rojo 2003). The EpUB study led to a "toolbox" for R&I policy evaluation based on European experiences, while at the same time 10 years of evaluation of the US American Advanced Technology Programme had led to a "toolkit" (Ruegg and Feller 2003). Some years later, the European Commission ordered the aforementioned study INNO-Appraisal to systematically take stock of R&I policy evaluation practices across European countries and at the European level (Edler et al. 2010).

The significance of those exercises is twofold. First, the R&I policy community had reached a state of maturity to come forward with major, almost standardised "toolkits" and recommendations as to when to use what approaches (Miles et al. 2006; Molas-Gallart and Davies 2006). Second, and maybe even more importantly, it signifies the self-awareness of the R&I community in Europe and the USA as an epistemic community, linked together by a convergent understanding of theoretical justifications for R&I policy interventions and a set of techniques to be shared and applied. This institutionalisation of the community was further supported by major

---

[8] EPuB: Socio-Economic Evaluation of Public Research, Technology and Development (RTD) Policies

[9] ASIF: Assessing the Socio-Economic Impact of the Framework Programme

international conferences initiated by the Austrian R&I policy evaluation community (2003 and 2006).

5 From the 2010s Until Today: Is Accountability Back? Evaluation Between Legitimacy and Learning

5.1 Political Ambitions and Developments

The years from 2010 onwards were characterised by a growing discussion on R&I policy as an instrument that should bring forth solutions for societal grand challenges such as climate change, demographic change or human diseases. Concepts such as "mission-orientation innovation policies" (Mazzucato 2018) and "transformative innovation policies" (Diercks et al. 2019) shifted the sphere of influence of R&I policies beyond the boundaries of the science system. Hence, research results were expected to influence non-research performing stakeholders and to have effects on non-research related sectors. As a consequence, demonstrating the impacts of research and innovation activities became increasingly important for R&I stakeholders (see discussion in the chapter by Lindner et al. 2024 in this anthology).

In parallel a, not entirely disconnected, discussion emerged, questioning the appropriateness of existing research assessment approaches and metrics ("the metric tide" (Wilsdon et al. 2015)) for assessing scientific output almost solely based on peer-reviewed journal publications. The unease towards the research assessment mechanisms (discussed already in the previous sections of this chapter) was reflected in two declarations stemming from the research community itself: The San Francisco Declaration on Research Assessment (DORA 2012) and the Leiden Manifesto for research metrics (2015)

In order to address the changes in the R&I policy landscape and methodological implications of the shift towards impact (namely the attribution challenge), the evaluation community directed its attention to developing and refining conceptual frameworks suitable to analyse the processes leading to longer-term impacts of research. It had to develop metrics capable of measuring those longer-term effects. Furthermore, the debate on the function of programme and policy evaluation and the roles evaluators opened up again in the discussions related to sustainability transition and responsible research and innovation (RRI).

5.2 Main Aims and Topics of Evaluation

As mentioned in the preceding chapter, societal impacts of funded research became a predominant topic at the turn of the decade. The observation that a longer-term impact of research had not been adequately addressed in evaluation studies of the early 2010s (Arnold 2012) is a result of several challenges which confront impact measurement of research:

First, there was no commonly shared definition of what impact of research means, the dividing lines reflecting the researchers' disciplinary backgrounds (Arnold 2011; Bornmann 2013; Donovan 2011; Flecha et al. 2018). In recent evaluation practice two main aspects have been emphasised with regard to impact, as e.g. put forward in the Glossary of the Evaluation Standards Kohlweg (2019): A) an impact can be positive or negative, primary or secondary, intended or not by the funded intervention. B) It is an effect that materialises on stakeholders not involved in the research processes and reaching out beyond the science system itself. [10] Building on earlier work mentioned above, the term "societal impact"[11] further grew in popularity in the last years, however, no commonly agreed definition exists for it until today (Bornmann 2013; Bührer et al. 2022a, b; Joly and Matt 2017; Muhonen et al. 2019; Reale et al. 2018).

A second challenge related to the attribution problems is the difficulty to establish a causal relation between a particular impulse such as a research result or a funding programme and a longer (societal) change. This was not a new discussion in the evaluation literature, but became more critical when focusing on longer-term outcomes and impacts of R&I funding. The consecutive challenges related to the evaluation of mission and challenge-oriented interventions have been raised by several authors lately: expected effects in various policy sectors (multiple impact dimensions (Bührer et al. 2022a, b); long-time horizons which effects would materialise that go beyond the evaluation exercise; changes that are beyond the control of the stakeholder initially involved in the research and innovation processes; interference and interaction of the research with other funding or sectoral policies (Amanatidou et al. 2014; Arnold 2018; Donovan 2011; Donovan and Hanney 2011; JIIP 2018; Molas-Gallart et al. 2021; Reale et al. 2014; Wittmann et al. 2022).

In consequence, the focus shifted away from traditional approaches measuring economic effects through well-established STI indicators. Rather, the concept of contribution as opposed to attribution has gained importance in the evaluation literature and the methodologies developed allow to analyse the contribution that research makes on broader socio-economic societal changes.

In line with the more ambitious attempts to capture impact, a concept that has grown in popularity in R&I programme evaluation in the last 10 years is one of the impact pathway modelling. Building on a "theory of change" (of the funded intervention) (Chen and Rossi 1983; Weiss 1993, 1997) idealised pathways are described showing how a stimulus in form of research results can contribute to longer-term changes. The pathways rely on a set of assumptions how effects can manifest

[10] fteval Evaluation Standards 2019; Glossary: "Positive and negative, primary and secondary long-term effects resulting directly or indirectly, intentionally or unintentionally from an intervention. OR effects beyond the intended reach of the evaluation object, i.e. on things (systems, institutions, organisations, individuals, outputs etc.) where the evaluated object has been a contributing factor".

[11] Also the term of "social return" is used (Donovan 2011) or "socio-economic impact" Reale et al. (2018).

themselves. Its particular strength lies in an explicit consideration of external conditions interfering with the initial research process and subsequent diffusion of research results. The focus of these approaches is on understanding the underlying mechanisms and processes leading (or not) to societal changes. Recent work of Belcher et al. (2020) extended the pathway logic by drawing attention to the stakeholders involved in the evaluated intervention and introduced the reasoning in spheres of dwindling control when moving up the impact pathway. The theory-based evaluation approaches thereby contrast the standard evaluation approaches which rely on experimental or quasi-experimental methods and aim at quantifying effects on the macro-economic level. However, they are not interested in how and why the measured effects develop ("black-box approach") (Chen and Rossi 1983).

An attempt to put into practice theory-based-evaluations analysing the changes along the theory of change is the method of contribution analysis (CA) (Mayne 2012). Studies applying this concept to R&I interventions remain scarce and raise doubts whether the idea of building a contrafactual situation relying on the theory of change can be used to measure contributions in complex systems such as the R&I field (Morton 2015). The effective use of CA in evaluation studies is hampered by the enormous costs that a rigorous analysis of processes implies (Delahais and Toulemonde 2012).

With the normative turn in innovation policy (Borrás and Edler 2020; Schot and Steinmueller 2018), whereby research and innovation policy is mobilised to support specific societal missions, most prominently in the field of sustainability and climate change mitigation, there has been again a growing interest in comprehensive formative evaluation frameworks (Amanatidou et al. 2014; Arnold 2018; Dinges et al. 2020a; JIIP 2018; Magro and Wilson 2013; Mickwitz et al. 2021; Molas-Gallart et al. 2021; Patton 2019; Wittmann et al. 2022). The focus is on the continuous improvement of the design and the implementation of an intervention. Capacity-building and learning of stakeholders involved in the evaluation becomes more important (Patton 2011). This impacts also the role of the evaluators themselves who become mediators with different interests but also co-producers in the evaluation process (Guba and Lincoln 1989). In addition, ex-ante impact assessments are receiving more attention using impact pathways to conceptualise possible future effects (JIIP 2018; Weber and Polt 2014; Wittmann et al. 2022).

5.3 Applied Methods and Concepts

With the shifts in R&I policy just described, a growing number of evaluations until today apply concepts of theory-based approaches and impact pathways to different research and innovation programme funding. Increasingly, studies across different fields and categories of policy measures apply the aforementioned approach of productive interaction (Molas-Gallart and Tang 2011; Kroll et al. 2022). Others further develop the impact pathways to their specific evaluation domain or combine both to evaluate process factors and measure impact at the same time (Muhonen et al. 2019). The mainstreaming of impact pathways reasoning for the evaluation in the

R&I field can be seen in the development of an assessment framework for the latest EU Research Funding Framework Programme, Horizon Europe, building "Key Impact Pathways" (Bruno and Kadunc 2019).[12] Further, the ASIRPA project has been a seminal project that used the impact pathways concept to trace the impacts of a research performing organisation by using case study methodology and cross-cases comparison (Joly et al. 2015). Indeed, the legitimacy of qualitative approaches as a method for impact evaluation studies is extended in the 2010s (Warta and Philipp 2014), a prominent example being the use of impact case studies and narrative in the Research Excellence Framework (REF) (Pinar and Horne 2022; Wilsdon et al. 2015). Nevertheless, it should be emphasised that aggregated econometric top-down models are also envisaged, at least for measuring the EU Key Impact Pathways (see, for example, the Nemesis model[13]).

To capture the increasing transformative ambition and pathway complexity of R&I policy intervention, evaluation practice more and more applies a mixed-method approach and uses triangulated data and analysis methods to capture societal impacts and transformation through R&I policies (Arnold 2018; Joly and Matt 2017; Magro and Wilson 2013; Seus and Bührer 2021; Warta and Philipp 2014).

The mid-2010s saw also a lively discussion on metrics used to measure research impact and the limitation of the traditionally used indicators (Hicks et al. 2015; Wilsdon et al. 2015). Although indicators measuring output and quality of research are, for the obvious reasons discussed in this chapter, only one type of assessment criteria used in evaluation studies, they nonetheless remain a fundament of most evaluations dealing with research funding. For example, the exploitation of bibliometric or patent databases measuring publication outputs and quality and cooperating patterns are continuously used in evaluation studies and the related methods of analysis are further fine-tuned. Today it is a standard tool in research funding evaluations (see, for example, the Evaluation of the Erwin Schrödinger Fellowship (Meyer and Bührer 2014)).

It is important to note that new metric avenues open particularly with regard to new data sources that can be summarised under the keyword "Big Data". This includes altmetrics or more general sources from the web or the use of quantitative text analysis methods to analyse big sets of documents. Examples of studies can be found in the SIPER repository (e.g. Bührer et al. 2021 and Feidenheimer et al. 2018). Furthermore, the combination of different data sets offer new possibilities for analysis. One example is the evaluation of the START-Wittgenstein programme (Seus et al. 2016) that used the bibliometric database Scopus to create a control group to be compared with the funded researchers. This was a methodological novelty, as evaluation studies usually only make use (if at all) of comparison groups drawing most of the time on non-successful applicants.

The last decade also saw a continued debate on the value of Randomised Controlled Trials (RCTs) in R&I policy evaluations. Those approaches are

[12] It must be noted that the programme theory approach is not new to the evaluation approaches of the EU, but a core approach in the European Structural Funds Evaluation and since 2017 also a standard component of the "EU better regulation guidelines".

[13] https://web.jrc.ec.europa.eu/policy-model-inventory/explore/models/model-nemesis

experimental in character which imply the inclusion of the RCT design right from the beginning of the intervention and, accordingly, have to be planned by the funding organisations before the launch of a programme (frontload of the evaluators work). Further, those approaches need a large number of funded entities and the possibility to experiment with randomised funding in the first place. The best known initiative promoting experimental innovation policy is the RCT Innovation Growth Lab (IGL) at Nesta's, which was founded in 2014 and its IGL Trails Database collects examples from experimentation and controlled trials.[14] The Austrian Research Promotion Agency (FFG) has engaged itself in this journey for three of its programmes (Landon and Hochreiter 2022). However, as pointed out by Firpo and Phipps (2019) RCTs have not been the main focus of R&I policy and methodological developments are rare with a few exceptions, such as the RCT+ approach (Bakhshi et al. 2015). Whether it will become an interesting method to fund and consecutively measure the impact of mission and transformation-oriented policies will be seen in the coming years. It will, in any case, be limited to very specific research questions and a very specific form of funding programmes, with a large number of funded and non-funded entities and the political opportunity for the experiments to be executed.

With the increasing importance of formative evaluation and ex-ante impact assessment, the demand has grown for broader data collection and analysis methods used in evaluation, and especially to include foresight methods, i.e. scenario building or Delphi surveys (Dinges et al. 2020b; JIIP 2018; Patton 2019; Weber and Polt 2014). These authors emphasised the fact that foresight methods can enrich the evaluation as they help explore the interaction between the different interventions and policies (policy mix) and acknowledge the complexity of longer-term change. Furthermore, inclusiveness, stakeholder involvement (opening up to non-traditional R&I stakeholders) and participation in evaluation design and implementation (co-creation) gained in importance as well as the evaluation of experiments and transdisciplinary research processes (Belcher and Hughes 2021).

Finally, this decade also saw a persistent interest in collating and learning from existing evaluation studies. One prominent example is the Handbook of Policy Impact (Edler et al. 2016), which, through a unified "evaluation synthesis approach", shows the diversity of effects generated by different policy instruments. A second example is the SIPER database (www.si-per.eu) which allows access to a wide variety of evaluation studies to be used for different analyses in terms of evaluation practice and innovation policy effects.

5.4 *Evaluation Community*

The R&I evaluation community has grown into a mature and stable community in which the changes in the R&I policy field resonate. Looking at the fteval's mission statement of 2014, one notices the shift from a narrow technology policy focus

[14] https://innovationgrowthlab.org/igl-database-v2

towards a broader definition of innovation (Warta and Philipp 2014). The field became further institutionalised. Most notably, between 2013 and 2022, the European R&I Policy Evaluation Conference (in 2013 still named "evaluation of STI policies, instruments and organisations") took place four times (in 2013, 2017, 2019 and 2022) and has established itself in a biennial rhythm. It is a forum which brings together researchers, evaluation practitioners and policy makers. The evaluation standards elaborated by the fteval members were updated for the last time in 2019, reflecting the changes in the R&I landscape and taking up the main discussion described in this chapter. In 2016 the Network for Advancing & Evaluating the Societal Impact of Science (AESIS) brought together the organisations involved in the evaluation of impact, research strategy and funding of research. Although the R&I evaluation community is today a consolidated community, it has nevertheless opened up to well-established evaluation practices used in other policy domains (especially social sector evaluation and development assistance) such as the concepts of theory-of-change modelling and applied them to R&I evaluations. As can be seen in numerous contributions of the latest European R&I evaluation conference (ftval-journal 2022), the evaluation community is reacting to these developments by discussing new approaches to measure effects and provide evidence for policy learning.

6 Summary

The evaluation practice in R&I policy has shown a remarkable development in terms of its very purpose, the questions asked, the methods used and the institutionalisation of the community. As we have seen, throughout the last decades, this development has naturally interacted with the development in R&I policy, practice and theory.

In sum, in the 1960s, 1970s and 1980s the purposive evaluation of public research and innovation activities and policies took shape. It covered a growing scope of topics and it gradually became more professional, from policy planning to policy implementation and policy impact evaluation; from science to technology, research and innovation processes as subjects of evaluation; from academic studies (sociology of science; economics; policy studies) to systematic professional research projects, often on behalf of governmental actors such as a service, nurtured by an internationally growing community of experts.

The 1990s took up and further developed the central approaches and work from the previous decades. However, the 1990s also saw important new developments in the field of policies (e.g. multi-actor, multi-measure programmes) as well as the strengthening of a national and international evaluation community. The decade was characterised by increased reflection on evaluation concepts and methods and the subsequent methodological and conceptual developments. On the one hand, these have brought the idea of evaluation as a core element of policy intelligence into focus, but on the other hand, they have also further developed numerous quantitative and indicator-based methods.

The decade between 2000 and 2010 was one of transition. R&I policy, while further differentiating, slowly moved from a system enhancement approach, focused mainly on economic and system effects, to an approach that was increasingly interested in societal impacts. In terms of strengthening the performance and structure of the science and innovation systems, policy measures targeting complex network structures led to improved evaluation methods to capture this complexity. Increasingly attempts were made to assess and benchmark entire systems and the role of R&I policies. At the same time, attempts intensified to capture impact and impact pathways of research, which foreshadowed a more ambitious R&I policy in the years to come. The policy and evaluation community also intensified stocktaking and learning exercises to reflect the opportunities and weaknesses of the evaluation practice and to further standardise the toolkit and strengthen the self-awareness of the evaluation practitioners as an epistemic community.

In line with the increased societal ambition of R&I policy, the years from 2010 to today have been marked by an intensified discussion on how to measure broader impacts of research and R&I policies. As a consequence, the evaluation community turns its attention to developing and refining conceptual frameworks suitable for analysing the processes leading to longer-term impacts of research and to develop metrics capable of measuring those long-term effects. Towards the end of the decade, the formative function of evaluation focusing on understanding funding mechanisms and distribution processes grew in importance reflecting the difficulties in measuring impacts and the need for advice how to govern the transition towards sustainability. The evaluation community in the R&I field has grown into an epistemic community using a wide range of qualitative and quantitate methods and exploring the possibilities of big data analysis.

7 Reflections and Outlook

7.1 Co-evolution of Policy and Evaluation: Interaction and Interdependencies

Our review of the emergence and development of evaluation concepts and efforts in the domain of research and innovation policy in the last 60 years suggests that, often, policy ambitions and designs, on the one hand, and evaluation methods and practices, on the other, go hand in hand; they co-evolve. Creative evaluation experts, based in academia or in consultancy, have developed mixed-method approaches to understand and model complex input/output relations of R&I policy interventions. They have advanced qualitative as well as quantitative data analysis methods, and they have explored new data sources. Vice versa, enlightened and creative policy-makers have used intelligence input from such evaluation studies to conceptualise and implement ever more sophisticated policy instruments that aim to stimulate effective and competitive research and innovation activities.

This co-evolution, though, occurs also with dialectic frictions: diverging policy targets and designs by competing policy actors (different ministries, agencies) interact with diverging evaluation concepts and designs by different experts, with diverse epistemological backgrounds. Evaluation results can and should inform the owner of an evaluated policy, however, evaluation results will also be perceived by other policy actors. This in turn can confuse, but also stimulate debate and competition between policy approaches and evaluation designs. Ideally, evaluation exercises would adopt a "moderation" function in policy arenas with competing actors. Evaluation can function as a strategic intelligence interface in a "multiple perspective framework" (Kuhlmann 1998a). Helping policy actors with the means of evaluation studies to reflect on the perspectives of other actors in the R&I policy arena can enrich and facilitate a re-framing (Schön and Rein 1994) of the orientations of corporatist and policy-administrative actors (Kuhlmann 1998a), a precondition for policy learning (Padilla 2016).

7.2 Critical Developments and Challenges

Over the past decades, the practice of evaluating publicly funded research and innovation has increased significantly. There is no doubt that evaluation exercises are necessary and justified. Funded projects must demonstrate that the use of taxpayers' money is efficient and meaningful. But what we also see today is that more and more evaluation exercises are done mechanistically by research or policy administrators or specialised service providers without a sound understanding of the scientific process or content, but rather following a given, more or less uniform set of technical criteria. This can lead to scientists performing to meet those criteria and stifle scientific creativity. The important search for evidence of public policy impacts can, if done mechanistically, revert and incite evaluated organisations to fabricate "policy-driven evidence" (Strassheim and Kettunen 2014).

Another worrying development in recent evaluation practice is the definition and application of a gold standard of evaluation in R&I policy which then puts pressure on the evaluation community and even on the design of programmes. More concretely, randomised controlled trials, as we have known them for decades, in health or education research policy have been increasingly asked for an innovation and entrepreneurship policy as well (Bravo-Biosca 2019; Landon and Hochreiter 2022; Roelandt and van der Wiel 2020). While those approaches have their merits for interventions in health and education, where the population is far more numerous and experiments can be on a large scale, and where large-scale intervention has an identifiable impact on individual units, their merit for science and innovation policy is limited. Nevertheless, the normative pressure to use RCTs has grown in a number of countries (Leko et al. 2016)[15] which—in our observation—puts pressure on policy makers to design programmes to fit the gold standard RCT or to apply RCTs even if, given the low numbers of units involved in trials, RCTs do not really fit the programme. Here, policy risks follow evaluation fashion rather than problem adequacy.

A further challenge is that the evaluation of research organisations and their performance, such as those carried out in the United Kingdom since the 2010s, but also in the Netherlands, have far-reaching implications not only for individual scientists and their institutions, but also for the entire research landscape. Many experts argue that research assessment procedures went out of control. In the Netherlands, for example, the initiative "Science in Change" (www.scienceintransition.nl/english) has addressed this prominently. The main concern of the initiative was that evaluation mechanisms used in academia are too mechanistic, if the quality of science is primarily measured by highly formalised performance indicators.

Such mechanistic, quantitative approaches can, inter alia, lead to an underestimation of the different dimensions of science as well as of societal relevance. Accordingly, there has recently been an increase in discourses on "responsible metrics", which primarily aim to make the existing procedures for evaluating science more responsible, as has been called for by numerous researchers in the context of DORA and the Leiden Manifesto, among others (Wilsdon et al. 2017; Hicks et al. 2015; DORA (San Francisco Declaration on Research Assessment 2012, (https://sfdora.org/read)). The current initiative at the EU level to improve research assessment is also part of this series of activities. More than 350 organisations from over 40 countries have signalled their interest in participating in this initiative and in July 2022 the final version of the agreement on the reforming research assessment was published (https://research-and-innovation.ec.europa.eu/news/all-research-and-innovation-news/reforming-research-assessment-agreement-now-final-2022-07-20_en).

7.3 Future Developments

Topics in evaluation research and practice for the years to come in particular support R&I policy in developing and implementing meaningful transformative or mission-oriented approaches. This will mean more formative work—including a link to foresight—and new attempts to understand, identify and monitor conditions for the pathways of impact. Additionally, the evaluation of action research and transdisciplinary research processes (Belcher and Hughes 2021) become important. At the same time, with the increased ambition of R&I policy to contribute to transformation and to achieve missions, there is more pressure to "deliver". Consequently, the

[15] See, for example, the broad internationally impactful activity by the think tank NESTA, https://www.nesta.org.uk/toolkit/running-randomised-controlled-trials-in-innovation-entrepreneurship-and-growth/

formative work will need a range of (new) summative endeavours to deliver on "measuring" the success of policy intervention and R&I activities in terms of transformative pathways, etc. There is also a need to open up the R&I evaluation community to experts in other policy areas (such as health or environment) to show effects of R&I on transformations. Also, those experts can improve the formative work.

In this context, it is increasingly challenging and important to address the attribution problem. This is why approaches like theory-based evaluations and modelling, already described above, are gaining in importance. This is because, first, the desired impacts of transformation-oriented R&I policies are by definition long-term effects and are thus often not yet available when evaluation is asked to measure impact. Second, any given research and innovation policy intervention is often only one factor among many influencing the form and speed of the transformation. Here, new approaches and new combinations are needed. The modelling of societal-impact scenarios can draw on a well-established approach of explicating assumptions that allows for a combination of R&I policy with other important determinants of impact. The plurality of scenarios then illustrates the future openness of the long-term impacts. Such an approach brings together qualitative bottom-up information with system analyses, i.e. it also addresses qualitative factors and does not intend to cover all relevant factors with one uno actu top-down (econometric) model. Thus, building on the experience with system-analytical modelling to deal with the complexity of impact relationships in the context of ex-ante sustainability impact assessments (e.g. in energy and mobility scenarios) can also be a suitable approach for making societal impacts tangible.

Further, there are increasing calls in evaluation research to conduct evaluations (even more so than in the past) with the active involvement of different stakeholder groups, a trend that is being intensified with the current claims of R&I policy to support transformation and thus to engage with a broader and more complex context and actor constellation. This stakeholder involvement can consist of asking non-academic user groups about their perception of the "impact" of research. It can also include a process component, namely that stakeholder groups and/or those affected are systematically involved during the design, implementation and validation phases of an evaluation. Such procedures are certainly easier to implement in formative evaluations than in summative ones, even if they are also possible in the latter.

Finally, our necessarily abridged and somewhat selective overview of five decades of evaluation practice in R&I policy has shown that we need continuous, in-built critical reflection on the (continuously changing) role of evaluation in R&I policy and what it can, and cannot, deliver.

Given the increased ambition of R&I policy, it remains critical that the practice of evaluation shall continue to interact productively with policy practice to co-develop both evaluation and policy in meaningful ways. Thus, it is more important than ever that evaluation exercises are not seen as necessary, mechanistic services following pre-defined scripts, but rather are understood as essential processes to learn and to further develop, apply and communicate methods and concepts.

References

Airaghi A, Becker C, Bobe B, Georghiou L, Kemper V, Rönström T, Wiegart N (1995) EUREKA evaluation report. Brussels

Airaghi A, Busch, Niels E, Georghiou L, Kuhlmann S, Ledoux MJ, van Raan AFJ, Baptista JV (1999) Options and limits for assessing the socio-economic impact of European RTD programmes. Report to the European Commission, DG XII, Evaluation Unit, Brussels/Luxembourg

Amanatidou E, Cunningham P, Gök A, Garefi I (2014) Using evaluation research as a means for policy analysis in a 'new' mission-oriented policy context. Minerva 52:419–438. https://doi.org/10.1007/s11024-014-9258-x

Arnold E (2004) Evaluating research and innovation policy: A systems world needs systems evaluations. Res Eval 13:3–17

Arnold E (2011) Understanding the long-term impacts of the EU framework programme of research and technological development. Inaugural lecture. School of Management and Governance of the University of Twente, Twente

Arnold E (2012) Understanding long-term impacts of R&D funding: The EU framework programme. Res Eval 21:332–343. https://doi.org/10.1093/reseval/rvs025

Arnold E (2018) How should we evaluate complex programmes for innovation and sociotechnical transitions? Technopolis. Checked on 21 May 2022

Arnold E, Kuhlman S, van der Meulen B (2001) A singular council: Evaluation of the Research Council of Norway. Technopolis, Brighton, England

Bach L, Lambert G (1992) Evaluation of the economic effects of large R&D programmes: The case of the European space programme. Res Eval 2:17–26

Bakhshi H, Edwards JS, Roper S, Scully J, Shaw D, Morley L, Rathbone N (2015) Assessing an experimental approach to industrial policy evaluation: Applying RCT+ to the case of Creative Credits. Res Policy 44:1462–1472. https://doi.org/10.1016/j.respol.2015.04.004

Becher G, Kuhlmann S (eds) (1995) Evaluation of technology policy programmes in Germany. Economics of science, technology and innovation, vol 4. Springer Dordrecht, Dordrecht

Belcher BM, Hughes K (2021) Understanding and evaluating the impact of integrated problem-oriented research programmes: Concepts and considerations. Res Eval 30:154–168. https://doi.org/10.1093/reseval/rvaa024

Belcher BM, Davel R, Claus R (2020) A refined method for theory-based evaluation of the societal impacts of research. MethodsX 7:100788. https://doi.org/10.1016/j.mex.2020.100788

Biegelbauer P (2007) Learning from abroad: the Austrian competence centre programme Kplus. Sci Public Policy 34:606–618

Bobe B, Bührer S, Chelli R, Eriksen H, Peksen Z, Nedeva M, Sand F (1999) The continuous and systematic evaluation of EUREKA. Annual impact report 1999, Brussels

Boekholt P (2010) The evolution of innovation paradigms and their influence on research, technological development and innovation policy instruments. In: Smits RE, Kuhlmann S, Shapira P (eds) The theory and practice of innovation policy. Edward Elgar, Cheltenham

Bornmann L (2013) What is societal impact of research and how can it be assessed? a literature survey. J Am Soc Inf Sci Tec 64:217–233. https://doi.org/10.1002/asi.22803

Borrás S, Edler J (2020) The roles of the state in the governance of socio-technical systems' transformation. Res Policy 49:103971. https://doi.org/10.1016/j.respol.2020.103971

Bozeman B, Rogers J (2001) Strategic management of government-sponsored R&D portfolios. Environ Plan C Governm Policy 19:413–442. https://doi.org/10.1068/c1v

Braun D, Hinze S, Hüsing B, Kuhlmann S, Menrad K, Peter V (1997) Interdisziplinäre Zentren für Klinische Forschung in der Frühphase der Bundesförderung—Vergleichende Analyse: Zwischenbericht zur Prozeßevaluation. Fraunhofer-IRB-Verl, Stuttgart

Bräunling G, Harmsen DM (1975) Die Förderungsinstrumente der Forschungs- und Technologiepolitik, Schriftenreihe der Kommission für wirtschaftlichen und sozialen Wandel, Bd. 85, Göttingen

Bravo-Biosca A (2019) Experimental innovation policy. MIT Press, Cambridge

Brisolla S (1996) Innovation indicators for less developed countries. Res Eval 6:187–192. https://doi.org/10.1093/rev/6.3.187

Bruno N, Kadunc M (2019) Impact pathways: tracking and communicating the impact of the European Framework Programme for research and innovation. fteval J Res Technol Policy Eval 62–71. https://doi.org/10.22163/fteval.2019.330

Bührer S (2001) Die Nutzung von Zukunftsstudien im Rahmen der Systemevaluation der Fraunhofer-Gesellschaft—neue Ansätze der Evaluation von Forschungseinrichtungen. In: Deutsche Gesellschaft für Evaluation e.V. (ed) Evaluation—Reformmotor oder Reformbremse? DeGEval, Köln, pp 155–160

Bührer S (2002) Network analysis. In: European Commission & Joanneum Reserach (ed) EPUB RTD Evaluation Toolbox, Sevilla, pp 163–169

Bührer S, Kuhlmann S (1999) Evaluation of science and technology in the new Europe. In: Proceedings of an international conference on 7 and 8 June 1999, Berlin,

Bührer S, Peter V (1999) Kommunikation & Kooperation in interdisziplinären Zentren für klinische Forschung. Fraunhofer IRB-Verlag, Stuttgart

Bührer S, Bierhals R, Hullmann A, Studer T, Erlinghagen R (2000) Begleitende Evaluation der Kompetenzzentren der Nanotechnologie. Statusanalyse, Karlsruhe

Bührer S, Bierhals R, Heinze T, Studer T, Erlinghagen R (2001) Begleitende Evaluation der Kompetenzzentren der Nanotechnologie in der Frühphase der Bundesförderung. Fraunhofer Institute for Systems and Innovation Research (ISI), Karlsruhe

Bührer S, Bierhals R, Heinze T, Hullmann A, Studer T, Erlinghagen R et al (2002) Die Kompetenzzentren der Nanotechnologie in der Frühphase der Bundesförderung : Ein Bericht der begleitenden Evaluation. Fraunhofer ISI, Karlsruhe

Bührer S, Edler J, Hinze S, Kuhlmann S, Braun D (2004) Förderung interdisziplinärer Zentren für klinische Forschung: Abschlussevaluation und Ausblick: Prozessevaluation der Fördermaßnahme des Bundesministeriums für Bildung und Forschung "Interdisziplinäre Zentren für Klinische Forschung an den Hochschulkliniken". Fraunhofer IRB-Verlag, Stuttgart

Bührer S, Wallwaey E, Seus S, Edler J (2021) Policy brief, issue 8/SIPER database, the increasing role of studies on social & environmental impacts

Bührer S, Feidenheimer A, Frietsch R, Karaulova M, Schubert T (2022a) Report on the impact of science. Current challenges, concepts and state of the art in impact assessment

Bührer S, Feidenheimer A, Walz R, Lindner R, Beckert B, Wallwaey E (2022b) Concepts and methods to measure societal impacts—an overview. Discussion papers innovation systems and policy analysis no. 74. Fraunhofer ISI, Karlsruhe

Bush V (1945) Science, the endless frontier. A report to the President by Vannevar Bush. Director of the Office of Scientific Research and Development. United States Government Printing Office, Washington

Cabo PG, Bijmolt THA (1992) International R&D networks: The Eureka map. Res Eval 2:161–168. https://doi.org/10.1093/rev/2.3.161

Cañibano C, Bozeman B (2009) Curriculum vitae method in science policy and research evaluation: the state-of-the-art. Res Eval 18:86–94. https://doi.org/10.3152/095820209X441754

Chen H-T, Rossi PH (1983) Evaluating with sense. Eval Rev 7:283–302. https://doi.org/10.1177/0193841X8300700301

Cozzens S (2010) Social impact assessment in U.S. S&T policy. In: Presentation to the SIAMPI workshop, 12 Mar 2010

Crozier M (1965) Pour une analyse sociologique de la planification française. Revue française de sociologie 6:147–163

Cunningham P, Nedeva M (1999) Towards a system of continuous evaluation and monitoring for European co-operation in scientific and technical research (COST). Res Eval 8:142–154. https://doi.org/10.3152/147154499781777487

Cunningham P, Miles I, Nedeva M, Stein J, Husson, JP, de Montgolfier P et al (1997) COST evaluation: main report, Brussels

Dale AJ, Barker KE (1994) The evaluation of EUREKA: a pan-European collaborative evaluation of a pan-European collaborative technology programme. Res Eval 4:66–74

Daniel H-D (1989) Ansätze zur Messung und Beurteilung des Leistungsstandes von Forschung und Technologie. Beiträge zur Hochschulforschung 223–231
Da Silva CM, Henriques L (1995) R&D evaluation in Portugal. Res Eval 5:89–97. https://doi.org/10.1093/rev/5.1.89
DeGEval—Gesellschaft für Evaluation e.V. (ed) (2016) Standards für Evaluation. Langfas-sung. Available online at https://www.degeval.org/degeval-standards-alt/download/
Dekker W et al (1991) Report of the EUREKA assessment panel. EUREKA Secretariat
Delahais T, Toulemonde J (2012) Applying contribution analysis: Lessons from five years of practice. Evaluation 18:281–293. https://doi.org/10.1177/1356389012450810
Deutsch KW (1973) Zum Verständnis von Krisen und politischen Revolutionen. In: Jänicke M (ed) Herrschaft und Krise. VS Verlag für Sozialwissenschaften, Wiesbaden
Diercks G, Larsen H, Steward F (2019) Transformative innovation policy: Addressing variety in an emerging policy paradigm. Res Policy 48:880–894. https://doi.org/10.1016/j.respol.2018.10.028
Dinges M, Meyer S, Brodnik C (2020a) Key elements of evaluation frameworks for transformative R&I programmes in Europe. fteval J Res Technol Policy Eval 26–40. https://doi.org/10.22163/fteval.2020.486
Dinges M, Wang A, Schuch K (2020b) Using the Delphi Method in evaluations—incorporating a future oriented perspective in evaluations. fteval J Res Technol Policy Eval 44–50. https://doi.org/10.22163/fteval.2020.469
Donovan C (2011) State of the art in assessing research impact: introduction to a special issue. Res Eval 20:175–179. https://doi.org/10.3152/095820211X13118583635918
Donovan C, Hanney S (2011) The 'payback framework' explained. Res Eval 20:181–183. https://doi.org/10.3152/095820211X13118583635756
Edler J, Lo V (2004) Assessment "Future of the competence centre programmes (Kplus und Kind/net) and future of the competence centres": executive summary. Fraunhofer Institute for Systems and Innovation Research (ISI); KMU Forschung Austria - Austrian Institute for SME Research, Karlsruhe
Edler J, Bührer S, Lo V, Rainfurth C, Kuhlmann S, Sheikh S (2004) Assessment "Zukunft der Kompetenzzentrenprogramme (K plus und K ind/net) und Zukunft der Kompetenzzentren". Fraunhofer Institut, Karlsruhe
Edler J, Joly PB, Kuhlmann S, Nedeva M, Propp T, Rip A et al (2006) Understanding "fora of strategic intelligence for research and innovation". The PRIME Forum Research Project, Report on major results, Strategic review. Manchester Institute of Innovation Research, Manchester
Edler J, Ebersberger B, Lo V (2008) Improving policy understanding by means of secondary analyses of policy evaluation. Res Eval 17:175–186. https://doi.org/10.3152/095820208X331667
Edler J, Amanatidou E, Berger M, Bührer S, Cunningham P, Daimer S et al (2010) INNO-appraisal. Perspectives on evaluation and monitoring. Final report: (contract number: 04377). Manchester Institute of Innovation Research, Manchester
Edler J, Cunningham P, Gok A, Shapira P (2016) Handbook of innovation policy impact. EU-SPRI Forum on Science, Technology and Innovation Policy. Edward Elgar, Northampton, MA
Etzioni A (1968) The active society, 2nd edn. Free Press, New York
European Commission, Directorate-General for Research and Innovation, Second European report on S&T indicators 1997 – Report, Publications Office, 1997
European Commission, Directorate-General for Research and Innovation (2006) Creating an innovative Europe : report of the independent expert group on R&D and innovation appointed following the Hampton Court Summit. Publications Office
Evaluierungskommission (ed) (1998) Systemevaluierung der Fraunhofer-Gesellschaft. Bericht der Evaluierungskommission, München
Fagerberg J, Martin BR, Andersen ES (2013) Innovation studies: evolution and future challenges. Oxford University Press, Oxford
Fahrenkrog G, Polt W, Rojo J, Zinöcker K, Tübke A (2002) RTD Evaluation Toolbox—assessing the socio-economic impact of RTD-Policies, Sevilla

Fayl G, Dumont Y, Durieux L, Karatzas I, O'Sullivan L (1998) Evaluation of research and technological development programmes: a tool for policy design. Res Eval 7:93–97. https://doi.org/10.1093/rev/7.2.93

Fayl G, Dumont Y, O'Sullivan L, Durieux L, Karatzas I, Teuber H (1999) Overview of the EC RTD programme monitoring and evaluation system. fteval J Res Technol Policy Eval 8:2–5

Feidenheimer A, Frietsch R, Schubert T, Neuhäusler P (2018) Intermediate report on the conceptual framework, project report. Big data approaches for improved monitoring of research and innovation performance and assessment of the societal impact in the Health, Demographic Change and Wellbeing Societal Challenge. Fraunhofer ISI, Karlsruhe

Feller I (2007) Mapping the frontiers of evaluation of public-sector R&D programs. Sci Public Policy 34:681–690. https://doi.org/10.3152/030234207X258996

Ferlie E, Ashburner L, Pettigrew A (1996) The new public management in action. Oxford University Press, Oxford

Firpo T, Phipps J (2019) New running experiments in innovation and growth policy: What can we learn from recent experience? fteval J Res Technol Policy Eval 47:46–50. https://doi.org/10.22163/fteval.2019.328

Flecha R, Radauer A, Besselaar P (2018) Monitoring the impact of EU framework programmes: expert report. Publications Office

Freeman C, Soete L (1997) The economics of industrial innovation. MIT Press, Cambridge, MA

Friedmann J (1987) Planning in the public domain: From knowledge to action. Princeton University Press, Princeton

ftval-journal (2022) Ausgabe Nr. 53 fteval Journal for Research and Technology Policy Evaluation conference proceeding: Proceedings of the revaluation conference 2021/22

Gabolde J (1998) New challenges for indicators in science and technology policy-making: a European view. Res Eval 7:99–104. https://doi.org/10.1093/rev/7.2.99

Georghiou L (1995a) Research evaluation in European national science and technology systems. Res Eval 5:3–10. https://doi.org/10.1093/rev/5.1.3

Georghiou L (1995b) Assessing the framework programmes. Evaluation 1:171–188. https://doi.org/10.1177/135638909500100204

Georghiou L (1999) Socio-economic effects of collaborative R&D—European experiences. J Technol Transf 24:69–79. https://doi.org/10.1023/A:1007724804288

Georghiou L, Meyer-Krahmer F (1992) Evaluation of socio-economic effects of European Community R&D programmes in the SPEAR network. Res Eval 2:5–15. https://doi.org/10.1093/rev/2.1.5

Georghiou L, Cameron H, Stein JA, Neveda M, Janes M, Yates J, Senker J (1993) The impact of European community policies for research and technological development upon science and technology in the United Kingdom. HMSO, London

Georghiou L, Rigby J, Cameron H, Vonortas NS, Prastacos G, Spanos Y, Guy K (2002) Assessing the socio-economic impacts of the framework programme. PREST, Manchester

Geuna A (2001) The changing rationale for European university research funding: are there negative unintended consequences? J Econ Issues 35:607–632

Grupp H, Hohmeyer O (1986) A technometric model for the assessment of technological standards and their application to selected technology-intensive products. Technol Forecast Soc Change 30:123–137

Grupp H, Schmoch U, Kuntze U (1995) New technology indicators for the evaluation of research and development programmes. In: Becher G, Kuhlmann S (eds) Evaluation of technology policy programmes in Germany. Springer, Dordrecht, pp 243–284

Guba EG, Lincoln YS (1989) Fourth generation evaluation, 13th edn. Sage, Newbury Park

Guy K, Polt W (1999) Strategic options for the evaluation of the R&D programmes of the European Union. fteval J Res Technol Policy Eval 8:5–9

Hage J, Jordan G, Mote J (2007) A theory-based innovation systems framework for evaluating diverse portfolios of research, part two: macro indicators and policy interventions. Sci Public Policy 34:731–741. https://doi.org/10.3152/030234207X265385

Hall PA, Soskice DW (2001) Varieties of capitalism: the institutional foundations of comparative advantage. Oxford University Press, Oxford
Hicks D (1991) A cautionary view of co-citation analysis. Res Eval 1:31–36. https://doi.org/10.1093/rev/1.1.31
Hicks D (2012) Performance-based university research funding systems. Res Policy 251–261
Hicks D, Wouters P, Waltman L, de Rijcke S, Rafols I (2015) Bibliometrics: The Leiden Manifesto for research metrics. Nature 520:429–431. https://doi.org/10.1038/520429a
Hills PV, Dale AJ (1995) Research and technology evaluation in the United Kingdom. Res Eval 5:35–44. https://doi.org/10.1093/rev/5.1.35
Holland D, Kuhlmann S (1995) Systemwandel und industrielle Innovation: Studien zum technologischen und industriellen Umbruch in den neuen Bundesländern. Technik, Wirtschaft und Politik, vol 16. Physica-Verlag HD, Heidelberg
Internationale Kommission (1999) Forschungsförderung in Deutschland. Bericht der internationalen Kommission zur Systemevaluation der Deutschen Forschungsgemeinschaft und der Max-Planck-Gesellschaft. Volkswagen-Stiftung, Hannover
JIIP (2018) Mission-oriented research and innovation : assessing the impact of a mission-oriented research and innovation approach: Final report. Publications Office, Brussels
Joly P-B, Matt M (2017) Towards a new generation of research impact assessment approaches. J Technol Transf. https://doi.org/10.1007/s10961-017-9601-0
Joly P-B, Gaunand A, Colinet L, Larédo P, Lemarié S, Matt M (2015) ASIRPA: A comprehensive theory-based approach to assessing the societal impacts of a research organization. Res Eval 24:440–453. https://doi.org/10.1093/reseval/rvv015
Jongbloed B, Lepori B (2015) The funding of research in higher education: mixed models and mixed results. In: Huisman J, De Boer H, Dill DD, Souto-Otero M (eds) The Palgrave international handbook of higher education policy and governance. Palgrave Macmillan, London, pp 439–462
Jongbloed B, Vossensteyn H (2001) Keeping up performances: an international survey of performance-based funding in higher education. J High Educ Policy Manag 23:127–145. https://doi.org/10.1080/13600800120088625
Kastrinos N, Katsoulacos Y (1995) Towards a national system of research evaluation in Greece. Res Eval 5:63–68. https://doi.org/10.1093/rev/5.1.63
Kingsbury N (1986) Coming together: evaluation network and evaluation research society share common business agendas at Evaluation '85 leading to the American Evaluation Association. Eval Pract 7:107–110
Kohlweg K (2019) Standards der Evaluierung in der Forschungs-, Technologie-, und In-novationspolitik. Technischer Bericht. fteval—Österreichische Plattform für For-schungs- und Technologiepolitikevaluierung, Vienna. Available online at https://repository.fteval.at/387/
Koschatzky K (1996) Technology transfer, innovation networking and regional development: Paper presented at the RESTPOR '96 conference on "global comparison of regional RTD & innovation strategies for development and cohesion", 19–21 Sept 1996, Brussels. Arbeitspapier Regionalforschung, Nr. 4. ISI, Karlsruhe
Koschatzky K (2000) The regionalisation of innovation policy in Germany—theoretical foundations and recent experience. Working Papers Firms and Regions, R1/2000, Karlsruhe
Krauch H (1970) Prioritäten für die Forschungspolitik. Hanser, München
Kroll H, Hansmeier H, Hufnagl M (2022) Productive interactions in basic research an enquiry into impact pathways at the DESY synchrotron. Technological forecasting and social change, vol 175
Krull W (1992) Perspektiven für Wissenschaft und Forschung in den neuen Lndern. Naturwissenschaften 79:49–59. https://doi.org/10.1007/BF01131803
Kuhlmann S (1995) Patterns of science and technology policy evaluation in Germany. Res Eval 5:23–33. https://doi.org/10.1093/rev/5.1.23
Kuhlmann S (1998a) Moderation of policy-making? Evaluation 4:130–148. https://doi.org/10.1177/13563899822208491

Kuhlmann S (1998b) Politikmoderation: Evaluationsverfahren in der Forschungs- und Technologiepolitik. Nomos Verlagsgesellschaft, Baden-Baden

Kuhlmann S (2009) Evaluation von Forschungs- und Innovationspolitik in Deutschland: Stand und Perspektiven. In: Widmer T, Beywl W, Fabian C (eds) Evaluation: Ein systematisches Handbuch. VS Verlag für Sozialwissenschaften, Wiesbaden, pp 283–294

Kuhlmann S, Arnold E (2001) RCN in the Norwegian research and innovation system: background report no 12 in the evaluation of the Research Council of Norway

Kuhlmann S, Holland D (1995) Erfolgsfaktoren der wirtschaftsnahen Forschung. In: Technik, Wirtschaft und Politik, Schriftenreihe des Fraunhofer-Instituts für Systemtechnik und Innovationsforschung (ISI), vol 15. Physica-Verlag HD; Imprint; Physica, Heidelberg

Kuhlmann S, Holland D, Grupp H (1995) Evaluation von Technologiepolitik in Deutschland: Konzepte, Anwendung, Perspektiven. Technik, Wirtschaft und Politik, vol 12. Physica-Verlag, Heidelberg

Kuhlmann S, Boekholt P, Georghiou L, Guy K, Héraud J-A, Larédo P, Lemola T, Loveridge D, Luukkonen T, Polt W, Rip A, Sanz-Menendez L, Smits R (1999) Improving distributed intelligence in complex innovation systems: final report of the Advanced Science & Technology Policy Planning Network (ASTPP), Karlsruhe

Kulicke M (1993) Chancen und Risiken Junger Technologieunternehmen: Ergebnisse des Modellversuchs "Förderung technologieorientierter Unternehmensgründungen" des Bundesforschungsministeriums. Technik, Wirtschaft und Politik, vol 4. Physica-Verlag, Heidelberg

Kulicke M (2002) Multi-Akteurs-/Multi-Maßnahmen-Wettbewerbe und ihre Anforderungen an die wissenschaftliche Evaluation—Das Beispiel EXIST. In: Wissenschaft und Wirtschaft im regionalen Gründungskontext, pp 45–66

Kulicke M, Görisch J, Tönniessen F (2000) EXIST Existenzgründer aus Hochschulen. Netzwerke für innovative Unternehmens-gründungen, Bonn

Kulicke M, Görisch J, Stahlecker T (2002) Erfahrungen aus EXIST. Querschau über die einzelnen Projekte, Bonn

Landon T, Hochreiter H (2022) Randomised controlled trials and other experimental approaches in the Austrian Research Promotion Agency. fteval J Res Technol Policy Eval 160–168. https://doi.org/10.22163/fteval.2022.554

Larédo P, Callon M (1990) L'impact des programmes communautaires sur le tissu scientifique et technique français: rapport du Centre de sociologie de l'innovation de l'Ecole des mines de Paris. Documentation française

Larédo P, Mustar P (1995) France, the guarantor model and the institutionalisation of evaluation. Res Eval 5:11–21. https://doi.org/10.1093/rev/5.1.11

Leko N et al (2016) How policy formation changes with evidence: Trends toward evidence-based policy in the UK, as verified by randomised controlled trials, and some implications for Japan (Mitsubishi UFJ Research and Consulting Policy Research Report). Available online at UFJ Research and Consulting Policy Research Report. http://www.murc.jp/uploads/2016/02/seiken_160212_1.pdf. Checked on 1 Mar 2017

Lepori B, Probst C (2009) Using curricula vitae for mapping scientific fields: a small-scale experience for Swiss communication sciences. Res Eval 18:125–134. https://doi.org/10.3152/095820209X441772

Lindner R, Edler J, Daimer S (2024) Understanding paradigm change in science, technology and innovation policy: Between science push and policy pull. In: Edler J, Walz R (eds) Systems and innovation research in transition: Research questions and trends in historical perspective. Springer, Heidelberg

Lipset SM (1959) Some social requisites of democracy: economic development and political legitimacy. Am Polit Sci Rev 53:69–105. https://doi.org/10.2307/1951731

Luukkonen T (1991) Citation indicators and peer review: their time-scales, criteria of evaluation, and biases. Res Eval 1:21–30. https://doi.org/10.1093/rev/1.1.21

Luukkonen T, Niskanen P (1998) Learning through collaboration: Finnish participation in EU framework programmes. VTT Technical Research Centre of Finland

Magro E, Wilson JR (2013) Complex innovation policy systems: Towards an evaluation mix. Res Policy 42:1647–1656. https://doi.org/10.1016/j.respol.2013.06.005
Mandel E (1986) In defence of socialist planning. New Left Rev 159:5–37
Martin BR (1996) Technology foresight: Capturing the benefits from science-related technologies. Res Eval 6:158–168. https://doi.org/10.1093/rev/6.2.158
Martinsen R (1992) Theorien politischer Steuerung—auf der Suche nach dem dritten Weg. In: Grimmer K, Häusler J, Kuhlman S, Simonis G (eds) Politische Techniksteuerung. VS Verlag für Sozialwissenschaften, Wiesbaden, pp 51–73
Mayne J (2001) Addressing attribution through contribution analysis: using performance measures sensibly. Can J Program Eval 16:1–24
Mayne J (2012) Contribution analysis: Coming of age? Evaluation 18:270–280. https://doi.org/10.1177/1356389012451663
Mayntz R (1979) Public bureaucracies and policy implementation. Int Soc Sci J 31:633–645
Mayntz R (1980) Implementation politischer Programme: Empirische Forschungsberichte. Verlagsgruppe Athenäum, Hain, Scriptor, Hanstein, Königsstein/Ts
Mayntz R, Scharpf FW (1975) Policy making in the German Federal Bureaucracy. Elsevier, Amsterdam
Mazzucato M (2018) Mission-oriented research & innovation in the European Union: A problem-solving approach to fuel innovation-led growth. Publications Office of the European Union, Brussels
Meyer N, Bührer S (2014) Impact evaluation of the Erwin Schrödinger fellowships with return phase. Final report for the Austrian Science Fund (FWF), Vienna
Meyer-Krahmer F (1981) The present status and problems of impact research in technology policy: A case study on the federal program for funding research and development personnel in Germany. Res Policy 10:356–366
Meyer-Krahmer F, Reiss T (1992) Ex ante evaluation and technology assessment—Two emerging elements of technology policy evaluation. Res Eval 2:47–54. https://doi.org/10.1093/rev/2.1.47
Meyer-Krahmer F, Gielow G, Kuntze U (1983) Impacts of government incentives towards industrial innovation: An analysis of the federal programme funding R&D personnel in the Federal Republic of Germany. Res Policy 12:153–169
Mickwitz P, Neij L, Johansson M, Benner M, Sandin S (2021) A theory-based approach to evaluations intended to inform transitions toward sustainability. Evaluation 27:281–306. https://doi.org/10.1177/1356389021997855
Miles I, Cunningham P, Cox D, Crowe C, Malik K (2006) Smart innovation: supporting the monitoring and evaluation of innovation programmes. European Commission: Enterprise and Industry
Moed H, Burger WJM, Frankfort JG, van Raan AFJ (1985) The use of bibliometric data for the measurement of university research performance. Res Policy 14:131–149
Molas-Gallart J, Davies A (2006) Toward theory-led evaluation. Am J Eval 27:64–82. https://doi.org/10.1177/1098214005281701
Molas-Gallart J, Tang P (2011) Tracing 'productive interactions' to identify social impacts: an example from the social sciences. Res Eval 20:219–226. https://doi.org/10.3152/095820211X12941371876706
Molas-Gallart J, Boni A, Giachi S, Schot J (2021) A formative approach to the evaluation of transformative innovation policies. Res Eval. https://doi.org/10.1093/reseval/rvab016
Morton S (2015) Progressing research impact assessment: A 'contributions' approach. Res Eval 24:405–419. https://doi.org/10.1093/reseval/rvv016
Muhonen R, Benneworth P, Olmos-Peñuela J (2019) From productive interactions to impact pathways: Understanding the key dimensions in developing SSH research societal impact. Res Eval. https://doi.org/10.1093/reseval/rvz003
Naschold F (1969) Systemsteuerung. Geschichte und Gegenwart, vol 2. Kohlhammer, Stuttgart
Noyons ECM, van Raan AFJ (1996) Actor analysis in neural network research: The position of Germany. Res Eval 6:133–142. https://doi.org/10.1093/rev/6.2.133

OECD (ed) (1997) Policy evaluation in innovation and technology: towards best practices. OECD, Paris
OECD (2002) Benchmarking industry-science relationships. Organisation for Economic Co-operation and Development, Paris
OECD Proceedings (1999) Boosting innovation: the cluster approach, Paris
Ormala E (1993) Evaluation of EUREKA industrial and economic effects, Brussels
Padilla P (2016) Policy learning through strategic intelligence: The American Small Business Innovation Research Program (SBIR) and British Small Business Research Initiative (SBRI). University of Twente, Enschede
Patton MQ (1997) Utilization-focused evaluation: the new century text. Sage, Thousand Oak
Patton MQ (2011) Developmental evaluation. Guilford Press, New York
Patton MQ (2019) Blue marble evaluation: premises and principles. Guilford, New York
Pavitt K (1998) Do patents reflect the useful research output of universities? Res Eval 7:105–111. https://doi.org/10.1093/rev/7.2.105
Picht G (1964) Die deutsche Bildungskatastrophe: Analyse und Dokumentation. Walter, Freiburg im Breisgau
Pinar M, Horne TJ (2022) Assessing research excellence: evaluating the research excellence framework. Res Eval 31:173–187. https://doi.org/10.1093/reseval/rvab042
Polt W (2003) New developments in evaluation methods and strategy at the European level—a short review of recent projects (ASIF, EPUB). fteval J Res Technol Policy Eval 11–12
Polt W, Ploder M, Breitfuss M, Daimer S, Jackwerth T, Zielinski A (2021) Politikstile und Politikinstrumente in der F&I-Politik. Studien zum deutschen Innova-tionssystem. EFI, Berlin
Pressman JL, Wildaysky A (1984) Implementation: How great expectations in Washington are dashed in Oakland; or, why it's amazing that federal programs work at all, this being a saga of the Economic Development Ad-ministration as told by two sympathetic observers who seek to build morals on a foundation, vol 708. University of California Press, Berkeley
Reale E, Nedeva M, Duncan T, Primeri E (2014) Evaluation through impact: a different viewpoint. fteval J Res Technol Policy Eval 36–41
Reale E, Avramov D, Canhial K, Donovan C, Flecha R, Holm P, Larkin C, Lepori B, Mosoni-Fried J, Oliver E, Primeri E, Puigvert L, Scharnhorst A, Schubert A, Soler M, Soòs S, Sordé T, Travis C, van Horik R (2018) A review of literature on evaluating the scientific, social and political impact of social sciences and humanities research. Res Eval 27:298–308. https://doi.org/10.1093/reseval/rvx025
Reger G, Kuhlman S (1995) European technology policy in Germany. Technology, innovation, and policy, vol 2. Physica; Springer series, Heidelberg
Reger G, Balthasar A, Bührer S, Bättig C (1998) Switzerland's participation in the European RTD framework programmes: a win–win game? Technovation 18:425–438. https://doi.org/10.1016/s0166-4972(98)00032-7
Removille J, Clarysse B (1999) Intra-European scientific co-operation: measuring policy impact. Res Eval 8:99–109. https://doi.org/10.3152/147154499781777586
Rip A, van der Meulen BJR (1995) The patchwork of the Dutch evaluation system. Res Eval 5:45–53. https://doi.org/10.1093/rev/5.1.45
Roelandt T, van der Wiel H (2020) The long-term impact of Dutch innovation vouchers: Back to the future with randomised controlled trials. https://www.innovationgrowthlab.org/blog/long-term-impact-dutch-innovation-vouchers-back-future-randomised-controlled-trials
Rogers JD, Jordan G (2010) New research evaluation frameworks and methods for systems level learning: introduction to a special section. Res Eval 19:235–237. https://doi.org/10.3152/095820210X12827366906526
Rojo J (2003) Evaluation of RTD policy foundations: the socio-economic dimension. fteval J Res Technol Policy Eval 18–24
Ruegg R (2006) Bridging from project case study to portfolio analysis in a public R&D program a framework for evaluation and introduction to a composite performance rating system (NIST GCR 06-891)

Ruegg RT (2007) Quantitative portfolio evaluation of US federal research and development programs. Sci Public Policy 34:723–730. https://doi.org/10.3152/030234207X259021
Ruegg R, Feller I (2003) A toolkit for evaluating public R&D investment models, methods, and findings from ATP's first decade (NIST GCR 03–857)
Sandström U (2009) Combining curriculum vitae and bibliometric analysis: mobility, gender and research performance. Res Eval 18:135–142. https://doi.org/10.3152/095820209X441790
Sanz-Menéndez L (1995) Research actors and the state: research evaluation and evaluation of science and technology policies in Spain. Res Eval 5:79–88. https://doi.org/10.1093/rev/5.1.79
Schmoch U (1999) Impact of international patent applications on patent indicators. Res Eval 8:119–131. https://doi.org/10.3152/147154499781777540
Schmoch U, Grupp H, Reiss T, Strauss E (1991) Monitoring science-based biotechnology by means of patent indicators. Res Eval 1:61–68. https://doi.org/10.1093/rev/1.2.61
Schofer E, Meyer JW (2005) The worldwide expansion of higher education in the twentieth century. Am Sociol Rev 70:898–920. https://doi.org/10.1177/000312240507000602
Schön DA, Rein M (1994) Frame reflection: Toward the resolution of intractable policy controversies. Basic Books, New York
Schot J, Steinmueller WE (2018) Three frames for innovation policy: R&D, systems of innovation and transformative change. Res Policy 47:1554–1567. https://doi.org/10.1016/j.respol.2018.08.011
Scriven M (1991) Evaluation Thesaurus, 4th edn. Sage, Newbury Park
Seus S, Bührer S (2021) How to evaluate a transition-oriented funding programme? Lessons learned from the evaluation of FONA, the German framework programme to promote sustainability research. fteval J Res Technol Policy Eval 10–18. https://doi.org/10.22163/fteval.2021.514
Seus S, Heckl E, Bührer S (2016) Evaluation of the START programme and the Wittgenstein Award. Available online at https://zenodo.org/record/50610/files/Eval-START-Witt_final_report.pdf. Checked on 13 Feb 2017
Shapira P, Kuhlmann S (eds) (2003) Learning from science and technology policy evaluation: Experiences from the United States and Europe. Edward Elgar, Cheltenham
Silvani A, Sirilli G (1995) R&D evaluation in Italy: A science and technology policy view. Res Eval 5:69–77. https://doi.org/10.1093/rev/5.1.69
Spaapen J, van Drooge L (2011) Introducing 'productive interactions' in social impact assessment. Res Eval 20:211–218. https://doi.org/10.3152/095820211X12941371876742
Spaapen J, van Drooge L, Propp T, van der Meulen B, Shinn T, Marcovich A (2011) Social impact assessment methods for research and funding instruments through the study of productive interactions (SIAMPI). Final report on social impacts of research. Royal Netherlands Academy of Arts and Sciences, Amsterdam
Steiner M, Sturn D (1995) Elements of evaluation of science and technology policy in Austria. Res Eval 5:98–108. https://doi.org/10.1093/rev/5.1.98
Stern E (1993) Ongoing and participative evaluation: Purpose, design and role in the evaluation of a large-scale R&D programme. Res Eval 3:75–82. https://doi.org/10.1093/rev/3.2.75
Stockmann R (ed) (2000) Evaluationsforschung. Leske+Budrich, Leverkusen
Strassheim H, Kettunen P (2014) When does evidence-based policy turn into policy-based evidence? Configurations, contexts and mechanisms. Evid Policy 10:259–277. https://doi.org/10.1332/174426514X13990433991320
Tomizawa H, Niwa F (1996) Evaluating overall national science and technology activity: general indicator of science and technology (GIST) and its implications for S&T policy. Res Eval 6:83–92. https://doi.org/10.1093/rev/6.2.83
van Raan AF (ed) (1988) Handbook of quantitative studies of science and technology. Elsevier, Amsterdam
van Raan AFJ (1993) Advanced bibliometric methods to assess research performance and scientific development: basic principles and recent practical applications. Res Eval 3:151–166. https://doi.org/10.1093/rev/3.3.151
van Steen J (1995) S&T indicators in science policy: how can they matter? Res Eval 5:161–166. https://doi.org/10.1093/rev/5.2.161

van Steen J, Eijffinger M (1998) Evaluation practices of scientific research in the Netherlands. Res Eval 7:113–122. https://doi.org/10.1093/rev/7.2.113

Vonortas NS, Stampfer M, Zinöcker K (2007) Evidence never lies: introduction to a special issue on new frontiers in evaluation. Sci Public Policy 34:679–680. https://doi.org/10.3152/030234207X273548

Wagner RM, Jordan GB (2009) Moving towards impact when evaluating research programs: introduction to a special section. Res Eval 18:339–342. https://doi.org/10.3152/095820209X483073

Warta K, Philipp S (2014) Was bringt die Plattform fteval und zu was hat sie es gebracht? Rückblick und Ergebnisse einer Mitgliederbefragung. fteval J Res Technol Policy Eval 15–24

Weber KM, Polt W (2014) Assessing mission-orientated R&D programs: combining foresight and evaluation. fteval J Res Technol Policy Eval 11–15

Weingart P, Sehringer R, Winterhager M (1988) Bibliometric indicators for assessing strengths and weaknesses of West German science. In: van Raan AF (ed) Handbook of quantitative studies of science and technology. Elsevier, Amsterdam, pp 391–430

Weiss CH (1993) Where politics and evaluation research meet. Eval Pract 14:93–106

Weiss CH (1997) Theory-based evaluation: past, present, and future. New Dir Eval 41–55

Widmer T (1996) Meta-Evaluation: Kriterien zur Bewertung von Evaluationen. Haupt, Bern

Wilsdon J, Allen L, Belfiore E, Campbell P, Curry S, Hill S et al (2015) The metric tide: report of the independent review of the role of metrics in research assessment and management. Checked on 22 May 2022

Wilsdon J, Bar-Ilan J, Frodeman R, Lex E, Peters I, Wouters P (2017) Next-generation metrics: responsible metrics and evaluation for open science. Report. European Commission, Brussels

Wissenschaftsrat (ed) (2000) Systemevaluation der Blauen Liste—Stellungnahme des Wissenschaftsrates zum Abschluss der Bewertung der Einrichtungen der Blauen Liste, Köln

Wissenschaftsrat (ed) (2001) Systemevaluation der HGF—Stellungnahme des Wissenschaftsrates zur Hermann von Helmholtz-Gemeinschaft Deutscher Forschungszentren, Köln

Wittmann F, Hufnagl M, Roth F, Lindner R, Kroll H (2022) A framework for formative evaluation and impact assessment of mission-oriented innovation policies. Final report of the Scientific Support Action to the German High-Tech Strategy 2025—volume 2. Checked on 17 May 2022

Wollmann H (1998) Evaluation research and politics: Between a science-driven and a pluralist controversy-responsive policy-making model. Potential and limitations. Paper given to the conference of the European Evaluation Society (EES). Rom

Worthen BR, Sanders JR, Fitzpatrick JL (1997) Program evaluation: alternative approaches and practical guidelines, 2nd edn. Longman, New York

Zellner C (2002) Evaluating the social economic benefits of publicly funded basic research via scientists' career mobility. Res Eval 11:27–35. https://doi.org/10.3152/147154402781776970

Zulueta MA, Bordons M (1999) A global approach to the study of teams in multidisciplinary research areas through bibliometric indicators. Res Eval 8:111–118

Susanne Bührer is coordinator of the Business Unit "Societal Change and Innovation" at the Fraunhofer Institute for Systems and Innovation Research. Her research areas are the development and implementation of sophisticated evaluation and impact analyses, the conceptualisation of impact-oriented monitoring systems, the topic of women in research and innovation systems (with special consideration of Gendered Innovations) as well as questions of responsible research and innovation.

Jakob Edler is Managing Director of the Fraunhofer Institute for Systems and Innovation Research ISI since October 2018. The Anglo-German is also Professor of Innovation Policy and Strategy at the Manchester Institute of Innovation Research (MIoIR), which he led from 2011 to September 2018. His research focuses on analysing and conceptualising governance, policy and management of innovation. J. Edler is Speaker of the Fraunhofer Group Innovation Research, member of the German Science Council, the German Academy of Science and Engineering (Acatech) and the Royal Society for the Encouragement of Arts, Manufactures and Commerce (UK).

Stefan Kuhlmann is a political scientist and emeritus professor "Science, Technology and Society" (STS) at the University of Twente (Netherlands). He works on research and technological innovation as social and political processes, focusing on governance and politics, and he publishes widely in the field of research and innovation policy studies. Prior he was professor of Innovation Policy at Utrecht University (Netherlands, 2001–2006), held leading positions at the Fraunhofer Institute for Systems and Innovation Research (Germany 1988–2006) and was researcher at the University of Kassel (Germany, 1979–1988). He graduated in History and Political Science from the University of Marburg (Germany, 1978).

Sarah Seus is a social scientist and a researcher in the Competence Centre Policy and Society of the Fraunhofer Institute for Systems and Innovation Research ISI. Her research areas are the evaluation of science and innovation policies and specifically research on methodologies for impact evaluation (incl. societal impacts) of research funding; the analysis of STI policies and governance at national and EU level; the analysis of transition processes within the science system (e.g. transdisciplinary research) and questions of Responsible Research and Innovation (RRI).

Assessing Technological Innovations: From Early Warning to the Governance of Socio-technical Transformations

Nils B. Heyen, Michael Friedewald, Eberhard Jochem, Thomas Reiß, and Axel Thielmann

With technology we want to secure prosperity and survival, but at the same time threaten our future to the extreme

Helmar Krupp (founding director of Fraunhofer ISI)

Abstract Technology assessment (TA) is an interdisciplinary field of problem-oriented research generating knowledge primarily for decision-making processes in politics, economy, and society with regard to the opportunities, risks, and challenges of emerging technologies and socio-technical change. This chapter outlines the development of the field since its roots in the 1960s to date with a focus on Germany, the USA, and Europe. Based on a rough temporal division into three phases representing an expert-based, participatory, and pragmatic TA approach, respectively, we illustrate the field's development by highlighting concrete projects and studies, reflecting on the (ever new) development of appropriate assessment methods and pointing to relevant thematic waves ("hype cycles") of technology groups, such as energy, genetic, or information technologies. Finally, we discuss the impact and future challenges of TA given current socio-political trends and upcoming socio-technical transformations. We conclude that there is a clear need for new approaches of TA so that TA will remain a valuable instrument providing guidance in a complex and uncertain technological environment in the coming decades.

N. B. Heyen (✉) · M. Friedewald · E. Jochem · T. Reiß · A. Thielmann
Fraunhofer Institute for Systems and Innovation Research ISI, Karlsruhe, Germany
e-mail: Nils.Heyen@isi.fraunhofer.de; Michael.Friedewald@isi.fraunhofer.de; Eberhard.Jochem@isi.fraunhofer.de; Thomas.Reiss@isi.fraunhofer.de; Axel.Thielmann@isi.fraunhofer.de

J. Edler, R. Walz (eds.), *Systems and Innovation Research in Transition*, Sustainability and Innovation, https://doi.org/10.1007/978-3-031-66100-6_6

1 Introduction

For a long time, until the middle of the twentieth century, technological progress was commonly equated with societal progress leading to economic growth, social wealth, and well-being. However, this technological optimism had already begun to crack as a result of the horrors of the atomic bomb in World War II and came also in civilian contexts under further pressure, for instance, with regard to the environmental and human risks of the use of pesticides in the 1960s (Carson 1962). As the awareness of the ambivalence of technologies increased in politics and public life (Daddario 1966), the strong need emerged to create robust knowledge on the unintended and potentially adverse effects of a specific technology entering markets and society and to contrast this with the technology's promises and potential benefits.

The world's first institution for such a task of "technology assessment" was launched by the US Congress in 1972. It was named *Office of Technology Assessment* (OTA) and installed as a science-based body for parliamentary policy advice (Kunkle 1995; Bimber 1996). Although the founding history of the *Fraunhofer Institute for Systems and Innovation Research* (ISI) is different (cf. Edler and Walz 2024 in this anthology), it is from a historical perspective certainly no coincidence that OTA and Fraunhofer ISI were founded in the same year (1972), given the broader societal context at that time with its increasing discomfort with new or more frequently applied technologies leading, for example, to severe environmental degradation such as intensive air and water pollution, related diseases, corrosion, fish kill, or high cost of river filtrate (Grandjean 1960; Baram 1970; Reimer 1971).

Since the foundation of OTA, technology assessment (TA) has developed, established, and diversified as an interdisciplinary field of problem-oriented research practices (Paschen et al. 1978; Grunwald 2019). This chapter aims to outline the field's development to date. In loose alignment with the work of Kuhn (1962), the Austrian sociologist Bogner (2021) distinguishes three "paradigms" of TA in its rather brief history.

In a first paradigm or phase ("politicisation"), beginning in the late 1960s, TA aimed at increasing the number of options for policy making, fighting the idea that everything is determined by the technology itself, and therefore turning technology into a political, debatable issue. Given the steering optimism of that time, the expertise of scientists was meant to help with creating a scientific knowledge base that gives orientation for political decision-making on new technologies (Coates 1974). This has been described as the expert model of "classic" TA (Grunwald 2002, pp. 123ff.).

In a second paradigm or phase ("democratisation"), which became dominant in the 1980s, TA aimed at increasing the inclusion of people and stakeholders affected by technological projects, striving to democratise (the assessment of) science and technology, a field that had long been considered as a matter for scientific experts alone. Given the large societal conflicts on technologies such as nuclear energy or genetic engineering, especially in Europe, the perspectives and (local) expertise of citizens and stakeholders were meant to widen the value (and knowledge) base for decision-making in the political and administrative institutions, increasing its social robustness, and to influence or even co-create the design of technologies. This has been described as the "participatory turn" of TA or simply as "participatory TA" (Joss and Bellucci 2002).

In a third paradigm or phase ("pragmatisation"), since the 2000s to date, TA has aimed at contributing to the responsible design, implementation, and governance of technological innovations. Given the "normative turn" (Daimer et al. 2012) in national and supranational science, technology and innovation (STI) policies towards the so-called "grand societal challenges", both technological and social innovations are increasingly seen as important means to address these challenges, be it demographic or climate change. In this context, TA has taken a rather "pragmatic" position using diverse forms of expertise and processes of knowledge production to reflect upon technologies and societal discourses on possible socio-technical futures.

Certainly, these three TA paradigms or phases are not to be understood as strictly separate from each other (Bogner 2021). Not only do they overlap temporally, but the diverse TA approaches developed in a specific phase (such as expert-based and participatory models) have also been taken and adapted to the changed context conditions in later phases. Especially the current "pragmatic" paradigm is characterised by a "peaceful coexistence" (Bogner 2021, p. 56) of diverse TA approaches that have proven to be valuable in the past. The paradigms also reflect different constellations in the relationship between TA and science in general. Wehling (2021), a sociologist of science, distinguishes four types of such constellations: a scientistic (dominant in the paradigm of "classic" TA; cf. Wynne 1975), a constructivist, a participative (both dominant in the second paradigm to date), and a normative-reflexive constellation. The latter represents a rather new and fuzzy development, since the ("classic") claim of neutrality and the question of TA's implicit or explicit normativity have recently become a central topic of self-reflexive debates within TA (Nierling and Torgersen 2020; Torgersen 2019)—with an open outcome.

In this chapter, we take the three TA paradigms as a rough temporal division and further outline the development of the field of technology assessment in a broad sense focusing on Germany, the USA, and Europe. In contrast to Bogner (2021), our aim is not to define and justify such paradigms or phases, but to illustrate the development in more detail and practice-oriented terms. We therefore focus on concrete projects and studies, on the (ever new) search for and development of appropriate assessment methods, and on the connection to the thematic waves ("hype cycles") of specific groups of technologies, such as energy, genetic, or information technologies. In addition, since Fraunhofer ISI has played an active and influential role in the field from the beginning to date, we highlight some of its activities to illustrate the field's development.

2 The Beginnings: Expert-Based TA as Policy Advice

In the late 1960s, concerns about technological innovations and their effects became increasingly known in the USA. They came from environmentalists, doctors, and psychotherapists and were put together as complaints to the legal system. Thus, it is surprising only at first glance that the demand for systematic forecasting of new (and known) technical developments appeared very early in jurisprudential publications (Baram 1970). Green (1967), for instance, states:

> The basic question is whether our legal system is capable of imposing effective social control over new technologies before they inflict very substantial, or even irreparable injury upon society. It seems clear that we cannot rely on the courts alone to protect society against fast-moving technological developments. Judge-made rules of law always come after, and usually long after, the potential for injury has been demonstrated. (Green 1967, cited in Baram 1970, p. 569)

The technical fields addressed by lawyers and courts at that time included motorisation, aviation, genetic engineering, and nuclear power. However, the reactive nature of the courts and the limited knowledge of judges of the various fields of technical, economic, and societal impacts made it obvious that they could not serve as society's primary instrument for TA. The limits to growth report to the Club of Rome (Meadows et al. 1972) added another global and far-reaching topic (on limited natural resources) to the discussion, using new types of simulation models (such as system dynamics by Forrester 1971) and the increasing capacity of computers.

In this challenging situation, it became very clear that there had to be both interdisciplinary scientific training and corresponding research funding for the development and application of projection methods and their interdisciplinary linkage. Both were achieved through corresponding initiatives by US universities (e.g., in the Boston area; cf. Baram 1970, pp. 576–578) and the establishment of the *Research Applied to National Needs Program* (RANN) of the *National Science Foundation* (NSF). Between 1970 and 1976, the NSF supported 43 TA studies and 23 methodological studies, surveys, and conferences (Coates 1978, pp. 54–59). This support of the NSF was an essential contribution to capacity building in interdisciplinary research, facilitating also the start of work of the *Office of Technology Assessment* (OTA) in the USA in 1973 (Coates 1976).

The OTA is the first institutional format of TA and represents a first benchmark or model for an "expert-based TA", building "own" expertise for the legislative power in distinction to the US Government. OTA started with a 10 million dollar programme on issues of energy, food, oceans, health, materials, transportation, and also on methodological developments and limits (Coates 1978, pp. 63–65). Similar considerations on methods, new technical developments, and the political process were published by Hetman (1973) to inform governments of the OECD countries. In this context (Hetman 1978), a booklet on methodological guidelines for TA, co-authored by F. Hetman, J. Coates, E. Jochem, and H. Paschen, was also produced (OECD 1975). This booklet is very balanced in its assessment of the various techniques and sensible in its awareness of their strengths and limitations. It avoids the temptation of staking everything on one particular technique or quantitative method, a fault which bedevils much of the literature at that time.

In Germany, first attempts to institutionalise TA at the Federal Parliament (Bundestag) started as early as the 1970s (TAB 2022). The establishment of a parliamentary "Enquête Commission" in 1985 finally spurred the debate. While the basic decision to establish a parliamentary TA institution was taken rather quickly, the discussion about the organisational form and mode of operation of this institution included a second parliamentary "Enquête Commission" and lasted until 1989 (Petermann 1994). The *Office of Technology Assessment at the German Bundestag* (TAB) was finally established in 1990 (Paschen and Petermann 2005). It was clearly inspired by the OTA and has given policy advice to the Parliament since then.

Before being involved in both parliamentary "Enquête Commissions" that paved the way for the foundation of TAB, Fraunhofer ISI completed a study (Krupp et al. 1978) suggesting a special research programme on TA at the *German Research Foundation* (DFG) and (1) recommending pertinent areas of TA research activities in Germany, (2) addressing the difficulties and challenges involved with interdisciplinary research, and (3) proposing organisational procedures for supporting TA research. However, the proposal was not implemented by the DFG.

Nevertheless, TA research was emerging in the 1970s, both internationally and also in Germany. One important research question, for instance, addressed the limitations of the expert-based TA concept with regard to forecasting, i.e. the methodological challenge to identify unintended impacts and to produce knowledge on their causal interrelationships, which might even change in the future. Three TA studies exploring the limitations of such analyses were conducted in the 1970s and mid-1980s.

First, a problem-oriented partial TA study (Denton et al. 1976) focused on the intended and unintended impacts of a further possible strong oil price increase (which became reality in 1979). The self-reflecting analysis brought up obviously lacking knowledge about the impact of the reduced demand for final energies produced from crude oil. In addition, there was little information on short-term options for the substitution of oil products in case that crude oil prices double or triple. Short-term elasticities of demand and substitution of energies were unknown at that time. Therefore, assumptions on data had to be made and used in the newly designed system dynamics or simulation models. Furthermore, the input–output table of the West German economy was projected according to the changing energy flows, demand, and investments.

Second, in contrast to the considerable uncertainties with regard to future developments, an ex-post TA study on motorisation in the former Federal Republic of (West) Germany for the period from 1953 to 1973 (Jochem et al. 1976) showed that if a historical analogy can be used, the predictions can be very accurate and complete. This was possible because of the comparable motorisation in the USA between 1919 and 1939 which could be used as a reference (for the approach of a retrospective TA, see Coates and Finn 1979).

Third, an ex-ante TA study on three different applications of solar energy (decentralised thermal solar energy, photovoltaics, and satellite photovoltaic use) (Jochem et al. 1988) demonstrated the limitations of TA in various aspects such as available time and budget, empirical data, controversial opinions or assumptions within the research team, lacking methods, etc. Although the research team was large, quite interdisciplinary, and working together for more than 2 years, the limitations of realising the TA concept were manifold (see the team's critical self-evaluation presented in Table 1). Major limitations were related to:

- the available time and budget (33% of all critical notes by team members collected during the TA process referred to that). The research team often had to stop looking for empirical data or new projection models and evaluation tools.
- missing data, methods, or knowledge (34%). This limitation of the analytical steps taken was a challenge and frustrating as the quantification of intended or unintended impacts was often not possible or only with high degrees of uncertainty.

Less importance was attributed to principal limitations of prognosis (15%), using data from other sources, although the authors were not convinced of their reliability (8%), and doubts about the own assumptions and results within the team (7%).

The low number on "controversial opinions within the team" (see Table 1) was in contrast to the political debate in the 1970s and early 1980s. An example is the labelling of renewable energies as "additive energies" by the German energy providers in the 1980s arguing that solving existing energy problems by "additive energies" was actually more wishful thinking than reality (Benz 1987).

As TA, in its beginnings, was explicitly understood as systemic analysis and projection, the new methods of system dynamics analysis (Forrester 1971) and graph methods (Boissevain 1979) were often applied in the 1970s. In the following decades, they were more and more substituted by other methods developed and used by various scientific disciplines. However, given that the role of scientific experts in policy advice and public debates in general was increasingly criticised (Nennen and Garbe 1996), new approaches of involving stakeholders and citizens in TA processes were also developed and tested (Várkonyi 2000). Naturally, expert-based TA approaches are still in use today, but they have been complemented by methods of integrating and dealing with different types of actors and their specific expertise and perspectives, as will be shown in the following sections.

Table 1 Critical self-evaluation during the TA process: Frequency and proportion of critical notes by members of the research team with respect to areas of limitation

	Critical notes	
Areas of limitation	Frequency	Proportion (%)
Lack of data, methods, or knowledge	47	34
Time and budget limits	45	33
Principal limitations of prognosis	20	15
Taking data from others, despite critiques	11	8
Questioning own assumptions and results	9	7
Personal and institutional limitations	3	2
Controversial opinions within the team	2	1
Sum	137	100

Source: Jochem et al. (1988, p. 353)

In other areas, such as Health Technology Assessment (HTA), the expert-based model has dominated until today. Although its origins lie in OTA activities as well, over the decades HTA has taken its own pathway of differentiation and institutionalisation (Banta 2003). Nowadays, HTA sees itself as an evidence-based instrument to support policy or management decisions within the healthcare system (e.g., whether the use of a new medical technology should be reimbursed by health insurances or not). At the forefront of the evaluation are, therefore, the efficacy and safety as well as the costs (or the cost-benefit ratio) of a new medical technology, while ethical and social aspects are dealt with rather rarely.

3 The Participatory Turn

Already in the 1960s and early 1970s, a wider public debate on science and technology was considered as an important element of TA. Accordingly, the critical public engagement in science and technology was seen as one of the driving forces for the institutionalisation of TA in general and also for the foundation of OTA (Joss 2002). However, the OTA actually developed mainly into an expert-driven institution. As Joss (2002) points out, a main reason for this development lies in the fact that OTA was founded to provide scientific and technological intelligence for the US Congress in order to counterbalance the respective expertise available at the White House. Since OTA was perceived as a role model for TA in political discussions worldwide, the expert-based mode of TA became prevalent in many countries in the 1970s and 1980s (Bimber and Guston 1997)—also in Germany.

A broader involvement of the public into TA emerged again on the political agenda in the 1980s when an increasing septicism and critical discussion of new technologies appeared. For example, in 1978, the human gene for insulin was first isolated and cultivated in bacteria, and in 1982, human insulin produced by genetically modified bacteria was introduced to the market (The 1989). Genetic engineering (for the production of pharmaceuticals or modification of food), plant biotechnology, reproduction medicine (e.g., in vitro fertilisation), other biomedical technologies (e.g., genetic testing), but also nuclear energy, automation in manufacturing, environmental pollution and related technologies steered public debates on risks and (to a lesser extent) chances of new technologies (for the intensive discussion on the humanisation of work, see Lerch and Jäger 2024 in this volume). Calls for stricter legislation of (perceived) risky technologies came along with these debates resulting in the first specific regulations of genetic engineering in some European countries such as Denmark where a genetic engineering act was issued in 1986 (Joss 2002).

Regulation of new technologies in general and biotechnology in particular became highly controversial issues, not only in the public domain but also in politics, industry, and science. One of the issues was to what extent legislation was hampering international competitive positions of key industries. A detailed analysis of genetic engineering regulations and their implementation in main world regions

(Europe, USA, Japan), for example, came to the conclusion that there was no systematic competitive disadvantage of European countries including Germany due to legislation compared to other world regions (Hohmeyer et al. 1994).

In the TA communities, such intensive discussions led to a renaissance of the idea of public participation in TA, democratisation of technology development and, in general, a stronger focus on affected stakeholder groups—participatory technology assessment (pTA) entered the stage.

The conceptual foundation of pTA is based on Habermas (1968) as discussed by Hennen (2012). Habermas (1968) elaborates on the relationship between scientific expertise and political decision-making and presented two ideal types of this relation: In the decisionistic model, policy makers use information from scientists, but power and interests finally shape the goals for which scientific information is employed. Here, scientific expertise could be considered as politically instrumentalised (Hennen 2012). On the other hand, in the technocratic model all political issues are reduced to factual ones assuming that decision-making issues can be resolved on the basis of science and technology. Here, political debate is replaced by expertise (Hennen 2012). Habermas (1968) realised that none of these extreme models provide an adequate description of political reality, and he proposed, as some kind of synthesis, a "pragmatist" model. In this model, normative claims in policy making have to be examined with regard to generalisability, feasibility, cost, and utility in the light of scientific and technological knowledge. At the same time, scientific and technological knowledge need to be assessed against normative and evaluative standpoints. According to Hennen (2012), this pragmatic approach forms the basis for pTA since the pragmatic discourse between science and policy making depends on an informed public debate.

The diffusion of pTA starting from the mid-1980s was fuelled by the development of an experimentation with new methods for public engagement in TA and by the further institutionalisation of TA with a specific focus on pTA. In Europe, Denmark became the forerunner of this movement. In 1985, the *Danish Board of Technology* (DBT) was set up by the Danish Parliament (Joss 2002). Main motives for its foundation were intensive political debates and public controversy about modern biotechnology and reproductive medicine (Klüver 2000). Two large Danish companies (Novo and Nordisk Gentofte) had announced their plans to produce human insulin using genetically modified bacteria. DBT developed and implemented new methods for stakeholder participation and public engagement in these controversial debates. These include in particular "consensus conferences", which became a kind of brand of the DBT, but also "voting conferences", "scenario workshops", and "future search conferences" (see, for instance, Slocum 2003).

Shortly after the setting up of the DBT, the *Netherlands Organisation for Technology Assessment* (NOTA; now *Rathenau Institute*) was created in 1986 (Joss 2002). The mission of NOTA was to broaden the basis for decision-making in science and technology by addressing social consequences and integrating different societal stakeholder into TA processes. NOTA also experimented with new participatory methods including "science shops". Other institutions and countries joined these trends, for example Switzerland and Germany. The *Swiss Science and Technology*

Council (now *TA-SWISS*) was founded in 1992 and, among others, developed the so-called "Publiforum" adopting experiences from the DBT. In Germany, the *Academy for Technology Assessment in Baden-Württemberg* was founded, also in 1992 (but closed in 2003), with a strong focus on exploring new methods for public participation such as the "Bürgerforum" (citizen forum) (Renn 2002).

The German TAB in contrast, although founded in 1990 (see Sect. 2), was largely rooted in the concept of "classic" TA. As an institution steered by a parliamentary committee, specific framework conditions and limitations arose. While Paschen (1999) stated that TAB has implemented many modern TA concepts, he also admitted that certain ideas were hardly feasible since, for example, many parliamentarians were critical of broad citizen participation activities because they see this as "questioning the decision-making sovereignty of MPs legitimised by elections" (Paschen 1999; own translation; see also Grunwald 2003). This was also reflected, at least in the early years of TAB, in the orientation of the studies, which mostly focused on large-scale and cutting-edge research. A distinguishing feature compared to other TA institutions of that time was the continuous monitoring of future technologies, but also of "soft" factors such as citizens' perceptions of technology (aka technology acceptance).

Another element of pTA comprises educational activities. For example, starting from the late 1990s, several such activities were initiated in Germany including the elaboration of specific curricula for debating biotechnology in classrooms or the production of movies on biotechnology for higher education and vocational training of teachers (Gaisser and Hüsing 2000). During that period, foresight approaches were increasingly integrated into TA not only focusing on future trends in science and technology but also on key societal issues (cf. Cuhls et al. 2024 in this volume).

Not only specialised institutions engaged in pTA but also "new entrants" such as museums (Joss 2002). For example, in the United Kingdom, the first consensus conference on plant biotechnology adopting the Danish model was organised and implemented at the *Science Museum* in London in 1994. In Germany, the first citizens' conference on human genetic testing was hosted by the *German Hygiene Museum* in Dresden in 2001 (Zimmer 2002). And in 2008, the *Boston Museum of Science* was partner in a consensus conference on nanotechnology in the USA (Guston 2023).

Although most pTA approaches were pursued in Europe, some more recent examples outside Europe are worth mentioning (Hahn et al. 2023): In South Korea, participative elements such as "citizen fora" are included in the parliamentary TA process; in South Africa, multi-stakeholder participatory assessments were implemented with a focus on evaluating developments in biotechnology; and Australia initiated participatory approaches on a regional level for environmental management involving experience and knowledge of indigenous communities.

The development in the USA since 1995 is most interesting considering the pioneering role of the OTA and its closure in that year (Guston 2023). In parallel to building up mainly expert-mode oriented TA capacities in the *Government Accountability Office* (GAO), pTA approaches were initiated in a less institutionalised way mainly by a group of academics called *Expert and Citizen Assessment of*

Science and Technology (ECAST). ECAST evolved in the aftermath of the above-mentioned consensus conference on nanotechnology conducted by the *Centre for Nanotechnology in Society at Arizona State University* (CNS-ASU). Recently, ECAST explored public perspectives of human gene editing based on CRISPR technology. As Guston (2023) points out, ECAST intensively strives for international networking and is collaborating among others with the DBT.

Along with the diffusion of pTA in the 1980s and 1990s, fundamental critique of this approach increasingly emerged in the science policy domain (Gethmann 2002). Mainly three critical points were raised (Hennen 2012): (1) lack of impact, (2) instrumentalisation, and (3) tampering of laypeople's perspectives by experts.

- Lack of impact refers to the observation that a direct influence of pTA on political decision-making is hardly detectable. However, as Hennen (2012) argues, this is not specific to pTA but can be observed for many types of scientific advice. Nevertheless, possible impact of pTA on the political decision-making process is hampered by a specific systematic feature of any TA process: TA has a systemic perspective and aims at exploring the full complexity of technical developments. Thereby, TA increases the complexity of decision-making processes at stake making it less likely that outcomes of the process are directly used by policy makers.
- Instrumentalisation is an inherent risk not only of pTA but also of many laypeople or expert-based consulting processes (Stirling 2008). A key issue to avoid or minimise instrumentalization is the institutional setting of pTA (Hennen 2012). Independent institutions with clear mission statements minimise this risk.
- Laypeople and experts can be perceived as complementary, without laypeople there are no experts. Accordingly, both play an important role in the pTA process. During the process, laypeople may change their view on a specific technological issue not least due to information provided by experts. This could be considered as tampering. On the other hand, as Hennen (2012) explains, this is also an indication of empowerment and learning in the pTA process: Minds are changing, new positions are taken.

In parallel with such critiques of pTA, evaluation activities were initiated. For example, the *European Commission* (EC) launched two large studies which analysed pTA processes in different countries and technological domains. The EUROPTA study compared 16 different pTA projects and found that mainly two factors are important for the visibility and resonance of pTA (Hennen 2002): the character and status of the public debate, and the institutional and political setting of the procedure. The ADAPTA project (Gaisser et al. 2001) explored pTA processes in several countries in three different technological domains: urban transport policy, genetically modified food, and genetic and predictive testing. The findings for the case of Germany were rather sobering as they detected only a very low impact of pTA activities on policy processes and public debate (Gaisser et al. 2001). One of the few other systematic evaluations of the impact of pTA activities concerns the above-mentioned citizens' conference on human genetic testing in the *German Hygiene Museum* in Dresden. In this case, the evaluation study was able to show

how the mentioned empowerment process of laypeople worked in practice (Zimmer 2002).

Aside from pTA, there is another prominent TA approach that has its roots in the 1980s: the concept of *Constructive Technology Assessment* (CTA) which had been developed in the Netherlands and Denmark (Rip et al. 1995). Even though the term CTA is used with different understandings, according to Schot and Rip (1997) there is a common feature of CTA, namely the modulation of ongoing technology development by all relevant stakeholders. As Schot and Rip point out, such a process can lead to new design practices of technologies which anticipate impacts and involve diverse societal groups from the beginning in a kind of societal learning. Although there is obviously some overlap with the idea of participation and inclusion in pTA, CTA has different theoretical foundations and focuses on the socially responsible design of technology ("better technology in a better society", Rip et al. 1995). It has also been taken up by more recent concepts of technology and innovation governance such as *Responsible Research and Innovation* (RRI). This plays an important role in the next ("pragmatic") phase of TA which is outlined in the following section. To begin with, the focus is on the development and discussion in Germany as it can be seen as paradigmatic for this phase.

4 Managing and Governing Technological Innovations

In the business world, TA used to have a notoriously bad reputation for being hostile to innovation. TA was considered to be critical of technological progress, which was not entirely unfounded in view of the failure of major large-scale research projects in the 1970s and 1980s (for example, next generation nuclear energy reactors or magnetic levitation train). For these reasons, in the USA, TA was sometimes denigrated as "technology arrestment" (Paschen and Petermann 1986, p. 22).

This does not mean, however, that there has not been a critical approach to technical progress in companies and among technicians. In Germany, the *Association of German Engineers* (VDI) had been working on a guideline on TA since 1976, which was rooted in its technology-reflecting tradition of engineering responsibility, but also took up the academic and political discussion of the time. This guideline—which was always controversial even within the association—was finally adopted in 1990 (VDI 1991). In terms of its character, it was not a "recipe book" for TA, but it had a considerable influence in industry and among engineers, mainly by raising awareness of the general TA discussion (Haberland 2016; König 2021). However, the guideline differed from other contemporary concepts of TA by its explicit orientation towards innovation processes in industrial contexts.

At the same time, some researchers in Germany noted a "crisis of the traditional TA concept" (Bröchler and Simonis 1998, p. 31). Apart from the orientation towards political and administrative decision-makers as TA's primary addressees they criticised the dominance of scientific experts and the focus on recommending options for policy making. According to these researchers, the "traditional" TA approach was based on the premises that scientific analyses can be translated into political decisions, that the state is capable of effectively steering technical development, and that this steering can be done by parliament (Bröchler and Simonis 1998, p. 34). They expressed doubts as to whether these premises were (still) valid, especially since the state technocratic approach had fallen into disrepute after the failure of many large-scale research projects. Instead, the example of Silicon Valley in the USA seemed to show that the market could produce innovations faster and more in line with demand: "TA is in danger of lagging behind the development of technology. TA is called upon to deal with this problem more intensively in conceptual and methodological terms" (Bröchler and Simonis 1998, p. 34; own translation). Moreover, social and technical developments could no longer be controlled by politics alone, but took place in networks of actors from the state, industry, science, civic associations and society. Finally, TA had to take into account the recognisable "change in consciousness [of companies] in their relationship to the social environment" and was called upon to "emphasise the non-technical factors in the process of shaping technology" (Bröchler and Simonis 1998, p. 35; own translation).

Such a turn towards TA as a "pragmatic innovation management" (Bogner 2021, pp. 51ff.; own translation) had parallels with the concept of innovation systems, which had gained popularity since the mid-1980s in the international community of innovation research and emphasised that the flow of knowledge and technology between people, companies, and institutions is the key to innovative processes (Fraunhofer ISI 2012; see also Frietsch et al. 2024 in this volume). Against this background, Meyer-Krahmer (1999, p. 214) pointed out at a conference on the occasion of "25 years of TA in Germany" that the contrast between problem-induced and technology-induced TA had to be overcome. In view of the increasing international and institutional integration of innovation actors, a development from a state-centred approach to a multi-actor-approach seemed necessary.

In order to remedy these weaknesses and to counter the perception of TA as technology-hostile and innovation-inhibiting, an innovation-oriented TA was called for, which could influence technology design through "organised innovation processes" (Tschiedel 1997, cited in Haberland 2016). Thus, in 2000, the German Government adopted central arguments for innovation-oriented TA, enriched them with considerations from social constructivist science and technology studies (STS) (Bode 2002), and finally presented a concept for an innovation and technology analysis (ITA) (Brüntink 2001). This was intended to complement and integrate existing TA measures and projects (Astor and Bovenschulte 2000), whereby "complement" meant in particular an increased addressing of businesses. According to Brüntink (2001, p. 8; own translation), the aim was "the promotion of cooperation between ITA and industry. [...] Innovation processes take place in companies, innovative companies change the economy and—more and more frequently—thereby also society". The proponents of this new concept assumed that companies do not act exclusively according to profitability criteria, but also take social needs into account in the sense of a collective responsibility for the common good, and concluded that "ITA is one, if not *the* tool of choice for companies" (Baron et al. 2003, p. 34; own translation, emphasis in the original). In that sense, it had a "problem-solving potential [...] for the German economy" and was a management tool that "could draw on mature methods and be useful for the economy" (Baron et al. 2003, p. 22, 24; own translation).

From the beginning, policy makers understood ITA as a strategic attempt to bring together the different TA traditions. First, policy makers should be supported by recommendations for science, technology, and innovation (STI) policy. Second, participatory approaches should be used to involve citizens and consumers in the development process in order to increase social acceptance. And finally, it was hoped to provide companies with knowledge about technical alternatives, foreseeable obstacles, and framework conditions to be considered (Astor and Bovenschulte 2000). For the established TA community, it was not so clear whether this approach was really new or just old wine in new skins as many of its elements had already been part of TA since the 1960s (Grunwald 2001). Others considered ITA mainly as a marketing attempt to extend the target group of TA to industrial actors and suspected that it was a strategy for business development and to increase technology acceptance (Haberland 2016, pp. 83f.).

Such a pragmatic-eclectic approach to TA was not limited to Germany, even if it was much less justified with a theoretical framework in most other countries. The ITA projects carried out in Germany since 2000 were very different in their character, but aimed, at least in terms of their claim, at a holistic assessment of technological developments and responsible innovation design. The strong role of industry, both in the conduct and in the exploitation of the studies, has admittedly not been fulfilled: ITA has primarily remained an instrument of government research planning. In many cases, the focus was on specific so-called "key technologies" such as information and communication technologies, nanotechnology, biotechnology, and genetic engineering, with studies focusing less on technology risks and more on their contribution to societal needs and global challenges. Research also included studies on the human factor in innovation, for instance on factors influencing technology perceptions and what role these perceptions play for the market success (Hüsing et al. 2002). Finally, there was an increasing number of studies focusing on foresight of scientific and technological developments and their innovative impact.[1]

The discussions about ITA also had an impact when the contract for the operation of the German TAB had to be renewed in 2001. TAB's operator, the *Institute for Technology Assessment and Systems Analysis* (ITAS), was requested to cooperate with Fraunhofer ISI and to supplement established areas of work by the so-called "future reports", "innovation reports", and "policy benchmarkings". Future reports aimed to analyse medium- and long-term fields of development. They were primarily intended to identify parliamentary need for action, while the innovation reports were meant to provide orientation knowledge about areas with high development dynamics (Cuhls et al. 2003; Petermann 2003). Although true foresight studies were not conducted as part of the TAB work programme, the new study formats gained

[1] In 2021, the German Government renamed ITA "Insight—Interdisciplinary perspectives on societal and technological change", yet without fundamentally changing the underlying concept.

much popularity, especially as they were also of interest to other parliamentary groups and committees. In particular the innovation reports addressed current and urgent issues relating to the competitiveness and innovative capability of German industry (Nusser et al. 2007; Thielmann et al. 2009; Gandenberger et al. 2012).

Internationally, with the new millennium the time of the big technology controversies was over. However, as in previous decades, some technological developments received particular attention, being often utopian visions of a technologically improved world. From 2000 onward, this was primarily nanotechnology, thus the use of materials on an atomic, molecular, and supramolecular scale for industrial purposes. Visionary publications, such as those of Drexler et al. (1991) and Joy (2000), initiated an intense debate not only about the potential but also about the risks of nanotechnology which led to a series of TA studies (Malanowski 2001; Paschen et al. 2003; Malsch et al. 2004; The Royal Society 2004). Some of these early studies were mainly concerned with the visionary aspects. These were linked to the notion that there is a convergence of Nano-Bio-Info-Cogno (NBIC) technologies, resulting in a fundamental boundary shift between the natural and the artificial, and with the goal of enhancing human (physical, sensory, and cognitive) capabilities (Roco and Bainbridge 2003; Beckert et al. 2007). Later, most TA studies on nanotechnology took a rather pragmatic approach and examined how much substance the promises of the technology visionaries actually had in specific application areas (e.g., NRM 2006; Möller et al. 2009). Starting around 2010, research focused on very specific problem areas such as nanotoxicology or product safety, with the goal of defining requirements that the new technology must meet in order to fulfil its promises. This has, for instance, resulted in a long-term activity like the "Nanotrust"[2] project which has continuously been investigating specific safety and risk-relevant aspects of nanomaterials and providing input for the regulation of innovative materials since 2007.

Another substantive strand of TA research took up the notion of a networked world (Castells 1996) that became popular with the advent of the Internet and was discussed as "ubiquitous computing", "ambient intelligence", or later "Internet of things". As in the case of nanotechnology, the first step was the analysis of certain technology visions that had a strong impact on politics. An early example was the scientific deconstruction of "ambient intelligence", a vision of the future information society where intelligent interfaces enable people and devices to interact seamlessly with each other and with the environment (Ducatel et al. 2001). Several TA studies showed how naive this idea actually was and highlighted the social and environmental risks involved (Hilty et al. 2003; Bizer et al. 2006; Wright et al. 2008). More recently, TA studies—especially those conducted on behalf of the European Parliament—have provided important input to the regulation of the connected world, from the General Data Protection Regulation (GDPR) in 2016 to the regulation of artificial intelligence (AI) at present (Boucher 2020; Christen et al. 2020; Kolleck and Orwat 2020).

---

[2] See https://www.oeaw.ac.at/en/ita/nanotrust/

Apart from issues of nanotechnology and digitisation, questions of energy, the environment, and sustainability became a new urgency in the context of the predicted climate change, but also changed the way TA was dealing with them. Whereas traditional TA focused primarily on the risks of single technologies to the natural environment, the focus has moved towards systemic interrelationships between technologies, society, and the environment (see also Hillenbrand et al. 2024 in this anthology). With this change in perspective and in view of the global challenges, TA has increasingly turned its attention to questions of management and governance of innovations, also bringing non-technological innovations into the focus (Howaldt et al. 2019; Ozoliņa et al. 2009). Accordingly, more research is being conducted into how socio-technical constellations should look like if they are expected to contribute to solving global challenges. The experts' knowledge of the technology's functions, effects, and unintended side effects plays an important role in this process, as does the identification and resolution of potential societal conflicts through the participation of as many stakeholders as possible.

In this context, at the EU level, TA concepts have received strong attention since around 2010 under the term *Responsible Research and Innovation* (RRI) or just *Responsible Innovation* (RI). Starting from debates on responsible development in the area of nanotechnologies in the early 2000s, RRI quickly attracted considerable attention in the academic discourse on the governance of research and innovation (Owen et al. 2021; Rip 2014). RRI is also an eclectic approach, building upon several earlier concepts (such as CTA, see Sect. 3), partially integrating and developing them further. Apart from TA in its numerous guises, it makes use of concepts and disciplinary contributions from STS, ethics of science and technology, ELSA/ELSI research (ethical, legal, social aspects/implications), sustainable technology development, value sensitive design, responsible development, participatory and transdisciplinary research, research integrity, responsible metrics, etc. (Lindner et al. 2016; Brundage and Guston 2019).

According to the definition and framework developed by Stilgoe et al. (2013) that has gained the most attention in academia, RI (or RRI) comprises four elements that are also found in various directions of TA: anticipation, reflexivity, inclusion, and responsiveness. Anticipation is about carefully considering both the intended and potential unintended consequences of research and innovation activities, covering elements of expert TA and, to some extent, foresight. Reflexivity is about reflecting on the motivations, assumptions, and commitments underlying technological developments. At the same time, reflexivity also means questioning the normative basis of the assessment (Hennen and Nierling 2019; Kollek 2019; Nierling and Torgersen 2019). Inclusion is closely related to public participation, which is the central element of pTA. It is not limited to citizens, but seeks to involve all relevant societal stakeholders (including businesses and politics) at an early stage in order to identify potential conflict fields and to reach a consensual design. Finally, responsiveness expresses that TA must not stop at the (ex-post) analysis and assessment of technologies, but must entail decisions and practical action. These decisions can then take the desirable and undesirable impacts of technology as well as the interests of citizens into account. Ideally, they result in solutions guided by values and norms in the interest of the common good.

As a concept, RRI was strongly promoted by powerful actors in the field of STI policy (cf. Lindner et al. 2024 in this anthology), particularly by the *European Commission* (EC), culminating in the integration of RRI as a crosscutting issue in the EC's Research Framework Programme "Horizon 2020" (2014–2020) (Blok 2023).[3] This development was also conducive to the uptake of RRI-related initiatives in a number of countries and organisations, which still continues (Wittrock et al. 2021).

5 Conclusion and Outlook

The description of the three TA paradigms or phases (expert-based, participatory, and pragmatic) shows that the paradigms are not strictly separated from each other. Rather, each highlights a specific, temporally dominant perspective on how to do TA. They all are still relevant today and continue to coexist. Figure 1 is an attempt to summarise and further illustrate the history of TA since the 1960s by roughly locating in time some of the technologies in focus, some of the methods introduced in TA, and the foundation of some TA institutions (see Fig. 1).

What can be said on the impact of TA (cf. Decker and Ladikas 2004), given the constant changes in objectives, concepts, addressees, and stakeholders involved? The most intensive interaction has been with politics, which gave the impetus for the institutionalisation of TA and has also financed much of its research. In this arena, TA has always had to strive for its independence and neutrality. TA has made numerous important contributions to evidence-based (technology) policy but had to be careful to fairly take into account the interests of all social groups. However, it cannot be denied that TA is also an instrument of power that can be—and actually is—used to justify political decisions. This can also be seen in the fact that politicians occasionally ignore even important TA results because they are inconvenient or do not fit into the current political agenda. One example is a TAB study from 2011 on the risks of a large-scale power blackout (Petermann et al. 2011), which only recently received greater attention in the context of the war in Ukraine.

Certainly, TA processes and results also have the potential to directly influence political decision-making. For example, in Denmark some restrictions of public research funding in biotechnology were informed by related TA activities of the DBT. In addition, in some areas of Europe such as Switzerland and again Denmark, TA has become an important instrument for informing citizens about technological developments and their impacts, or even for participation of citizens in technological processes. In contrast, the function of TA as an instrument of welfare-oriented innovation management has not yet been able to establish itself broadly in industry.

[3] However, in the subsequent Research Framework Programme "Horizion Europe" (2021–2027) the significance of RRI has strongly declined.

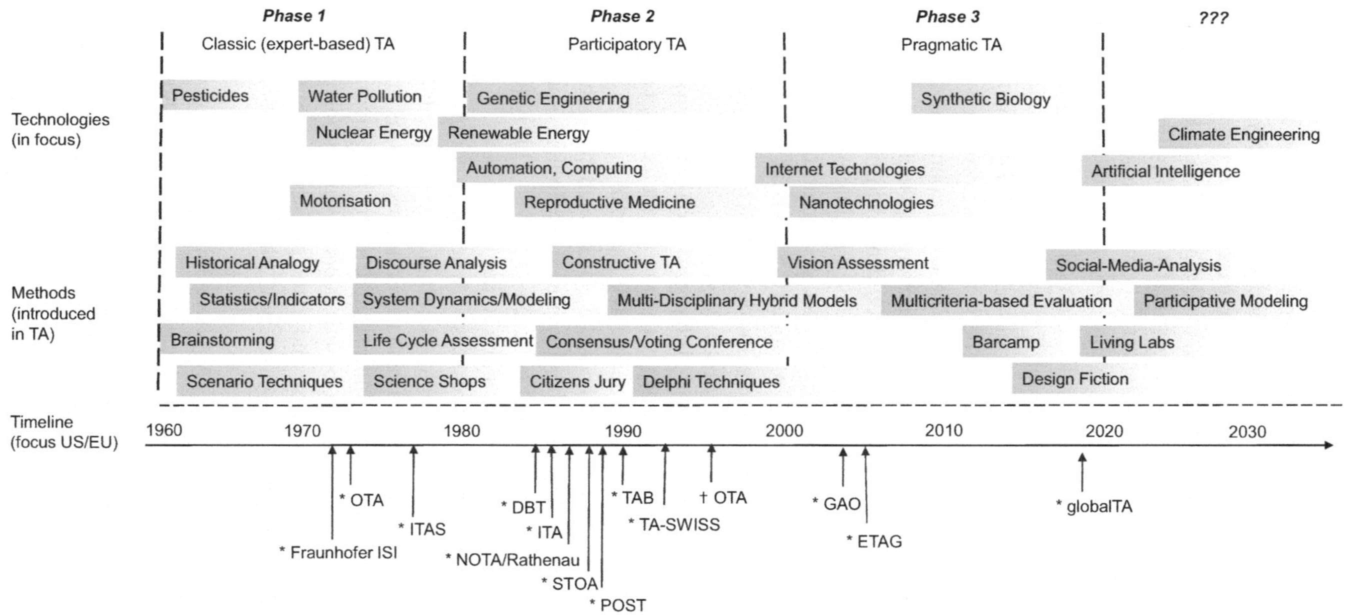

Fig. 1 The assessment of technological innovations over time; own illustration, based on three phases by Bogner (2021). See appendix for further details and explanations regarding the TA institutions

All in all, it seems that the impact of TA is characterised by a similar diversity as TA activities themselves, ranging from almost undetectability to direct influence on political decision-making.

Looking at the future of TA, problem-orientation is needed more than ever. The problem-orientation of TA had been discussed from the very beginnings (Enzer 1974) and had been, already some time ago, expected to gain importance in the future in order to address the grand societal challenges (Decker and Fleischer 2010). Today, after the "normative turn" (Daimer et al. 2012), STI policies have to a certain extent internalised such problem-orientation, following a so-called mission-oriented approach (see also Lindner et al. 2024 in this anthology), with the goal of overcoming grand societal challenges, such as demographic or climate change. We expect TA to increasingly contribute to this goal. TA would then need to focus not only on individual technologies, but rather on socio-technical transitions and the governance of urgent system transformations, be it the transformation of the mobility, energy, healthcare, or nutrition system. Here, TA could contribute to finding appropriate system solutions as combinations of (converging) technologies, innovations, and also non-technological approaches. It would need to be embedded in knowledge and decision-making processes characterised by intense cooperation between science, industry, policy, administration, and society including NGOs, citizens, consumers, and users. The involvement of society here is at least bi- if not multi-directional, on the one hand with regard to the development of products and technical solutions and their societal (consumer/user) acceptance, and on the other hand with regard to increasingly required behavioural changes in consumption and resource use.

Of course, such an orientation comes with several challenges. One is to mediate complex impact dependencies and to decide and navigate through innovation pathways whose differences are no longer to be determined by techno-economic or socio-technical criteria alone, but rather by ethical, ecological, societal, geopolitical, and other criteria, which we might not even know yet. This requires new instruments (and indicators) for assessment and a sound factual basis.

In addition, with regard to its addressees, TA needs to communicate its (generally complex) findings in a way that reduces complexity and produces transparency through the explanation of assumptions and uncertainties. This is important because also governments, parliaments, and administrative actors in democratic societies are required to explain their (possibly TA-based) decisions in a transparent way, since they are exposed to the public discourse and to the problem that both decision criteria and scientifically derived assumptions will likely be questioned by parts of society.

This entails another challenge for TA. Given the present crisis of confidence in science and scientific policy advice, not least obvious in the dispute on COVID-19 vaccination, TA has an ambivalent position like never before between claiming neutrality, on the one hand, and representing normative perspectives, on the other (Hennen and Nierling 2019; Nierling and Torgersen 2020; Torgersen 2019). First of all, legitimate questions may be raised regarding the weighting criteria used in TA processes, especially if TA commits to contributing to specific goal-oriented socio-technical transformations. What is more, in an ever more complex, both techno-scientifically driven and democratic society, the question arises to what extent and for what reasons the various societal groups trust or distrust scientists and their research results. And if they distrust, to what extent and with what means is it possible to build up trust into TA processes and results (e.g., in terms of quality, correctness, independency) and to reach acceptance by all societal groups including those following right-wing populist views? If TA aims to contribute, for instance, to the transformation towards a more sustainable society, how can there be trust if such a goal is politically not shared? And with regard to an increasingly mission-oriented STI policy, does it not automatically make itself vulnerable to discussions in social media and fake news when apparently established certainties and common assumptions are doubted?

A further challenge is the fact that technologies are at a specific stage of development and maturity when they become the centre of public or political attention at a given time. There is not only the well-known dilemma (Collingridge 1980) that at an early point in time little is known about the impacts and unintended side effects of a technology, although the possibilities to control and shape the development are high, whereas at a later stage much more is known, but control is more difficult to achieve. In addition, technologies are subject to socially conditioned cycles of attention in research funding, in the media, and in the public and political discourse. It is therefore obvious to assume that one technology at other times, in different societal contexts, with different research efforts may be evaluated with different results leading to different decisions, selections, and design mechanisms. If so, may then the push of certain technologies at a certain time possibly hinder the development of better, alternative solutions? This opens up a field of conflict with questions about how long and intensively specific and alternative technologies (as well as non-technical innovations) should be researched and evaluated before a societal decision can be made about their significance and use.

Finally, TA increasingly needs to face the global dimension of technologies and socio-technical change including geopolitical aspects. The foundation of the *globalTA network*[4] in 2019 can only be a very first step in this regard (Hennen et al. 2023). Given the increasing importance and development of both sustainability goals (e.g., compliance with planetary boundaries, global justice, global health; cf. the UN Sustainable Development Goals, SDGs) and sustainability criteria (e.g., carbon footprint, water and energy consumption, living and health conditions), there is the need to establish local and regional structures of circular economies and value creation structures with reduced transport routes in light of current global trade. Moreover, with supply bottlenecks (e.g., of semiconductor chips and dependent products such as vehicles) in the COVID-19 pandemic and dependencies on energy (gas) and other raw materials from Russia in light of the war in Ukraine, the discussion on technology sovereignty, raw materials, and technology dependencies on other countries (including China) have attained highest actuality (Edler et al. 2020). The assessment of stability and trust in countries and regions as trading partners is becoming a new and critical parameter in decision-making processes.

---

[4] See https://globalta.technology-assessment.info/.

In addition, there is the question of sufficiency in the (global) consumption of goods and mobility in today's societies. This connects to more fundamental questions of how our economies could and should work, to what extent the development towards a post-growth society could be an option for achieving the sustainability goals, and what role technological and social innovations (Sartorius et al. 2022; Heyen et al. 2024) as well as TA (Grunwald 2018) might play in this regard.

With all those challenges, TA is more than ever asked to take the "bigger picture" into account and not to focus too much on an individual technology alone. This implies questioning current hypes on the potential of new technologies, such as hydrogen or electric cars, to solve the grand societal challenges of our time. Certainly, TA will be needed for emerging technologies such as climate engineering and for the further digitisation of our societies. With such new technologies and the challenges mentioned above comes the need for new approaches of TA so that TA will remain a valuable instrument providing guidance in a complex and uncertain technological environment in the coming decades.

Appendix

Year dates and abbreviations used in Fig. 1 (* = foundation; † = closure):

1972: * Fraunhofer ISI—Fraunhofer Institute for Systems and Innovation Research ISI (originally founded as Fraunhofer Institute for System Technology and Innovation Research, re-named in 2004)

1972/73: * OTA—Office of Technology Assessment at the US Congress; closure in 1995

1977: * ITAS—Institute for Technology Assessment and Systems Analysis, at Karlsruhe Institute of Technology KIT (foundation of predecessor institution in 1977, as ITAS since 1995)

1985: * DBT—Danish Board of Technology

1985: * ITA—Institute of Technology Assessment, at the Austrian Academy of Sciences (originally as working group of a pre-existing research institute; as ITA since 1994)

1986: * NOTA/Rathenau—Netherlands Organisation for Technology Assessment, re-named in Rathenau Institute in 1994

1987: * STOA—European Parliament Office for Scientific and Technological Option Assessment

1989: * POST—Parliamentary Office of Science and Technology, United Kingdom

1990: * TAB—Office of Technology Assessment at the German Bundestag

1992: * TA-SWISS—Foundation for Technology Assessment, at the Swiss Academies of Arts and Sciences (originally founded as Swiss Science and Technology Council)

2004: * GAO—US Government Accountability Office (originally founded in 1921, but no TA relation before the closure of OTA, re-named in 2004)

2005: * ETAG—European Technology Assessment Group (network of TA institutions)

2019: * globalTA—global Technology Assessment Network (network of TA institutions)

References

Astor M, Bovenschulte M (2000) Innovations- und Technikanalyse. Zukunftschancen erkennen und realisieren [Innovation and technology analysis: identifying and realizing future opportunities]. BMBF, Bonn

Banta D (2003) The development of health technology assessment. Health Policy 63:121–132. https://doi.org/10.1016/S0168-8510(02)00059-3

Baram MS (1970) The social control of science and technology. Denver Law Rev 47:567ff

Baron W, Häußler S, Luther W, Zweck A (2003) Innovations- und Technikanalyse: Chancen und Barrieren betrieblicher Integration [Innovation and technology analysis: opportunities and barriers of integration into companies]. Campus, Frankfurt

Beckert B, Blümel C, Friedewald M (2007) Visions and realities in converging technologies: exploring the technology base for convergence. Innov Eur J Soc Sci Res 20:375–394. https://doi.org/10.1080/13511610701767932

Benz E (1987) Umweltschutz im Verständnis eines Energieversorgungsunternehmens [Environmental protection in the understanding of an energy supply company]. In: Henn R (ed) Technologie, Wachstum und Beschäftigung: Festschrift für Lothar Späth. Springer, Berlin, pp 923–932

Bimber B (1996) The politics of expertise in congress: the rise and fall of the Office of Technology Assessment. State University of New York Press, Albany, NY

Bimber B, Guston DH (1997) Introduction: The end of OTA and the future of technology assessment. Technol Forecast Soc Chang 54:125–130. https://doi.org/10.1016/S0040-1625(97)81491-6

Bizer J, Spiekermann S, Günther O, Dingel K, Fabian B, Hansen M, Klafft M, Möller J (2006) Technikfolgenabschätzung Ubiquitäres Computing und Informationelle Selbstbestimmung [Technology assessment ubiquitous computing and informational self-determination]. Unabhängiges Landeszentrum für Datenschutz Schleswig-Holstein & Institut für Wirtschaftsinformatik der Humboldt-Universität zu Berlin, Kiel

Blok V (ed) (2023) Putting responsible research and innovation into practice: a multi-stakeholder approach, vol 40. Springer, Cham

Bode O (2002) Die ITA der Gesellschaft: Praxisbeobachtungen zur Innovations- und Technikanalyse auf der Grundlage der Theorie sozialer Systeme [The ITA of society: practical observations on Innovation and Technology Analysis based on social systems theory]. Dev Perspect 2:35–68

Bogner A (2021) Politisierung, Demokratisierung, Pragmatisierung. Paradigmen der Technikfolgenabschätzung im Wandel der Zeit [Politicization, democratization, pragmatization. Paradigms of technology assessment in the course of time]. In: Böschen S, Grunwald A, Krings B-J, Rösch C (eds) Technikfolgenabschätzung: Handbuch für Wissenschaft und Praxis. Baden-Baden, Nomos, pp 43–58

Boissevain J (1979) Network analysis: a reappraisal. Curr Anthropol 20:392–394

Boucher P (2020) Artificial intelligence: How does it work, why does it matter, and what can we do about it? STOA Study, Brussels. Available at https://www.europarl.europa.eu/thinktank/en/document/EPRS_STU(2020)641547. Accessed 23 Feb 2024

Bröchler S, Simonis G (1998) Konturen des Konzepts einer innovationsorientierten Technikfolgenabschätzung und Technikgestaltung [Contours of the concept of innovation-oriented technology assessment and technology design]. TA-Datenbank-Nachrichten 7:31–40. https://doi.org/10.14512/tatup.7.1.31

Brundage M, Guston DH (2019) Understanding the movement(s) for responsible innovation. In: von Schomberg R, Hankins J (eds) International handbook on responsible innovation: a global resource. Edward Elgar, Cheltenham, pp 102–121

Brüntink C (2001) Zum Konzept der Innovations- und Technikanalyse des BMBF [On the concept of the BMBF's innovation and technology analysis]. TATuP 10:6–9. https://doi.org/10.14512/tatup.10.2.6
Carson R (1962) Silent spring. Houghton Mifflin, Boston, MA
Castells M (1996) The rise of the network society. Blackwell, Oxford
Christen M, Mader C, Čas J, Abou-Chadi T, Bernstein A, Binder NB, Dell'Aglio D, Fábián L, George D, Gohdes A, Hilty L, Kneer M, Krieger-Lamina J, Licht H, Scherer A, Som C, Sutter P, Thouvenin F (2020) Wenn Algorithmen für uns entscheiden: Chancen und Risiken der künstlichen Intelligenz [When algorithms decide for us: Opportunities and risks of artificial intelligence]. vdf Hochschulverlag, Zürich. https://doi.org/10.3218/4002-9
Coates JF (1974) Some methods and techniques for comprehensive impact assessment. Technol Forecast Soc Chang 6:341–357. https://doi.org/10.1016/0040-1625(74)90035-3
Coates JF (1976) The role of formal models in technology assessment. Technol Forecast Soc Chang 9:139–190
Coates VT (1978) A handbook of technology assessment. Final report. School of Engineering and Applied Science, George Washington University, Washington, DC
Coates VT, Finn B (eds) (1979) A retrospective technology assessment: submarine telegraphy—the transatlantic cable of 1866. San Francisco Press, San Francisco, CA
Collingridge D (1980) The social control of technology. St. Martin's Press, New York
Cuhls K, Edler J, Lay G, Meyer-Krahmer F (2003) Zukunftsreports—Politik-Benchmarking—Innovationsreports: Neue Analyseansätze im TAB [Future reports—policy benchmarking—innovation reports: new analysis approaches at TAB]. TAB-Brief 25:7–11
Cuhls K, Dönitz E, Erdmann L, Gransche B, Kimpeler S, Schirrmeister E, Warnke P (2024) Foresight: Fifty years to think your futures. In: Edler J, Walz R (eds) Systems and innovation research in transition: Research questions and trends in historical perspective. Springer, Heidelberg
Daddario E (1966) Inquiries, legislation, policy studies review: science and technology review and forecast. US Congress, House. Committee on Science and Astronautics. Second Progress Report of the Subcommittee on Science, Research and Development, Washington, DC
Daimer S, Hufnagl M, Warnke P (2012) Challenge-oriented policy-making and innovation systems theory: reconsidering systemic instruments. In: Fraunhofer ISI (ed) Innovation system revisited: Experiences from 40 years of Fraunhofer ISI research. Fraunhofer, Stuttgart, pp 217–234
Decker M, Fleischer T (2010) When should there be which kind of technology assessment? A plea for a strictly problem-oriented approach from the very outset. Poiesis Praxis Int J Ethics Sci Technol Assess 7:117–133. https://doi.org/10.1007/s10202-010-0074-6
Decker M, Ladikas M (eds) (2004) Bridges between science, society and policy: technology assessment—methods and impacts, Wissenschaftsethik und Technikfolgenbeurteilung, vol 22. Springer, Berlin
Denton R, Jochem E, Bossel G, Brand P, Herz H, Legler H (1976) Entwicklung einer Methode zur Durchführung einer Technikfolgen-Abschätzung mittels dynamischer Simulation unter der Annahme eines veränderten Mineralölangebotes [Development of a method for conducting a technology assessment study by means of dynamic simulation assuming a changed supply of crude oil]. Fraunhofer ISI Report, Karlsruhe
Drexler KE, Peterson C, Pergamit G (1991) Unbounding the future: The nano-technology revolution. William Morrow, New York
Ducatel K, Bogdanowicz M, Scapolo F, Leijten J, Burgelman J-C (2001) Scenarios for ambient intelligence in 2010. Institute for Prospective Technological Studies (IPTS), Sevilla
Edler J, Walz R (2024) Introduction. In: Edler J, Walz R (eds) Systems and innovation research in transition: Research questions and trends in historical perspective. Springer, Heidelberg
Edler J, Blind K, Frietsch R, Kimpeler S, Kroll H, Lerch C, Reiss T, Roth F, Schubert T, Schuler J, Walz R (2020) Technology sovereignty. From demand to concept. Fraunhofer ISI, Karlsruhe. Available at https://www.isi.fraunhofer.de/content/dam/isi/dokumente/publikationen/technology_sovereignty.pdf. Accessed 23 Feb 2024
Enzer S (1974) Assessing a problem-oriented social technology. Futures 6:486–498. https://doi.org/10.1016/0016-3287(74)90032-9

Forrester JW (1971) World dynamics. Wright-Allen Press, Cambridge, MA
Fraunhofer ISI (ed) (2012) Innovation system revisited: Experiences from 40 years of Fraunhofer ISI research. Fraunhofer, Stuttgart. Available at http://publica.fraunhofer.de/dokumente/N-215972.html. Accessed 23 Feb 2024
Frietsch R, Reiß T, Schmoch U (2024) Development of innovation monitoring and innovation indicators in the past 50 years. In: Edler J, Walz R (eds) Systems and innovation research in transition: Research questions and trends in historical perspective. Springer, Heidelberg
Gaisser S, Hüsing B (2000) Dokumentation der Lehrerfortbildung "Gentechnik bei Pflanzen (Grüne Gentechnik)" am 11./12. Januar 2000 in Stuttgart: Im Rahmen des Vorhabens "Gesellschaftlicher Diskurs zur Bio- und Gentechnik in Baden-Württemberg" [Documentation of the teacher training course "Genetic engineering in plants (green genetic engineering)" on January 11/12, 2000 in Stuttgart: As part of the project "Social discourse on biotechnology and genetic engineering in Baden-Württemberg"]. Fraunhofer ISI, Karlsruhe
Gaisser S, Bührer S, Menrad K, Zimmer R (2001) Assessing debate and participative technology assessment (ADAPTA): National report of Germany. Fraunhofer ISI, Karlsruhe
Gandenberger C, Glöser-Chahoud S, Marscheider-Weidemann F, Ostertag K, Walz R (2012) Die Versorgung der deutschen Wirtschaft mit Roh- und Werkstoffen für Hochtechnologien—Präzisierung und Weiterentwicklung der deutschen Rohstoffstrategie. Innovationsreport [The supply of raw and other materials for high technologies to the German economy—specification and further development of the German Raw materials strategy. Innovation report]. Büro für Technikfolgen-Abschätzung beim Deutschen Bundestag, Berlin. https://doi.org/10.5445/IR/1000085570
Gethmann CF (2002) Participatory technology assessment: some critical questions. Poiesis Praxis Int J Ethics Sci Technol Assess 1:151–159. https://doi.org/10.1007/s10202-002-0015-0
Grandjean E (1960) Luftverschmutzung und Gesundheit [Air pollution and health]. In: Demoll R (ed) Im Schatten der Technik: Beiträge zur Situation des Menschen in der modernen Zeit. Bechtle, München
Green HP (1967) The new technological era: a view from the law. Bull At Sci 23:12–18. https://doi.org/10.1080/00963402.1967.11455128
Grunwald A (2001) Arbeitsteilige Technikgestaltung und verteilte Beratung: TA zwischen Politikberatung und Technikbewertung in Unternehmen [Work-sharing technology design and distributed consulting: TA between policy advice and technology assessment in companies]. TA-Datenbank-Nachrichten 10:61–71. https://doi.org/10.14512/tatup.10.2.39
Grunwald A (2002) Technikfolgenabschätzung—eine Einführung [Technology assessment—an introduction]. Ed. Sigma, Berlin
Grunwald A (2003) Technology assessment at the German Bundestag: 'expertising' democracy for 'democratising' expertise. Sci Public Policy 30:193–198. https://doi.org/10.3152/147154303781780515
Grunwald A (2018) Diverging pathways to overcoming the environmental crisis: A critique of eco-modernism from a technology assessment perspective. J Clean Prod 197:1854–1862. https://doi.org/10.1016/j.jclepro.2016.07.212
Grunwald A (2019) Technology assessment in practice and theory. Routledge, London
Guston D (2023) Technology assessment activities in the United States of America. In: Hennen L, Hahn J, Ladikas M, Lindner R, Peissl W, van Est R (eds) Technology assessment in a globalized world: facing the challenges of transnational technology governance. Springer, Cham, pp 266–268
Haberland M (2016) Technology Arrestment oder Innovationsmanagement? Technikfolgenabschätzung und Industrie in Deutschland in den Jahren 1970 bis 2000 [Technology arrestment or innovation management? Technology assessment and industry in Germany from 1970 to 2000]. Dissertation, Technische Universität Berlin
Habermas J (1968) Verwissenschaftlichte Politik und öffentliche Meinung [Scientificated politics and public opinion]. In: Habermas J (ed) Technik und Wissenschaft als Ideologie. Suhrkamp, Frankfurt, pp 120–146
Hahn J, Heyen NB, Lindner R (2023) Tracing technology assessment internationally—TA activities in 12 countries across the globe. In: Hennen L, Hahn J, Ladikas M, Lindner R, Peissl W,

van Est R (eds) Technology assessment in a globalized world: facing the challenges of transnational technology governance. Springer, Cham, pp 17–29

Hennen L (2002) Impacts of participatory technology assessment on its social environment. In: Joss S, Bellucci S (eds) Participatory technology assessment: European perspectives. Center for the Study of Democracy, London, pp 257–275

Hennen L (2012) Why do we still need participatory technology assessment? Poiesis Prax 9:27–41. https://doi.org/10.1007/s10202-012-0122-5

Hennen L, Nierling L (2019) The politics of technology assessment: Introduction to the special issue. Technol Forecast Soc Chang 139:17–22. https://doi.org/10.1016/j.techfore.2018.07.048

Hennen L, Hahn J, Ladikas M, Lindner R, Peissl W, van Est R (eds) (2023) Technology assessment in a globalized world: facing the challenges of transnational technology governance. Springer, Cham

Hetman F (ed) (1973) Society and the assessment of technology: premises, concepts, methodology, experiments, areas of application. OECD, Paris

Hetman F (1978) Social assessment of technology and some of its international aspects. Technol Forecast Soc Chang 11:303–313. https://doi.org/10.1016/0040-1625(78)90015-X

Heyen NB, Zenker A, Aichinger H, Bratan T, Kaufmann T, Schnabl E (2024) Innovation without growth? Exploring the (in)dependency of innovation on economic growth. In: Discussion papers innovation systems and policy analysis no. 83, Karlsruhe: Fraunhofer ISI. https://doi.org/10.24406/publica-2916

Hillenbrand T, Hiessl H, Marscheider-Weidemann F, Niederste-Hollenberg J, Sartorius C, Walz R (2024) Environmental and sustainability research using water as an example: Systemic, integrated approach in research and policy for the sustainable use of water. In: Edler J, Walz R (eds) Systems and innovation research in transition: Research questions and trends in historical perspective. Springer, Heidelberg

Hilty LM, Behrendt S, Binswanger M, Bruinink A, Erdmann L, Frölich J, Köhler A, Kuster N, Som C, Würtenberger F (2003) Das Vorsorgeprinzip in der Informationsgesellschaft: Auswirkungen des Pervasive Computing auf Gesundheit und Umwelt [The precautionary principle in the information society: implications of pervasive computing for health and the environment.]. TASwiss, Bern

Hohmeyer O, Hüsing B, Maßfeller S, Reiß T (1994) Internationale Regulierung der Gentechnik: Praktische Erfahrungen in Japan, den USA und Europa [International regulation of genetic engineering: Practical experience in Japan, the USA and Europe]. Physica, Heidelberg

Howaldt J, Kaletka C, Schröder A, Zirngiebl M (eds) (2019) Atlas of social innovation: 2nd volume: a world of new practices. Atlas of social innovation, 2nd vol. oekom verlag, München

Hüsing B, Bierhals R, Bührlen B, Friedewald M, Kimpeler S, Menrad K, Wengel J, Zimmer R (2002) Technikakzeptanz und Nachfragemuster als Standortvorteil [Technology acceptance and demand patterns as a location advantage]. Fraunhofer ISI, Karlsruhe

Jochem E, Bossel G, Herz H, Hoeflich M (1976) Die Motorisierung und ihre Auswirkungen. Untersuchung zur Frage der Realisierbarkeit der Technikfolgenabschätzung anhand von Ex-Post-Projekten [The motorization and its impacts: Analysis of the possibilities to realise technology assessment by means of ex-post projects]. Schwartz, Göttingen

Jochem E, Garnreiter F, Gruber E, Hohmeyer O, Jäger F, Kober W, Mannsbart W, Poppke H, Reichert J (1988) Technikfolgen-Abschätzung am Beispiel der Solarenergie Nutzung [Technology assessment—the example of solar energy use]. Peter Lang, Frankfurt

Joss S (2002) Toward the public sphere—reflections on the development of participatory technology assessment. Bull Sci Technol Soc 22:220–231. https://doi.org/10.1177/02767602022003006

Joss S, Bellucci S (eds) (2002) Participatory technology assessment: European perspectives. Center for the Study of Democracy, London

Joy B (2000) Why the future does not need us. Wired 8:238–262

Klüver L (2000) The Danish board of technology. In: Vig NJ, Paschen H (eds) Parliaments and technology: The development of technology assessment in Europe. State University of New York Press, Albany, NY, pp 173–197

Kolleck A, Orwat C (2020) Mögliche Diskriminierung durch algorithmische Entscheidungssysteme und maschinelles Lernen—ein Überblick [Possible discrimination by algorithmic decision systems and machine learning—an overview]. Büro für Technikfolgen-Abschätzung beim Deutschen Bundestag, Berlin. https://doi.org/10.5445/IR/1000127166
Kollek R (2019) Implizite Wertbezüge in der Technikfolgenabschätzung: Plädoyer für eine Praxis der reflexiven Normativität [Implicit value references in technology assessment: a plea for a practice of reflexive normativity]. TATuP 28:15–20. https://doi.org/10.14512/tatup.28.1.15
König W (2021) VDI-Richtlinie zur Technikbewertung [VDI Guideline on technology assessment]. In: Grunwald A, Hillerbrand R (eds) Handbuch Technikethik. Metzler, Stuttgart, pp 447–451
Krupp H, Jochem E, Harmsen D-M, Legler H (1978) Konzeptionsstudie zur inhaltlichen Abgrenzung und zur Organisation eines Schwerpunktprogramms "Technikfolgen-Abschätzung" der DFG [Conceptual study on the content and organization of a special research program "Technology Assessment" at the German Research Foundation (DFG)]. Fraunhofer ISI, Karlsruhe. Available at http://fachliteratur.isi.fraunhofer.de/isipublica-archiv-extern/Konzeptionsstudie-Technikfolgen-Abschaetzung.pdf. Accessed 23 Feb 2024
Kuhn TS (1962) The structure of scientific revolutions. University of Chicago Press, Chicago
Kunkle GC (1995) New challenge or the past revisited? Technol Soc 17:175–196. https://doi.org/10.1016/0160-791X(95)00007-E
Lerch C, Jäger A (2024) Industry in a changing era: Production paradigms during the last 50 years. In: Edler J, Walz R (eds) Systems and innovation research in transition: Research questions and trends in historical perspective. Springer, Heidelberg
Lindner R, Goos K, Güth S, Som O, Schröder T (2016) "Responsible Research and Innovation" als Ansatz für Forschungs-, Technologie- und Innovationspolitik—Hintergründe und Entwicklungen: TA-Vorstudie ["Responsible research and innovation" as an approach to research, technology and innovation policy—background and developments: TA Pre-Study]. Büro für Technikfolgen-Abschätzung beim Deutschen Bundestag, Berlin. https://doi.org/10.5445/IR/1000133371
Lindner R, Edler J, Daimer S (2024) Understanding paradigm change in science, technology and innovation policy: Between science push and policy pull. In: Edler J, Walz R (eds) Systems and innovation research in transition: Research questions and trends in historical perspective. Springer, Heidelberg
Malanowski N (2001) Vorstudie für eine Innovations-und Technikanalyse zur Nano-technologie [Preliminary study for an innovation and technology analysis on nano-technology]. VDI-Technologiezentrum, Düsseldorf
Malsch I, Gleiche M, Hoffschulz H, Bøgedal M, Locatelli S, Nicollet C, Guibert JC, Denis M, Morrison M, Grunwald A, Dupuy J-P, Parr D, Speller S, Oud M (2004) Fourth nanoforum report: benefits, risks, ethical, legal and societal aspects of nanotechnology. European Nanotechnology Gateway
Meadows DH, Meadows DL, Randers J, Behrens WW (1972) The limits to growth: a report for the club of Rome's project on the predicament of mankind. Universe Books, New York
Meyer-Krahmer F (1999) Technikfolgenabschätzung im Kontext von Innovationsdynamik und Globalisierung [Technology assessment in the context of innovation dynamics and globalization]. In: Petermann T, Coenen R (eds) Technikfolgen-Abschätzung in Deutschland: Bilanz und Perspektiven. Campus, Frankfurt/Main, pp 197–215
Möller M, Eberle U, Hermann A, Moch K, Stratmann B (2009) Nanotechnology in the food sector. STOA Study, Brussels
Nennen H-U, Garbe D (eds) (1996) Das Expertendilemma: Zur Rolle wissenschaftlicher Gutachter in der öffentlichen Meinungsbildung [The expert dilemma: On the role of scientific experts in the formation of public opinion]. Springer, Berlin
Nierling L, Torgersen H (2019) Normativität in der Technikfolgenabschätzung [Normativity in technology assessment]. TATuP 28:11–14. https://doi.org/10.14512/tatup.28.1.11
Nierling L, Torgersen H (eds) (2020) Die neutrale Normativität der Technikfolgen-abschätzung: Konzeptionelle Auseinandersetzung und praktischer Umgang [The neutral normativity of tech-

nology assessment. Conceptual debate and practical handling]. Nomos Verlagsgesellschaft, Baden-Baden
NRM (2006) Road maps for nanotechnology in energy. NanoRoadMap (NRM), Stirling
Nusser M, Wydra S, Hartig J, Gaisser S (2007) Forschungs- und wissensintensive Branchen: Optionen zur Stärkung ihrer internationalen Wettbewerbsfähigkeit. Innovationsreport [Research- and knowledge-intensive industries: options for strengthening their international competitiveness. Innovation Report]. Büro für Technikfolgen-Abschätzung beim Deutschen Bundestag, Berlin. https://doi.org/10.5445/IR/1000132381
OECD (1975) Methodological guidelines for social assessment of technology. OECD, Paris
Owen R, von Schomberg R, Macnaghten P (2021) An unfinished journey? Reflections on a decade of responsible research and innovation. J Responsible Innov 8:217–233. https://doi.org/10.1080/23299460.2021.1948789
Ozoliņa Z, Mitcham C, Stilgoe J, Andanda P, Kaiser M, Nielsen L, Qiu R-Z, Stehr N (2009) Global governance of science. European Commission, Luxembourg
Paschen H (1999) Technikfolgenabschätzung in Deutschland—Aufgaben und Herausforderungen [Technology assessment in Germany—tasks and challenges]. In: Petermann T, Coenen R (eds) Technikfolgen-Abschätzung in Deutschland: Bilanz und Perspektiven. Campus, Frankfurt/Main, pp 77–93
Paschen H, Petermann T (1986) Technikfolgen-Abschätzung: Ein strategisches Rahmenkonzept für die Analyse und Bewertung von Techniken [Technology assessment: a strategic framework for the analysis and evaluation of technologies]. In: Petermann T (ed) Technikfolgen-Abschätzung als Technikforschung und Politikberatung. Campus, Frankfurt, pp 19–42
Paschen H, Petermann T (2005) Die Institutionalisierug der Technikfolgenabschätzung beim Deutschen Bundestag—Ein kurzer Blick zurück [The institutionalization of technology assessment at the German Bundestag—a brief glance back]. In: Petermann T, Grunwald A (eds) Technikfolgen-Abschätzung für den Deutschen Bundestag: Das TAB—Erfahrungen und Perspektiven wissenschaftlicher Politikbera-tung. Ed. Sigma, Berlin, pp 11–18
Paschen H, Gresser K, Conrad F (1978) Technology assessment—Technologie-Folgenabschätzung. Campus, Frankfurt
Paschen H, Coenen C, Fleischer T, Grünwald R, Oertel D, Revermann C (2003) TA-Projekt Nanotechnologie: Endbericht [TA project nanotechnology: final report]. Büro für Technikfolgen-Abschätzung beim Deutschen Bundestag, Berlin. https://doi.org/10.5445/IR/1000103212
Petermann T (1994) Historie und Institutionalisierung der Technikfolgenabschätzung [History and institutionalization of technology assessment]. In: Bullinger H-J (ed) Technikfolgenabschätzung (TA). Teubner, Stuttgart, pp 89–113
Petermann T (2003) Neue Eckpunkte und Inhalte für das TAB [New key points and contents for TAB]. TAB-Brief 25:5–6. https://doi.org/10.5445/IR/1000131593
Petermann T, Bradke H, Lüllmann A, Poetzsch M, Riehm U (2011) What happens during a blackout: Consequences of a prolonged and wide-ranging power outage. Büro für Technikfolgen-Abschätzung beim Deutschen Bundestag, Berlin. https://doi.org/10.5445/IR/1000103292
Reimer H (1971) Müllplanet Erde [Waste planet earth]. Hoffmann & Campe, Hamburg
Renn O (2002) 10 Jahre Akademie für Technikfolgenabschätzung [10 years academy for technology assessment]. TATuP 11:93–96. https://doi.org/10.14512/tatup.11.2.93
Rip A (2014) The past and future of RRI. Life Sci Soc Policy 10:17. https://doi.org/10.1186/s40504-014-0017-4
Rip A, Misa TJ, Schot J (eds) (1995) Managing technology in society: The approach of constructive technology assessment. Pinter, London
Roco MC, Bainbridge WS (2003) Converging technologies for improving human performance: nanotechnology, biotechnology, information technology and cognitive science. Kluwer, Dordrecht
Sartorius C, Dütschke E, Hansmeier H, Heyen NB, Preuß S, Warnke P, Zenker A (2022) The role of innovation in a postgrowth society. In: Leal Filho W, Azul AM, Doni F, Salvia AL (eds) Handbook of sustainability science in the Future. Springer, Cham, pp 1–17

Schot J, Rip A (1997) The past and future of constructive technology assessment. Technol Forecast Soc Chang 54:251–268. https://doi.org/10.1016/S0040-1625(96)00180-1
Slocum N (2003) Participatory Methods Toolkit: A practitioner's manual. King Baudouin Foundation, Brussels
Stilgoe J, Owen R, Macnaghten P (2013) Developing a framework for responsible innovation. Res Policy 42:1568–1580. https://doi.org/10.1016/j.respol.2013.05.008
Stirling A (2008) "Opening up" and "closing down": power, participation, and pluralism in the social appraisal of technology. Sci Technol Hum Values 33:262–294. https://doi.org/10.1177/0162243907311265
TAB (2022) History of TA in the German Bundestag. Available at https://www.tab-beim-bundestag.de/english/history.php. Accessed 23 Feb 2024
The M-J (1989) Human insulin: DNA technology's first drug. Am J Health Syst Pharm 46:S9–S11. https://doi.org/10.1093/ajhp/46.11_Suppl_2.S9
The Royal Society (2004) Nanoscience and nanotechnologies: opportunities and uncertainties, London. Available at https://royalsociety.org/-/media/Royal_Society_Content/policy/publications/2004/9693.pdf. Accessed 23 Feb 2024
Thielmann A, Zimmermann A, Gauch S, Nusser M, Hartig J, Wydra S, Blümel C, Blind K (2009) Blockaden bei der Etablierung neuer Schlüsseltechnologien. Innovationsreport [Blockades in the establishment of new key technologies. Innovation report]. Büro für Technikfolgen-Abschätzung beim Deutschen Bundestag, Berlin. https://doi.org/10.5445/IR/1000133263
Torgersen H (2019) Three myths of neutrality in TA—how different forms of TA imply different understandings of neutrality. Technol Forecast Soc Chang 139:57–63. https://doi.org/10.1016/j.techfore.2018.06.025
Tschiedel R (1997) Neun Thesen zu einem neuen Leitbild: Innovationsorientierte Technikfolgenabschätzung und Technikgestaltung [Nine theses on a new role model: Innovation-oriented technology assessment and technology design]. VITA-Newsletter 1/97:9–11
Várkonyi L (2000) Participative approach in technology development. Danish experience from a Hungarian aspect. University of Technology and Economics, Budapest
VDI (1991) Technikbewertung. Begriffe und Grundlagen [Technology assessment: concepts and foundations]. VDI-Richtlinie, Düsseldorf
Wehling P (2021) Technikfolgenabschätzung und Wissenschaft [Technology assessment and science]. In: Böschen S, Grunwald A, Krings B-J, Rösch C (eds) Technikfolgenabschätzung: Handbuch für Wissenschaft und Praxis. Nomos, Baden-Baden, pp 178–190
Wittrock C, Forsberg E-M, Pols A, Macnaghten P, Ludwig D (2021) Implementing responsible research and innovation: organisational and national conditions. Springer, Cham
Wright D, Gutwirth S, Friedewald M, Vildjiounaite E, Punie Y (2008) Safeguards in a world of ambient intelligence. Springer, Dordrecht
Wynne B (1975) The rhetoric of consensus politics: a critical review of technology assessment. Res Policy 4:108–158. https://doi.org/10.1016/0048-7333(75)90028-1
Zimmer R (2002) Begleitende Evaluation der Bürgerkonferenz "Streitfall Gendiagnostik" [Accompanying evaluation of the citizens' conference "Dispute over genetic diagnostics"]. Fraunhofer ISI, Karlsruhe

Nils B. Heyen is the Coordinator for Technology Assessment of the Fraunhofer Institute for Systems and Innovation Research ISI and a Senior Researcher at the institute's Competence Centre Emerging Technologies (since 2011). From 2023 to 2025, he is also a Deputy Professor for Socio-Economic Innovation Design at Darmstadt University of Applied Sciences, Faculty of Social Sciences. Heyen studied Sociology and Psychology and did his postgraduate studies at the Institute of Science and Technology Studies (IWT) at Bielefeld University. His main research areas are social and technical innovations in medicine and healthcare, technology assessment, citizen science, and the relationship between innovation and post-growth.

Michael Friedewald is a Senior Research Fellow at Fraunhofer ISI in the Competence Centre Emerging Technologies since 1999 and has been coordinating the ICT Research Group since 2009. He holds degrees in computer engineering and in economics from RWTH Aachen University. His current work focuses on assessing the impacts of digitisation on society and the economy, as well as the challenges for privacy and data protection in emerging digital systems. M. Friedewald has been coordinating the (German) Platform Privacy and Self-Determined Life in a Digital World for over 10 years.

Eberhard Jochem is a Senior Scientist at the Fraunhofer Institute of Systems and Innovation Research ISI. Internationally acknowledged expert in technical and socio-economic research, mainly in the fields of energy, material efficiency, and climate change. Major contributions to the development and implementation of a sustainable energy and resources policy. Member of various national and international scientific organisations and advisory committees during the last five decades. Author of several books and numerous publications in the fields of energy, resource efficiency, and environment with regard to R&D, innovation, obstacles and policy issues, and energy economics. Promoter of energy efficiency and climate protection networks in industry.

Thomas Reiß is a Senior Scientist at the Fraunhofer Institute for Systems and Innovations Research ISI. He studied Biology and Chemistry at Freiburg University, Germany, and holds a doctoral degree in Biology. Following a post-doctoral fellowship at the DOE Plant Research Laboratory, Michigan State University (USA) he joined Fraunhofer ISI in 1987. From 1996 to 2004 he was the Head of the department Innovations in Biotechnology, from 2005 to 2020 Head of the Competence Centre Emerging Technologies. From August 2006 to March 2007 Thomas Reiß was the Deputy Head of Fraunhofer ISI.

Axel Thielmann studied Physics at the Friedrich-Alexander University (FAU) Nuremberg-Erlangen (Diploma in Physics 2001). He was Research Assistant at the Institute of Nanotechnology (INT) of the Research Centre Karlsruhe (FZK) and at the University of Karlsruhe until 2005 (Doctoral degree in Physics 2005). He worked as Patent Engineer at the Firm Dr. Gassner & Partner, Patent Attorneys, Erlangen, from 2006 to 2007. Since 2007 he is Researcher and Project Manager at the Fraunhofer Institute for Systems and Innovation Research ISI. From 2010 to 2020 he was the Deputy Head and since 2021 he is the Head of the Competence Centre Emerging Technologies.

Industry in a Changing Era: Production Paradigms During the Last 50 Years

Christian Lerch and Angela Jäger

Abstract Germany has always been known for its strong industrial base. Although the process of deindustrialisation started in the 1970s, manufacturing is still considered a driver for the high innovation and competitiveness of the German economy. Nevertheless, several paradigm shifts in manufacturing have been observed since that time, changing not only the production philosophy but also predominant ways of thinking. These production paradigms are influenced not only by global events and external developments, but also by an advanced innovation and technology policy of Germany. Likewise, they are not only subject to technological innovations, as suggested, for example, by the concept of the four industrial revolutions (Industry 1.0 to 4.0). Rather, the role of humans and work in production has received increasing attention since the 1970s, and revolutionary organisational concepts have also shaped industrial development in Germany. In order to highlight these lines of development, we present in this article what we consider to be the central production paradigms over the last 50 years and highlight the major guiding issues of German manufacturing. Furthermore, we would like to give an outlook on upcoming production paradigms, as well as possible future developments in manufacturing industries.

1 Introduction

Germany's economy has always been known for its strong industrial core. The high quality of industrial products, coupled with constant productivity gains, means that manufacturing in Germany enjoys great recognition worldwide. This is also expressed by the "Made in Germany" label, which not only highlights the high quality and reliability of German products, but also ascribes a uniquely high level of competitiveness to Germany as a production location in general (BMWi 2019). The

---

C. Lerch (✉) · A. Jäger
Fraunhofer Institute for Systems and Innovation Research ISI, Karlsruhe, Germany
e-mail: Christian.Lerch@isi.fraunhofer.de; Angela.Jaeger@isi.fraunhofer.de

J. Edler, R. Walz (eds.), *Systems and Innovation Research in Transition*, Sustainability and Innovation, https://doi.org/10.1007/978-3-031-66100-6_7

strength of the manufacturing sector as part of the German economy continues to this day (BDI 2013). Nevertheless, industrial production has also been subject to major change over the last half century.

At the beginning of the 1970s, industrial production in most leading economies was considered to be a discontinued model. In the USA and the UK in particular, the process of deindustrialisation was already taking place. In Germany, too, the tertiary sector overtook the manufacturing sector for the first time in the early 1970s, both in terms of gross value added and the number of employees (Krupp 1984). Hopes for the second half of the twentieth century centred on services (Fourastié 1949). In the following decades, consequently, a decline of the industrial sector and a growth of the service sector were observed in all leading economies (Eickelpasch 2012, 2014). Despite concerns about growing competition from Asia and Eastern Europe and the relatively high volume of production relocation abroad during the 1990s and in the years around the millennium, a rethink did not take place until 2009, when the financial crisis escalated into a global economic crisis and there was a refocus on physical production. Under the term "reindustrialisation", more importance was now to be attached to industrial production again (Capello and Cerisola 2022; Prisecaru 2015).

Nevertheless, during the last half century, roughly since the beginning of deindustrialisation, various paradigm shifts have taken place in manufacturing that has changed not only the production philosophy but also the predominant way of thinking and has also been reflected in the development of the research field. These production paradigms are influenced not only by global events and external developments, but also by an advanced German innovation and technology policy. Likewise, they are not only subject to technological innovations, as suggested, for example, by the concept of the four industrial revolutions (Industry 1.0 to 4.0). Rather, the role of humans and work in production has received increasing attention since the 1970s, and revolutionary organisational concepts have also shaped the industrial development in Germany (Ernst 2009). Moreover, these developments co-evolved with the research system. On the one hand, research about changes in manufacturing resulted in different paradigms brought forward as an output of research. On the other hand, the ongoing development in manufacturing posed new challenges for research and influenced topics but also methodologies used in research. Thus, looking at the changing paradigms also yields insights into the development of the research field.

Therefore, in this article, we would like to highlight the major developments in manufacturing industries during the last half century with a special view on Germany. For this purpose, we not only draw on relevant literature but also on discussions with experts. In order to provide broad empirical evidence of industrial developments, we also draw on the analysis from the representative *German Manufacturing Survey* (GMS), which has been conducted regularly in Germany's manufacturing sector since 1993 (see Box 1). Moreover, we use text boxes to refer by way of example to two major funding programme lines that at the time shaped innovation and technology policy in Germany fundamentally. Lastly, we give an outlook on upcoming production paradigms, as well as possible future developments in manufacturing industries.

Box 1: German Manufacturing Survey (GMS) in the Context of Empirical Research

Developments in empirical methods, the emergence of computers and the resulting possibilities for the collection and processing of more extensive data, and the proliferation of analytical software encouraged the spread of large, standardised surveys from the early 1990s onward. Building on the first experiences (Meyer-Krahmer 1984), several large, standardised, long-term surveys were developed in order to systematically monitor key economic issues such as innovation, value creation, or employment in the company context and to collect data for differentiated analysis. These data make it possible to systematically analyse complex interrelationships and context-specific developments in manufacturing; thus, they do not only contribute to the understanding of the research topics, but also provide empirical evidence for policy and business. In addition to the *GMS*, two further examples illustrate this development. The IAB-Betriebspanel, an employer survey on company and workplace factors determining employment, was first launched in 1993/1996. The Community Innovation Survey (CIS), the reference survey on innovation in enterprises that is coordinated by Eurostat, has been conducted in the European Union and selected other countries, since the first pilot CIS 1 in 1993 (Arundel and Smith 2013).

Based on a survey carried out in 1993 in East Germany in the context of the evaluation of a support programme for East German manufacturing companies promoting computer integrated manufacturing (CIM) (Wengel 1999), the German Manufacturing Survey of Fraunhofer ISI, previously known as the Manufacturing Innovation Survey, has been conducted regularly since 1995 (Lay 1999). The *GMS* initially addressed Germany's capital goods industry; since 2006, a representative sample of production locations in Germany has been surveyed for all manufacturing sectors. From the outset, the focus of the survey has been on the diffusion of new technologies and new organisational concepts, issues relating to personnel deployment and qualification and questions about the decision to locate production. Indicators on performance such as productivity, flexibility, and quality are also surveyed. In 1997, questions on product-related services and subsequently service-based business models were added. Since 1999 aspects of sustainability and the circular economy have been an integral part of this survey.

The focus of the *German Manufacturing Survey* is to cover process innovation and value creation processes in manufacturing in particular. The specialisation in production makes it possible to capture production structures and modernisation processes based on facts and thus to reflect the diversity of small- and medium-sized production companies. Since 1995, between 1300 and 1600 randomly selected companies have taken part in the survey, enabling a representative data set in terms of industry, regional, and size structure. In 2001, this survey became international and developed into the *European*

Manufacturing Survey (EMS). The aim of the EMS is to use a set of common core questions in the respective national language in all partner countries and thus to collect internationally comparable data and enable internationally comparative analyses, e.g., on backshoring (Dachs et al. 2019) or servitisation (Dachs et al. 2014). In 2022, 19 European countries participated in the EMS.

2 Production Paradigm Shifts in the Last 50 Years

Within the last five decades, fundamental changes in industrial manufacturing can be observed. These paradigm shifts have changed not only the basic framework conditions for industrial companies, but also the way of thinking in research, politics, and practice. To identify the production paradigms that showed up in the research, we used triangulation. We therefore draw on the existing literature on the one hand, and on the other hand, we worked through the central developments of industrial production in discussions with experts. Subsequently, we combined and mirrored the results of both parts. In particular, we focused on the trends in Germany, which, however, are strongly embedded in the international context. By a paradigm shift, we understand the change of fundamental framework conditions in industrial production, which is also paired with a renewal of the production philosophy and triggered by technological or organisational disruptions.

If we relate to these assumptions, we can identify five production paradigms. These production paradigms are characterised by either technological or organisational disruptions, and we find four paradigm shifts during the period of the last five decades. Paradigms that emerge from technological disruption (technology-driven paradigm) correspond to those of the concept of the four industrial revolutions. Industry 3.0 and 4.0 fall within the period of the last 50 years, while the paradigm "mass production and division of work" (I2.0) also marks the beginning of the deindustrialisation in the leading economies and runs out at the beginning of the early 1970s. However, we were also able to identify three paradigms that emerge from organisational disruptions (organisational-driven paradigms). From this, in turn, a cycle can be derived in the paradigm shifts. On the one hand, this cycle can be seen in the fact that a paradigm lasts about 10 to 15 years before it is overlapped and dominated by another paradigm. On the other hand, it becomes clear that the paradigms alternately follow technological and organisational disruptions. Therefore, we assume that a new paradigm not only brings fundamental innovations, but also forms a kind of counter-design to the previous paradigm.

The first organisational-driven paradigm "quality of working life" focuses on humans in production and aims to improve their working conditions. It thus represents a counter-design to the monotonous work on the assembly line system during Industry 2.0. The second organisational-driven paradigm consists of the idea of "lean production and lean thinking" originating from Japan and revolutionising the

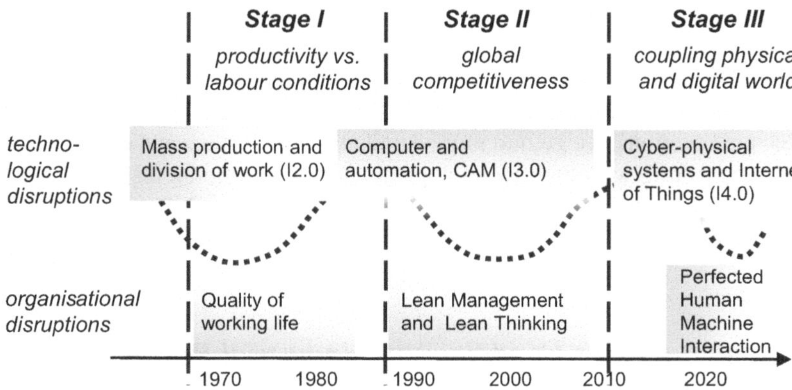

Fig. 1 Production paradigm cycle over the last five decades. Source: own illustration

predominant production principles. "Perfected human–machine interaction" represents the most recent concept of the organisational-driven paradigms. Here, the interface of humans with intelligent machines based on Artificial Intelligence (AI) is considered in particular.

The production paradigms as well as their shifts can be summarised again into three general stages (see Fig. 1):

- Stage I covers the trade-off between industrial productivity and working conditions and extends from the early 1970s to the late 1980s.
- Stage II focuses on the question of Germany's global competitiveness as a production location and extends from the late 1980s to the early 2010s.
- Stage III aims to couple physical objects in production with the digital world and has been prevalent since the early 2010s.

These stages are characterised, on the one hand, by specific developments in manufacturing and, on the other hand, by particular political challenges and a changing understanding of how the economy, industry, and society as a whole can be shaped. Research acted in this field of tension, seized opportunities, offered contexts of understanding and developed into a strong partner, which over time also became anchored in newly institutionalised structures, e.g. the special non-university research institutions in Germany. The interplay between research and practice, as well as the creative will of technology and innovation policy, is also expressed by the so-called "indirect-specific measures", which led to unique funding programmes in Germany. Two of the most important research programmes in this context are explicitly highlighted in the boxes of this study.

The aim of policy in these indirect-specific funding programmes is to enable industry in Germany, through cooperation with applied research institutions, to catch up in key areas with global competition. Since the research institutions were also able to help shape the funding programmes, they closed the gap between policy-makers, on the one hand, and industrial practitioners, on the other. As a result, three central players have emerged in recent decades that shape industrial development: innovation and technology policy, applied research institutions, and industry or rather practitioners.

In the next sections, we go into more detail about each stage and describe its key developments along with their technological and organisational disruptions. To do so we also address the respective paradigms and paradigm shifts during the individual stages as well as the respective political challenges and research agendas.

2.1 Stage I: Labour Conditions and Industrial Productivity (Early 1970s Until the Late 1980s)

Until the 1970s, manufacturing was strongly characterised by standardised mass production using highly specialised and monofunctional machines such as the assembly line system (Kagermann et al. 2011). The organisational form followed the principle of division of labour, according to which employees performed recurring subtasks. This makes it possible to realise a high number of units at low cost, which leads to high productivity in production and work processes. However, this form of organisation leads to a monotonous way of working, since the same activity is performed again and again at regular intervals. For employees, this led to a high workload, one-sided qualifications and increased work-related health risks.

Simultaneously, this period was marked by numerous economic disruptions, as e.g. the collapse of the world currency system, the oil crisis, or the decline in employment and demand, which required changes from a political point of view. Policy-makers searched for answers for a global positioning of the domestic economy and saw the need for a more active research and technology policy. The need to achieve high productivity, on the one hand, and to improve working conditions, on the other, created a field of tension at this stage, which was also evident in research. This is because the development and implementation of new technologies and organisational concepts should succeed not only in achieving further advances in productivity, but also in improving the quality of working life. At this time, industry experienced the decline of lead sectors, but also saw the opportunities of a development towards more flexible production with rapidly changing product lines and greater product diversity.

This initial situation led to a fundamental debate about working conditions in a changing working world in the early 1970s, both in society and in the scientific community, and appeared omnipresent (Kleinöder 2016). In this decade, fundamental regulations of employee co-determination in the workplace were renewed. Moreover, new regulations were found in the area of occupational health and safety. After the foundations for modern accident prevention had been laid in the 1960s, the 1970s saw a revolution in occupational health and safety regulations and new protective provisions for working in production. Consequently, the quality of working life thus moved to the centre of innovative organisational concepts in industrial production (Davis and Cherns 1975). In Germany at this time, numerous programmes and projects were set up to help improve working conditions in manufacturing companies (compare Box 2), with the aim of reducing stress while at the same time expanding employees' scope for action. Support measures addressed, e.g. semi-autonomous group work, the further qualification of employees and the democratisation of work structures (Seibring 2011).

Box 2: Funding Programme "Humanisierung der Arbeitswelt" (from 1974 to 1989)

Until the 1960s, it was hoped that with increasing process automation, the monotonous workloads in production would also be reduced. However, these hopes were not met. Likewise, a growing discrepancy between the personal standard of living and the standard of working conditions became increasingly apparent (Kleinöder 2016). As a result of the pressure to act, various programmes were launched that came to be known as "Humanisierung der Arbeitswelt" (HdA). These were intended to help improve working conditions in companies, reduce stress, and expand employees' scope for action. Another focus was on promoting the industrial application of new upcoming technologies in the fields of data processing, telecommunications, and semiconductor technology.

Until 1980, not only the state and the scientific community were represented, but also employers and labour unions. However, political disputes led to a variety of changes and restrictions, for example the participatory programme organisation was gradually dissolved and the democratisation approaches of the first programme phase (1974–1980) were stopped. This programme comprised around 1500 projects with a total funding volume of 550 million euros and ran in two phases from 1974 to 1989 (Ernst 2009). Despite its clear title, this funding programme can be seen as a politically controversial project in which the relationship between rationalisation and humanisation had to be constantly reassessed. Entrepreneurs hoped for savings in labour costs, while those who believed in progress saw the end of the monotonous everyday life on the assembly line. In addition to many fundamental findings and a scientific foundation that is still received today, the programme polarises to this day and triggered critical debates early on due to continuous problems with its implementation and organisation (Kleinöder 2016; Raehlmann 1981). In terms of content, too, the progress of the programme, measured in terms of its financial and organisational size, has been questioned from various sides. However, due to its political weight and its gradual but nonetheless long-term progress, the programme is a persistent myth to this day (Kleinöder 2016).

At this time, efforts were made to push forward alternative technology development together with alternative forms of work organisations (Raehlmann 1981). However, the hope that the quality of working life would go in line with the continuous improvement of technological progress was soon destroyed by actual developments. It also became clear that an organisation solely based on the division of labour and productivity no longer meets the requirements of flexible production with high quality standards (Kleinöder 2016). These findings ultimately led to the problem of the incompatibility of productivity and flexibility, creating a field of tension between human-centred working conditions and industrial productivity (Matthöfer 1980).

Figure 2 illustrates the need for this debate at this time. As an example, it shows the prevalence of two organisational measures in terms of quality of working life as well as two automation technologies as part of I3.0 among Germany's manufacturing companies: Continuous Improvement Processes (CIP) in manufacturing and assembly aim to ensure that workers continuously improve both products and processes through their experience and ideas. This process thus creates spaces for participation, manifests the view on workers as valuable knowledge carriers, and enables creative participation even in highly complex processes. Flattened hierarchies are structures with fewer management levels. As a result, individual employees have more responsibility and decision-making autonomy. With the introduction of the first CNC machine tools, programming shifted to the work preparation department; skilled workers at the machine lost their responsibility of converting a drawing into working movements of the machine. CAD/NC programming then refers to the linking of machine programming with solutions from the CAD system. A shop floor control system aims at a dynamic, flexible, and resource saving control of the entire production. It refers to precise scheduling and permanent transparency of whole production processes as well as the optimum capacity utilisation on the shop floor.

In Fig. 2, we see that manufacturers hardly made use of these new work concepts or automation technologies until the late 1980s—after technological opportunities

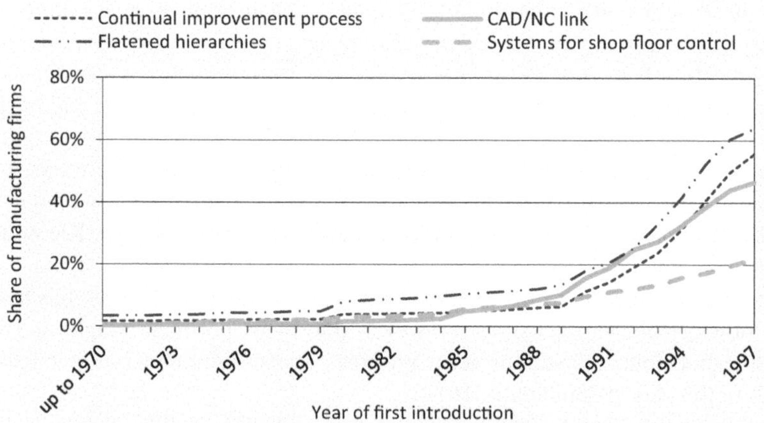

Fig. 2 Diffusion of automation technologies and labour-saving measures. Source: *Innovationen in der Produktion* 1997—Germany, N = 1.329. Fraunhofer ISI, Germany

arose and organisational concepts became known. While the user rate in the 1970s was almost zero percent, the shares increased slightly to just under 10 percent in the 1980s. It was not until the end of the 1990s that user rates in these new work concepts and automation technologies of between 20 and 60 percent were achieved.

Consequently, both organisational concepts for improving the working conditions of employees and CAM-driven automation processes have diffused into the breadth of German manufacturing only since the 1990s. Therefore, this empirical finding shows that there is a very large time lag between the political and scientific discussion on the one hand, and the actual changes in the production companies on the other. Interestingly, these time lags between research and practice can also be observed later in paradigm shifts, although not necessarily in the same order (compare, e.g., stage III, the I4.0-paradigm).

2.2 *Stage II: Global Integration and Competition (Late 1980s Until the Early 2010s)*

The second stage of industrial developments at the end of the 1980s is largely characterised by German reunification and the resulting European reorganisation. The elimination of borders also opens up completely new opportunities for global value creation for product manufacturers. In the 1990s, the People's Republic of China became a serious competitor and interesting market, inducing further productivity pressure. Furthermore, industrial development in emerging countries rapidly picked up speed, which melted the lead of the traditional industrialised countries in terms of production (Gornig and Schiersch 2015). At the same time, novel production concepts from Japan, based on different principles and ways of thinking (Womack et al. 1990), spilled over into the USA, Europe including Germany and revolutionised the predominant production principles (Womack and Jones 1997). From the late 1980s onward, the German economy has been significantly shaped by globalisation and with it the related question of how German industry can survive in global competition. The political challenges and research agendas revolve around how value chains can be expanded internationally while Germany remains attractive as a production location at the same time. Funding programmes at this time were justified by competition policy and aimed primarily at promoting economic modernisation.

After the German reunification, production in the new German states faced structural economic difficulties for a long time, confronting the consequences of the currency conversion and market restructuring on the one hand, but also had to deal with ailing infrastructure and on average low international competitiveness (BMWi 2015). In view of these dramatic changes in the economic parameters, manufacturers in the new German states were faced with the task of quickly adapting their structures to the requirements of the market economy (Lay 1995).

A central phenomenon of this development discussed in the literature is the productivity gap (Lay 1996). In the mid of the 1990s, East German firms achieved less than two-thirds of the productivity of their Western counterparts (Kinkel 2001).

Consequently, a major challenge for Germany at this time was to promote the industrial development of eastern German states, to handle the dramatic declines in the industrial bases of these regions and to bring the companies located there up to global competition.

As a result of the global political changes, a new wave of globalisation started, which in turn also had an impact on the organisation of production. Frequently, parts of production were outsourced to Eastern Europe or the People's Republic of China. As a result, competitive pressure increased and innovation and product life cycles became shorter. Global competition became increasingly evident, and many companies see the international configuration of value chains as an opportunity or a necessity. Empirical analyses also confirm that globalised value creation was particularly dynamic in this period (Kinkel 2001; Kinkel et al. 2002; Zanker et al. 2013). The upper part of Fig. 3 shows this development in Germany from 1995 to 2018. The share of manufacturing firms that relocate parts of their production abroad is displayed: Between 1995 and 2006, one in four or rather five German manufacturers relocated parts of their production abroad, a share that dropped significantly in subsequent years. Especially larger manufacturers from Germany massively expanded their global production activities around the turn of the millennium,

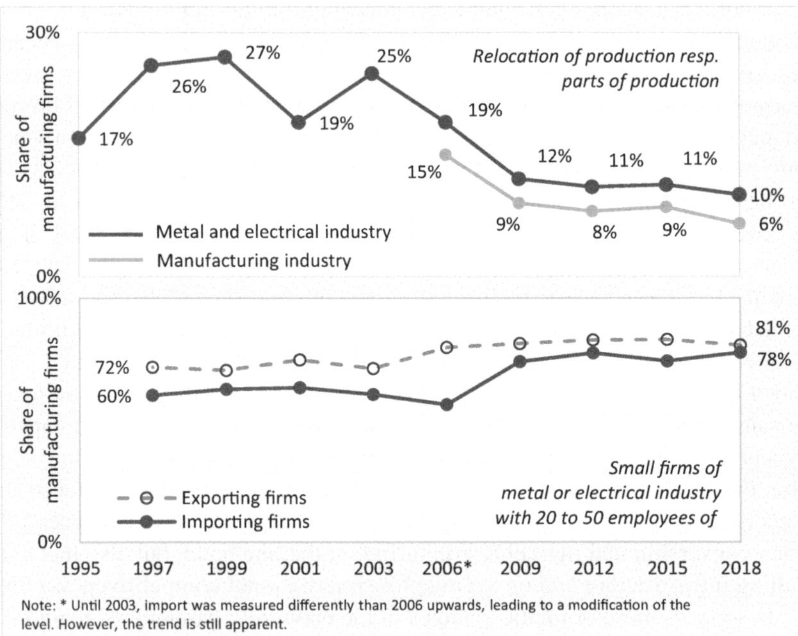

Fig. 3 Production relocations abroad and small exporting and importing firms. Source: *German Manufacturing Survey* 1995–2018, Fraunhofer ISI, Weighted data

not only to produce more cost-effectively but also to open up new markets (Kinkel et al. 2002).

For the same period of time, it is known that the share of exporting manufacturers hardly increased. German manufacturing is traditionally highly export oriented, with around nine out of ten manufacturers exporting their products since the mid-1990s. However, during this period of time, the share of manufacturers buying parts of their inputs abroad increased, catching up with the share of exporters. As can be seen in the lower part of Fig. 3, mainly small firms started internationalising their supply chains more in the 1990s and 2000s in order to obtain higher quality inputs and thus made their own products more competitive. The combination of both empirical findings clearly shows that production relocations do not replace product exports, but complement them (Lerch and Jäger 2021). Hence, globalised value chains represent a value-adding opportunity for manufacturers.

Nevertheless, globalisation of production is characterised not only by the international networking of value creation structures, but also by the internationalisation of thinking approaches and production principles. Around 1990, lean production was introduced as a guiding principle for the creation of value in larger companies (Womack et al. 1990). In the literature, the predominant view is that this principle is responsible for the superiority of Japanese automobile production over American and European manufacturers and enables value creation without waste. The goal is to optimally coordinate all value creation activities and to avoid superfluous activities (waste). To achieve this, two perspectives must be integrated: the view of the customer, whose requirements for availability, quality, and a low price must be met as optimally as possible, and the view of the manufacturer itself, which must produce profitably and improve its competitiveness. Five core principles must be taken into account and implemented using new methods: value, value stream, flow, pull, and perfection (Womack et al. 1990). Hereby, Lean offers an answer to the question of how the organisation of complex production can succeed due to increased automation and mass production. Automated production proved to be particularly fragile in the case of unforeseen disruptions. Therefore, Lean aims to strengthen the potential of people compared to machines, but without losing the productivity advantages of standardised processes.

Lean Thinking has spread rapidly worldwide since the 1990s (Womack and Jones 1997). German manufacturers also introduced the new production concepts at this time (Pfeiffer and Weiß 1994). Since the topic of Lean Thinking has been of greater interest, empirical studies are being conducted to estimate the spread of Lean Concepts in manufacturing. In 1995, for example, a quarter of all capital goods manufacturers in Germany were already using Kanban (Lay 1997), while at the same time demonstrating significantly higher labour productivity than other companies (Lay et al. 1996). The success story of Lean Thinking has persisted for decades and defines the prevailing production paradigm, resulting in the core principles of manufacturers being considered in production to this day (Lerch et al. 2021).

2.3 *Stage III: Entering Cyber-Physical Production (Since the Early 2010s)*

The third stage is initially still characterised by the economic crisis in 2009. In Europe, but especially in Germany, there is a renewed focus on the manufacturing industry (Capello and Cerisola 2022; Prisecaru 2015), which proves to be particularly resistant to the crisis (BMBF 2014). However, it is not only the industrial core and its economic significance that is receiving new attention; a broader understanding of manufacturing and the meaning of industrial value creation is also established (BMBF 2014) (compare Box 3). This new understanding is shaping not only policies but also research activities for the coming years. This includes in particular the linking of production and service in order to develop customer-specific offerings. Role models from the service sector with innovative business concepts are also leading to new business models beyond pure product sales which are discussed for manufacturers (Lay 2014; Lerch and Maloca 2020). Terms such as "Hybrid Value Creation" (Velamuri et al. 2011) or "Product-Service System" (Tukker 2004) emerge, describing the merging of previously separate disciplines. Furthermore, the coupling between production and the digital economy is also becoming increasingly important (Kagermann 2015).

In 2011, finally, manufacturing experienced a definite renaissance with the vision of Industry 4.0, which is perceived worldwide as a new production paradigm and a mode of production to strive for in the future (Kagermann et al. 2013). The vision of Industry 4.0 describes the idea of a comprehensive digitisation of production and an associated interlinking of production technologies and information and communication technologies. The technical basis is formed by the so-called cyber-physical systems that communicate with each other via the (Industrial) Internet of Things (IIoT) (Kagermann et al. 2011). With their help, largely self-organised production should become possible, ultimately leading to individualised production with batch size 1, but under the conditions of mass production. This technological foundation can also be used to develop new types of digital business models based on IIoT-platforms (Lerch and Jäger 2020a; Plattform Industrie 4.0 2019a). During this time, funding policy is attempting to use digitisation to keep the domestic production location competitive and to locate digital value creation in Germany and Europe. Integrating digital technologies with conventional physical technologies also creates completely new fields of research for engineers, computer scientists, and economists. Due to this high impact, the vision of Industry 4.0 is shaping funding programmes and innovation activities for the next 10 years.

Box 3: Funding Programme "Zukunft der Wertschöpfung" (since 2021)
In 2021, the new funding programme "Zukunft der Wertschöpfung" replaced the existing programme "Innovationen für die Produktion, Dienstleistung und Arbeit von morgen" (2014–2021). This one started in 2014, integrated research on production, services, and work for the first time, and emphasised the central role of industry as the backbone of the German economy, as well as the interlinking of product and service innovations. Moreover, the digital refinement of production and ICT as an innovation driver for production and services were central aspects of the programme. It thus responds to the increasing fusion of the various disciplines and sectors in practice as well as the need for a holistic innovation perspective (BMBF 2014).

This trend is continued in the new programme "Zukunft der Wertschöpfung" that was opened in 2021. The three former main topics production, services, and work are discarded as separate guidelines and united in an overall approach to the term value creation. It is mainly influenced by Industry 4.0, digitalisation, and the data and platform economy and is intended to address in particular the increasing complexity and dynamics of value creation processes. Artificial intelligence is seen as a central link for the interaction of humans and machines. It continues the programme line's tradition of harmonising digital technologies and the world of work and life. The programme defines six fields of action: Dynamics of value creation systems, people in value creation, business models and value propositions, resources, socio-technical and methodological innovations, networking and collaboration. However, the fields of action are not understood as rigid blocks, but are part of a continuous further development of the programme and an expression of its ability to learn, with the intention to make the new programme more flexible (BMBF 2021).

However, while research and policy-makers are setting up new funding programmes, manufacturers are already in a real digital transformation phase (see Fig. 4). As in the 1970s, a time lag between research and practice can again be observed. However, this time the development in production is ahead of the political and scientific discussion (compare stage I, Fig. 1). This can be demonstrated with the help of an I4.0-index that uses seven digital production technologies and addresses different technological fields. Hereby it is possible to show the digital progress of production in Germany over time (Lerch et al. 2015). Therefore, manufacturers are divided into three groups: Non-users, who do not apply any of the seven digital production technologies; Basic users, who apply at least one of these technologies; Top users, who are extensively digitised and have at least two cyber-physical systems in use.

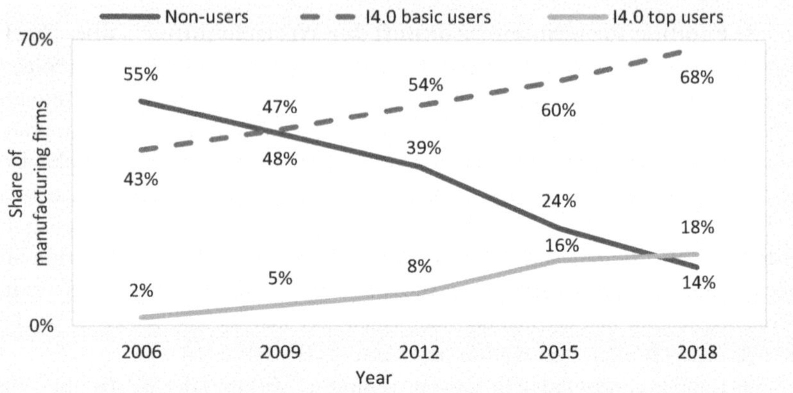

Fig. 4 Development of I4.0-readiness groups. Source: *German Manufacturing Survey* 2015 und 2018, Fraunhofer ISI

As Fig. 4 shows, significant changes took place between the years 2006 and 2018 with regard to the I4.0-readiness level of companies. Whereas in 2006, non-users still accounted for more than half of all manufacturers in Germany, in 2018 this group only made up the minority of manufacturers at 14 percent. The group of I4.0 basic users in particular benefited from this. The vast majority of companies have therefore begun to digitise their production step by step over these 12 years. During this period, more and more companies also reached the group of I4.0 readiness top users, who is larger than the group of non-users for the first time in 2018. Just under one in five industrial enterprises in Germany are already prepared to use I4.0 applications and are already able to apply technologies as, e.g., near real-time production control systems, digital data exchange systems with customers and suppliers, automated internal logistics, and digital management in 2018 (Lerch and Jäger 2020b).

In particular, the dramatic decline in non-users already heralded the digital transformation phase towards the end of the 2000s. Nevertheless, I4.0 basic users still represented by far the largest group in production in 2018. Moreover, this group was still growing at this point, which means that more non-users became basic users than basic users became top users. As a result, at the end of the 2010s, the transformation phase was far from complete and a digital maturity phase will not be reached for years, possibly even decades.

However, towards the end of the 2010s, a broader understanding of Industry 4.0 was developing. The point of criticism here was an overly strong focus on technological development. Once again, a counter-design developed that no longer pursued the goal of a factory devoid of people, but rather the approach was one of human-centred production. In addition to this, topics such as sustainability and resilience were also to be increasingly taken into account (Bendig et al. 2021). The initiator of this concept was the European Commission, which in some cases was already proclaiming the fifth industrial revolution (Renda et al. 2021). While

Industry 4.0 at least partially called into question the role of humans in factories, the EU's human-centric view described a fundamentally different paradigm: In the future, new technologies are to be adapted in such a way that they ideally complement the work of humans, and not vice versa. In this perfected version of human–machine interaction, not only productivity increases are promised, but ideal working conditions are also created for employees (Bendig et al. 2021). This concept was driven also by the search for the future role of employees in Europe and the question of how much and what kind of work manufacturing will offer in future. Studies focusing on the humanisation of work, which are discussed today under the keyword "good work", re-enter into political and scientific discussion. Consequently, the questions that arose in the early 1970s are also returning 50 years later.

2.4 *The Future of Manufacturing and the Search for New Production Paradigms*

As the timeline of industrial developments shows, global trends play a crucial role in the origin of production paradigms during the last decades. We can therefore also assume that the coming production paradigms will be determined by the future trends and thus also the related research activities in the field of innovation and production research. It is therefore useful to take a look at the emerging trends in order to obtain indications about upcoming production paradigms.

At present, we are (still) at the stage of linking the physical and digital world of production, whereby the human being seems to be moving back into the centre of investigation. This is already pointing to the first key trend for the future of manufacturing. Because this means, that the long-awaited tension between humans and intelligent machines in production is on the horizon. This raises the question of what role humans will play in the smart factory, to what extent intelligent machines will take over production work and, above all, what collaboration between humans and intelligent machines will look like (Plattform Industrie 4.0 2019b; Renda et al. 2021). Interestingly, the key question that arises is similar to the one from the early 1970s. In future, production will continue to be driven by the question of the role of human beings under new technological conditions, in this latest case the clash of natural and artificial intelligence.

A second key trend is certainly characterised by climate change and resource scarcity (BMBF 2021). Not only new production processes, but also new product and service offerings will be focused more than ever on decarbonisation. This development brings not only threats to industrial production, but also opportunities for new environmentally friendly production processes and product portfolios and hence, new value creation potential. The central question for German and European production will be to what extent will industrial firms be able to exploit the opportunities of decarbonisation and resource efficiency and take on a global leadership role. This challenge has also been occupying policy-makers and researchers for a long time. Nevertheless, this new paradigm will be less about exploiting the saving potentials of existing production systems and more about developing a completely new value creation model that combines prosperity and sustainable development in equal measure.

A third trend that has emerged in recent years is the so-called glocalisation of production. For decades, the internationalisation of value chains has been seen as a guarantor of value creation and prosperity. However, more recent developments, such as new protectionism of markets, the Covid crisis as well as a new war in Eastern Europe show that they can affect the industrial core hard. These developments raise the question of a reorganisation of global production chains and the need for strengthening the resilience of manufacturing. Manufacturing of the future will therefore have to consider not only economic but also strategic aspects, such as which value networks are to be localised and which are to be globalised, for creating a new mode of production and strengthening the resilience of manufacturing at the same time (Plattform Industrie 4.0 2022). This will also raise the question of how deeply politics intervene in strategic decisions of (systemically relevant) large enterprises or support them with state funds. Accordingly, the design of a future-oriented industrial strategy for Germany and Europe will also have to be addressed (BMWi 2019).

Beyond these foreseeable trends, the question remains as to what a fifth industrial revolution (I5.0) might look like in the future. This will require a new basic technology that finds its way into companies' production structures. There are currently no signs of a breakthrough for widespread use of any emerging technology. Nevertheless, new production principles are being discussed that could be based on nanotechnology or quantum computing, for example. A biologisation of manufacturing industries, which would entail a stronger infiltration of biotechnologies and biological principles within production, is also currently under discussion (Bauernhansl et al. 2019). When and in what form this fifth industrial revolution will take place, however, still seems to be an open question. Looking at the periods of the four previous industrial revolutions, this could be many decades.

In order to counter and master these potential key trends, the successful interplay of technology and innovation policy, applied research, and industry will continue to be crucial. Only in this way will the German and European industrial landscape continue to play a leading role, but will also be able to shape upcoming production paradigms in a significant way. As the past shows, industrial key trends can arise in a variety of ways: through the continuous rise of a new basic technology (Industry 4.0 in the 2010s), political agendas (quality of working life in the 1970s) through economic recessions (financial crisis 2008), and social crises or global shocks (climate change, unexpected outbreak of war in Eastern Europe 2022). While unforeseeable, disruptive events can hardly be predicted and occur at short notice, they can at best be mitigated by resilience mechanisms. It is important to recognise the signs of the times as early as possible in the case of long-term developments and to create visions for new value creation models, what we call the production paradigms in this study (cf. Cuhls et al. 2024).

The task of applied research will therefore remain to notice trends at an early stage, to discuss them and to provide contexts of understanding policy and industry.

At the same time, research is also needed as an evaluator of policy measures in order to draw conclusions for future activities. Visions and funding programmes must be designed not only by policy-makers but also by researchers in order to provide industry and practitioners with guidance on how to master future developments and challenges.

The principle of indirect-specific measures, in particular through funding programmes, will continue to play a key role in the industrial landscape in future. Since 2020, these are also supplemented with "direct non-specific" instruments, such as state R&D funding, anchored by laws. However, the search for suitable funding measures for industrial modernisation and the realisation of new production paradigms will continue to occupy policy-makers and researchers.

3 Conclusions: Manufacturing at a Crossroads Again

In this article, we look back at major changes in manufacturing over the last 50 years. In order to capture the predominant philosophy of production in a given period over the last half century, we outline five paradigms, which shaped production as well as research about it. These extend the concept of the four industrial revolutions (here I3.0 and I4.0), which we believe were designed from a purely technology-driven perspective. We complement these paradigms with organisational-driven paradigms and classify the resulting paradigm shifts into three major stages. Thereby, each stage is characterised by a central guiding question that has been groundbreaking for policy-makers, researchers, and industrial practitioners: while the first stage is driven in particular by the tension between high productivity and improved working conditions, the second stage consists of the question of the competitiveness and integration of German manufacturing in a globalised world. The third and current stage is dominated by the coupling of physical production with the digital world.

More conclusions can be drawn regarding the interplay of the three central players of industrial development, as well as the cycle of production paradigms: (i) a paradigm is only replaced or renewed by a paradigm of the same type. (ii) However, paradigms of a different type overlap in time and thereby interact in terms of research questions and the political agenda. Moreover, they represent a kind of counter-design to the previous paradigm. (iii) On the one hand, this results in a sequential flow of the same paradigm types, but on the other hand, it also creates a paradigm cycle through the interaction and counter-design of different paradigm types. (iv) In order to design a production paradigm, applied research must identify trends at an early stage and, together with policy-makers, design a vision that results in funding programmes. In turn, the cooperation between applied research and industry through these indirect-specific measures leads to the implementation of the vision and the rise of a new production paradigm.

At the moment, society and the economy are shaken as seldom before by several disruptive events that also influence the manufacturing industry: The UK's exit from the EU and increasing market protectionism in the USA and China are challenging the free movement of goods. The Covid-19 pandemic, triggered by a coronavirus in 2020, also led to restrictions in production and supply chain disruptions for several years. A new war in Eastern Europe not only causes established economic relations to be abandoned, but also leads to rapidly rising energy prices and supply bottlenecks. All these developments bring new challenges for production, which are associated with further structural change, new risks, but also opportunities. New production paradigms must provide solutions to these challenges and show ways for production in future. Once again, manufacturing is at a crossroads 50 years after the start of deindustrialisation.

References

Arundel A, Smith K (2013) History of the community innovation survey. In: Gault F (ed) Handbook of innovation indicators and measurement. Edward Elgar Publishing, pp 60–87

Bauernhansl T, Brecher C, Drossel W-G, Gumbsch P, ten Hompel M, Wolperdinger M (eds) (2019) Biointelligenz: Eine neue Perspektive für nachhaltige industrielle Wertschöpfung: Ergebnisse der Voruntersuchung zur biologischen Transformation der industriellen Wertschöpfung (BIOTRAIN). Fraunhofer Verlag, Stuttgart

BDI (2013) Industrielle Wertschöpfungsketten: Wie wichtig ist die Industrie? Online: https://bdi.eu/media/presse/publikationen/BDI_Industrielle_Wertschoepfungsketten.pdf. Accessed 16.07.2022

Bendig D, Lau K, Schulte J, Endriß S (2021) Industrie 5.0 – Die Europäische Kommission auf den Spuren der nächsten industriellen Revolution? I40M 2021:20–22. https://doi.org/10.30844/I40M_21-6_S20-22

BMBF (2014) Innovationen für die Produktion, Dienstleistung und Arbeit von morgen. Online: https://www.bmbf.de/bmbf/de/forschung/forschung-fuer-arbeit/innovationen-fuer-die-produkti-leistung-und-arbeit-vonmorgen/innovationen-fuer-die-produkti-leistung-und-arbeit-von-morgen_node.html. Accessed 13.05.2022

BMBF (2021) Zukunft der Wertschöpfung – Forschung zu Produktion, Dienstleistung und Arbeit. Online: https://www.bmbf.de/SharedDocs/Publikationen/de/bmbf/5/31662_Zukunft_der_Wertschoepfung.html. Accessed 13.5.2022

BMWi (2015) Industrielle Entwicklung in den neuen Bundesländern

BMWi (2019) Industriestrategie 2030: Leitlinien für eine deutsche und europäische Industriepolitik. Online: https://www.bmwk.de/Redaktion/DE/Downloads/Monatsbericht/Monatsbericht-Themen/2019-03-nationale-industriestrategie-2030.pdf?__blob=publicationFile&v=4. Accessed: 8.7.2022

Capello R, Cerisola S (2022) Regional reindustrialization patterns and productivity growth in Europe. Reg Stud 57:1–12. https://doi.org/10.1080/00343404.2022.2050894

Cuhls K, Dönitz E, Erdmann L, Gransche B, Kimpeler S, Schirrmeister E, Warnke P (2024) Foresight. In: Edler J, Walz R (eds) Innovation and sustainability research in transition: research questions and trends in historical perspective. Springer, Heidelberg

Dachs B, Biege S, Borowiecki M, Lay G, Jäger A, Schartinger D (2014) Servitisation of European manufacturing: evidence from a large scale database. Serv Ind J 34:5–23. https://doi.org/10.1080/02642069.2013.776543

Dachs B, Kinkel S, Jäger A, Palčič I (2019) Backshoring of production activities in European manufacturing. J Purch Supply Manag 25(3):100531. https://doi.org/10.1016/j.pursup.2019.02.003

Davis LE, Cherns AB (1975) The quality of working life: cases and commentaries, vol 2. Free Press; Collier Macmillan, New York, London

Eickelpasch A (2012) Industrienahe Dienstleistungen: Bedeutung und Entwicklungspotenziale; Expertise im Auftrag der Abteilung Wirtschafts- und Sozialpolitik der Friedrich-Ebert-Stiftung. Arbeitskreis Nachhaltige Strukturpolitik, Bonn. Online: http://library.fes.de/pdf-files/wiso/09101.pdf. Accessed 17.07.2022

Eickelpasch A (2014) Funktionaler Strukturwandel in der Industrie: Bedeutung produktionsnaher Dienste nimmt zu. DIW Wochenbericht. Online: https://www.diw.de/documents/publikationen/73/diw_01.c.478717.de/14-33-1.pdf. Accessed 8.7.2022

Ernst G (2009) Von der Humanisierung zu Innovationen für die Arbeit von morgen: 40 Jahre Arbeitsforschung

Fourastié J (1949) Die grosse Hoffnung des 20. Jahrhunderts. Online: https://www.europa.clio-online.de/quelle/id/q63-28281. Accessed 8.7.2022

Gornig M, Schiersch A (2015) Perspektive der Industrie in Deutschland. Vierteljahrshefte zur Wirtschaftsforschung 84:37–54. https://doi.org/10.3790/vjh.84.1.37

Kagermann H (2015) Change through digitization—value creation in the age of industry 4.0. In: Albach H, Meffert H, Pinkwart A, Reichwald R (eds) Management of permanent change. Springer Gabler, Wiesbaden, pp 23–45. https://doi.org/10.1007/978-3-658-05014-6_2

Kagermann H, Lukas W-D, Wahlster W (2011) Industrie 4.0: Mit dem Internet der Dinge auf dem Weg zur 4. industriellen Revolution. VDI nachrichten 13(1):2–3

Kagermann H, Wahlster W, Helbig J (2013) Recommendations for implementing the strategic initiative INDUSTRIE 4.0. Securing the future of German manufacturing industry. Final report of the Industrie 4.0 Working Group. Hg. v. Forschungsunion und acatech. Online: https://www.din.de/resource/blob/76902/e8cac883f42bf28536e7e8165993f1fd/recommendations-forimplementing-industry-4-0-data.pdf. Accessed 28.7.2021

Kinkel S (2001) Auf den Spuren der Produktivitätslücke ostdeutscher Betriebe: Das Beispiel der Investitionsgüterindustrie Sachsens. Mitteilung aus der Produktionsinnovationserhebung 21. Karlsruhe: Fraunhofer ISI. Online: https://www.isi.fraunhofer.de/content/dam/isi/dokumente/modernisierung-produktion/erhebung1999/pi21.pdf. Accessed 17.7.2022

Kinkel S, Jung Erceg P, Lay G (2002) Auslandsproduktion—Chance oder Risiko für den Produktionsstandort Deutschland? Stand, Entwicklung und Effekte von Produktionsverlagerungen im Verarbeitenden Gewerbe. Mitteilung aus der Produktionsinnovationserhebung 26. Karlsruhe: Fraunhofer ISI. Online: https://www.isi.fraunhofer.de/content/dam/isi/dokumente/modernisierung-produktion/erhebung2001/pi26.pdf. Accessed 17.7.2022

Kleinöder N (2016) Humanisierung der Arbeit: Literaturbericht zum "Forschungsprogramm zur Humanisierung des Arbeitslebens". Hans Böckler Stiftung, Düsseldorf

Krupp H (1984) Werden wir‘s erleben? Mitteilungen aus der Arbeitsmarkt- und Berufsforschung (MittAB) 17:5–8

Lay G (ed) (1995) Strukturwandel in der ostdeutschen Investitionsgüterindustrie. Technik, Wirtschaft und Politik, vol 17. Physica-Verl, Heidelberg

Lay G (1996) Sachsens Investitionsgüterindustrie im Regionalvergleich: Trotz Modernisierungsschub unterdurchschnittliche Produktivität. Mitteilung aus der Produktionsinnovationserhebung 3. Karlsruhe: Fraunhofer ISI. Online: https://www.isi.fraunhofer.de/content/dam/isi/dokumente/modernisierung-produktion/erhebung1995/pi3.pdf. Accessed 13.7.2022

Lay G (1997) Neue Produktionskonzepte und Beschäftigung: Kann die Reorganisation einen Beitrag zur Schaffung neuer Arbeitsplätze leisten? Mitteilung aus der Produktionsinnovationserhebung 8. Karlsruhe: Fraunhofer ISI. Online: https://www.isi.fraunhofer.de/content/dam/isi/dokumente/modernisierung-produktion/erhebung1995/pi8.pdf. Accessed 13.7.2022

Lay G (ed) (1999) Innovation in production: the adoption and impacts of new manufacturing concepts in German industry; with 6 tables. Technology, innovation and policy, vol 8. Physica-Verl, Heidelberg

Lay G (ed) (2014) Servitization in industry. SpringerLink Bücher. Springer, Cham

Lay G, Dreher C, Kinkel Steffen (1996) Neue Produktionskonzepte leisten einen Beitrag zur Sicherung des Standorts Deutschland. Mitteilung aus der Produktionsinnovationserhebung 1. Karlsruhe: Fraunhofer ISI. Online: https://www.isi.fraunhofer.de/content/dam/isi/dokumente/modernisierung-produktion/erhebung1995/pi1.pdf. Accessed 13.7.2022

Lerch C, Jäger A (2020a) Digitale Plattformen auf dem Vormarsch?: Verbreitung und Umsatzeffekte des Plattformgeschäfts im Verarbeitenden Gewerbe. Mitteilung aus der Erhebung Modernisierung der Produktion 77. Karlsruhe: Fraunhofer ISI. Online: https://www.isi.fraunhofer.de/content/dam/isi/dokumente/modernisierung-produktion/erhebung2018/PI_77_B2B_Plattformoekonomie.pdf. Accessed 13.7.2022

Lerch C, Jäger A (2020b) Industrie 4.0 quo vadis? Neuere Entwicklungen der vierten industriellen Revolution im Verarbeitenden Gewerbe. Mitteilung aus der Erhebung Modernisierung der Produktion 76. Karlsruhe: Fraunhofer ISI. Online: https://www.isi.fraunhofer.de/content/dam/isi/dokumente/modernisierung-produktion/erhebung2018/PI_76_Industrie_4-0_quo_vadis.pdf Accessed 13.7.2022

Lerch C, Jäger A (2021) Export und Globalisierung in Zeiten des digitalen Wandel: Wie Produkthersteller beim Exportgeschäft von der Digitalisierung profitieren können. Mitteilung aus der Erhebung Modernisierung der Produktion 79. Karlsruhe: Fraunhofe ISI. Online: https://www.isi.fraunhofer.de/content/dam/isi/dokumente/modernisierung-produktion/erhebung2018/PI_76_Industrie_4-0_quo_vadis.pdf. Accessed 13.7.2022

Lerch C, Maloca S (2020) Digitale Geschäftsmodelle—Modethema oder Wettbewerbsfaktor? Mitteilung aus der Erhebung Modernisierung der Produktion 75. Karlsruhe: Fraunhofer ISI. Online: https://www.isi.fraunhofer.de/content/dam/isi/dokumente/modernisierungproduktion/erhebung2015/PI_75_Digitale%20Gesch%C3%A4ftsmodelle.pdf Accessed 13.7.2022

Lerch C, Jäger A, Maloca S (2015) Wie digital ist Deutschlands Industrie wirklich? Mitteilung aus der Erhebung Modernisierung der Produktion 71. Karlsruhe: Fraunhofer ISI. Online: https://www.isi.fraunhofer.de/content/dam/isi/dokumente/modernisierung-produktion/erhebung2015/pi71_readiness_i4-0.pdf. Accessed 17.7.2022

Lerch C, Jäger A, Heimberger H (2021) Lean 4.0: Smart und schlank produzieren: Potenziale und Effekte einer neuen Managementdisziplin. Mitteilung aus der Erhebung Modernisierung der Produktion 80. Karlsruhe: Fraunhofer ISI. Online: https://www.isi.fraunhofer.de/content/dam/isi/dokumente/modernisierung-produktion/erhebung2018/PI80_Lean40_Management.pdf. Accessd 13.72022

Matthöfer H (1980) Humanisierung der Arbeit und Produktivität in der Industriegesellschaft, 3rd edn. Bund-Verlag, Köln

Meyer-Krahmer F (1984) Recent results in measuring innovation output. Res Policy 13:175–182. https://doi.org/10.1016/0048-7333(84)90025-8

Pfeiffer W, Weiß E (1994) Lean management. Grundlagen der Führung und Organisation lernender Unternehmen. 2., überarb. und erw. Aufl. Schmidt, Berlin

Plattform Industrie 4.0 (2019a) Digitale Geschäftsmodelle für die Industrie 4.0. Online: 2022 Online: https://www.plattformi40.de/IP/Redaktion/DE/Downloads/Publikation/digitale-geschaeftsmodelle-fuer-industrie-40.html. Accessed 13.7.2022

Plattform Industrie 4.0 (2019b) Technologieszenario—„Künstliche Intelligenz in der Industrie 4.0". Online: Online: https://www.plattform-i40.de/IP/Redaktion/DE/Downloads/Publikation/KI-industrie-40.html. Accessed 15.7.2022

Plattform Industrie 4.0 (2022) Resilienz im Kontext von Industrie 4.0. Online: https://www.plattformi40.de/IP/Redaktion/DE/Downloads/Publikation/Resilienz.pdf?__blob=publicationFile&v=1. Accessed 15.7.2022

Prisecaru P (2015) EU reindustrialization on the coordinates of scientific and technical progress. Procedia Econ Financ 22:485–494. https://doi.org/10.1016/S2212-5671(15)00243-9

Raehlmann I (1981) Technokratie statt Reformen: Strukturprobleme in der Forschungsverwaltung. https://doi.org/10.20378/IRBO-54251

Renda A, Schwaag Serger S, Tataj D, Morlet A, Isaksson D, Martins F, Mir Roca M, Hidalgo C, Huang A, Dixson-Declève S, Balland P-A, Bria F, Charvériat C, Dunlop K, Giovannini E (2021) Industry 5.0, a transformative vision for Europe: governing systemic transformations towards a sustainable industry. ESIR Policy Brief, No. 3. Publications Office of the European Union, Luxembourg

Seibring A (2011) Die Humanisierung des Arbeitslebens in den 1970er-Jahren: Forschungsstand und Forschungsperspektiven. In: Andresen K, Bitzegeio U, Mittag J (eds) "Nach dem Strukturbruch"? Kontinuität und Wandel von Arbeitsbeziehungen und Arbeitswelt(en) seit den 1970er-Jahren. Dietz, Bonn, pp 107–126

Tukker A (2004) Eight types of product–service system: eight ways to sustainability? Experiences from SusProNet. Bus Strat Env 13:246–260. https://doi.org/10.1002/bse.414

Velamuri VK, Neyer A-K, Möslein KM (2011) Hybrid value creation: a systematic review of an evolving research area. J Betriebswirtsch 61:3–35. https://doi.org/10.1007/s11301-011-0070-5

Wengel J (1999) Promotion of CIM in the new Länder. In: Innovation in production. Physica-Verl., Heidelberg [u.a.]

Womack JP, Jones DT (1997) Lean thinking—banish waste and create wealth in your corporation. J Oper Res Soc 48:1148. https://doi.org/10.1057/palgrave.jors.2600967

Womack JP, Jones DT, Roos D (1990) The machine that changed the world: how Japan's secret weapon in the global auto wars will revolutionize Western industry. HarperPerennial, New York, NY. Online: http://www.loc.gov/catdir/description/hc044/91055106.html

Zanker C, Kinkel S, Maloca S (2013) Globale Produktion von einer starken Heimatbasis aus: Verlagerungsaktivitäten deutscher Unternehmen auf dem Tiefstand. Mitteilung aus der Erhebung Modernisierung der Produktion. 63. Karlsruhe: Fraunhofer ISI. Online: https://www.isi.fraunhofer.de/content/dam/isi/dokumente/modernisierung-produktion/erhebung2012/pi63.pdf. Accessed 17.07.2022

Christian Lerch is the head of the Business Unit Industrial Change and New Business Models as well as coordinator of the cross-cutting topic "Transformation to a new added value paradigm" at the Fraunhofer Institute for Systems and Innovation Research ISI in Karlsruhe. He studied Economics Engineering at the Karlsruhe Institute of Technology KIT and holds a PhD in innovation management from the Freie Universität Berlin. For more than 15 years, he deals with topics on innovation and industrial change, where he analyses innovation activities from the company level to the sectoral and macroeconomic level and advises policy-makers and practitioners.

Angela Jäger studied Social Sciences with a focus on methods of empirical social research, Sociology, and Contemporary History at the University of Mannheim. Afterwards she worked at the Mannheim Centre for European Social Research (MZES) in the field of egocentric network analysis. In 2006, she moved to the Fraunhofer Institute for Systems and Innovation Research ISI as an expert in empirical methods of social and economic research. Her work focuses on research design, firm surveys, and quantitative analyses. Among other things, she coordinates the *European Manufacturing Survey* (EMS).

From Niche to Mainstream: Exploring Innovation and Progress of Renewable Energy Development

Julia Panny, Anne Held, Jenny Winkler, Barbara Breitschopf, Eberhard Jochem, and Rainer Walz

Abstract This article explores the evolution of research in the field of renewable energy over the past five decades, tracing its development through various phases. Initially sparked by the 1970s energy crises and growing environmental consciousness, the journey began with a focus on technological solutions for renewables. The article highlights the shift over time away from purely technology-driven research to a broader, interdisciplinary orientation. Following the first phase of exploring technology solutions, we discuss the market expansion phase of renewables, their market integration as well as the current speeding up of the transition towards a more and more renewable electricity system. We highlight the evolution of support mechanisms and concomitant scientific debate that accompanied the move from quota obligations to feed-in tariffs. With renewables now a key element in achieving climate neutrality, research has expanded to include market and system integration, the socio-economic impacts of the renewable energy expansion, and systems transformation perspectives. The article underscores the contribution of different types of institutions and players in shaping renewable energy research and policy, emphasising the increasing importance of a systemic and interdisciplinary approach to address current energy and sustainability challenges in a holistic manner.

1 Introduction

This article looks at the development of renewable energies over the last 50 years to identify the main research topics in this field and how these relate to the "major fault lines" of energy policy during this period. We reveal the major development lines of

J. Panny (✉) · A. Held · J. Winkler · B. Breitschopf · E. Jochem · R. Walz
Fraunhofer Institute for Systems and Innovation Research ISI, Karlsruhe, Germany
e-mail: julia.panny@isi.fraunhofer.de; anne.held@isi.fraunhofer.de; Jenny.Winkler@isi.fraunhofer.de; Barbara.Breitschopf@isi.fraunhofer.de; Eberhard.Jochem@isi.fraunhofer.de; Rainer.Walz@isi.fraunhofer.de

J. Edler, R. Walz (eds.), *Systems and Innovation Research in Transition*, Sustainability and Innovation, https://doi.org/10.1007/978-3-031-66100-6_8

renewables research by assessing project types, key methods, and research approaches and how these have changed over time.

As topical as it may seem, the societal, political, and scientific discourse regarding renewable energy technologies and policies has been shaped by ideological, political, and cultural dynamics for decades. The 1970s energy crises as exogenous events, however, marked the beginning of a new era, making Germany, Europe, and the rest of the industrialised world start to question their energy mix and its reliance on fossil fuels as well as shaping a lasting awareness of global interdependencies (Mittlefehldt 2018). At about the same time, Meadows et al. (1972) published their seminal work on the *"Limits to Growth"*, which highlighted unsustainable patterns of energy consumption. In Germany, the anti-nuclear movement started to become a force of growing importance, eventually ending the pro-nuclear consensus in the aftermath of Chernobyl (1986) (Hake et al. 2015). Since then, renewable energies research has continuously made key contributions to the debate surrounding the energy transition and energy security, with the topic becoming ever more important, not least due to the rise of climate policy and the urgency of limiting global warming.

In the European Union (EU) and many other countries across the globe, there is now little doubt that renewables have a key role to play in reaching climate neutrality and future-proofing our energy system. The topic's relevance is undisputed in light of national and global developments, and accelerated deployment of renewables is generally regarded as a necessity. In addition to being a pillar of the energy transition, developments in the renewable energy sector can also provide interesting lessons with respect to innovation, transformation, and market diffusion. Likewise, they are a showcase for the development of policy support and adapting this in line with technology and market maturity. Finally, it cannot be underscored enough that knowledge about the history of the research and thinking on renewable energies can help us tackle current energy and sustainability challenges. A large number of research institutions, including Fraunhofer ISI, have been actively involved in these research and policy areas over the last 50 years and have helped to shape their development.

Today, renewables hold an important place in multiple sectors, most notably electricity, heating, and cooling as well as transport. Fig. 1 depicts the development of renewables in the different sectors in Germany until 2021. The graph clearly shows that even if the debate regarding renewable sources of energy started much earlier than 1990, renewable shares in the electricity mix were negligible until then. In addition, growth rates were moderate at first and expansion was slow until 2000. This was followed by increased growth, but progress has been slowing again in recent years. Higher expansion rates than those currently seen will be required to reach future targets.

In terms of technology, hydropower was dominant to start with, but soon complemented by onshore wind, which started to develop in Germany at the beginning of the 1990s and then underwent rapid growth. The expansion of solar PV took place slightly later, but then with similar growth rates. The deployment of offshore wind is a relatively recent phenomenon, but rapid growth is expected in the coming years. Renewable support policies have been a crucial driver of the renewable

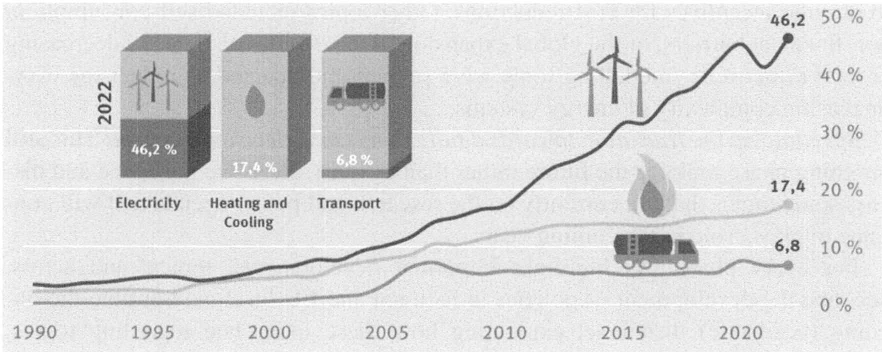

Fig. 1 Shares of RES in Germany in different sectors 1990–2022. Source: Umweltbundesamt, https://www.umweltbundesamt.de/en/topics/climate-energy/renewable-energies/renewable-energies-in-figures

capacities in Europe and Germany from 1990 until today, although their main design elements have changed considerably over this period. When it comes to sectors other than electricity, renewable heating, in particular, is central to the debate.

Based on the development of capacities and policies, we can identify four phases of renewable energy expansion in Germany, which are briefly characterised below. The main dynamics, scientific debate, and methods applied during each of these phases are described in more detail in the respective sections.

Exploring technology solutions: In this phase, which lasted from about 1970 to 1990, the focus was on technological developments, including solar thermal plants in Spain, for example, but also the big wind turbine, "Growian", in Germany (*Gro*ße *Wi*ndenergie*an*lage, commissioned in 1983 by the energy industry). Attention was also paid to renewable heating technologies due to the perceived necessity to divest from oil as a result of the global oil crises. The end of this phase was marked by the first expansion of renewables.

Market expansion: In this phase, which lasted from about 1990 to 2010, the growth of renewables increased substantially, especially in the electricity sector. Germany introduced feed-in tariffs to support them. There was a heated debate at EU level about market-based or state support systems during the second half of this phase. This resulted in Member States having a high degree of freedom in the choice of support systems. At the end of this phase, the rising support costs, mainly for solar PV, triggered a debate about possible cost savings. The boom in PV expansion and the associated sharp rise in subsidy costs were due to the administratively determined level of remuneration (fixed tariffs), on the one hand, and the rapidly falling technology costs, on the other.

Market integration: In the 2010s, the market and system integration of renewables became more important given the rising shares of renewables in the electricity system. Key developments included the introduction of premium schemes instead of fixed tariffs as well as the use of auctions for allocating support. One major goal was to introduce a measure capable of controlling the costs of support while trying to provide incentives for cost reductions. Other topics included self-consumption, non-financial barriers, or the global expansion of renewables also due to decreasing costs. Furthermore, modelling tools were refined and adapted to reflect the ever-increasing complexity of energy systems.

Speeding up the transition towards a fully renewable electricity system: This still on-going phase looks to the future rather than the past. Here, we introduce and discuss some topics that are currently on the research and policy agenda and will continue to play a role in the coming years.

For every phase, we highlight important developments, topics, and actors, describe the development of policies at national and EU level, and outline the on-going (academic) discourse, explaining how these interrelate regarding topics, methodologies, and research actors. The chapter concludes with a summary of important developments across phases and lessons learned about the development of the research field.

2 Exploring Technological Solutions and Dealing with Resistance from Incumbent Electricity Suppliers (1970–1990)

2.1 Starting Research on Technological Solutions

In the years after the first oil price crisis in 1973, increasing attention was given to renewable energies, in particular in US research laboratories. In his book "Man, Energy, Society" based on the analysis of 165 citations, Cook (1976) concluded "that man must ultimately rely on renewable energy forms, but there is no promise that these energy sources can be made available at costs low enough to ensure man's survival". Many energy technologists started applied research activities, supported by government funding, and focused on specific renewable energies like wind power, low and high temperature use of solar thermal, geothermal energy, use of wood and biogas, and photovoltaics. This was also observed in Germany, where the Ministry of Research and Technology launched dedicated research and development funding programmes for energy, which also focused on specific renewable technologies. Up until 1982, a total of 150 million Deutsche Mark was spent on renewables research (Hake et al. 2015).

This development phase was characterised by the emergence and consolidation of several new actors in energy research, but already established institutions also reinforced their research into renewable energies. Many of these actors, of which we will only mention a few, are still an integral part of the German and European energy and renewables research landscape today.[1] Especially at the beginning of the phase,

[1] For a comprehensive account of the individual actors in the German renewable energy landscape as well as their role and evolvement, see Stadermann (2021).

there was a clear focus on exploring and analysing renewable technology solutions including socio-economic aspects and less emphasis on policies and instruments. The most relevant actors in Germany were the German Aerospace Centre DLR (formerly DFVLR), Forschungszentrum Jülich (formerly Kernforschungsanlage Jülich, KFA), and the Helmholtz Association of German Research Centres, all of which had energy-focused institutes or units, e.g. the DLR site in Stuttgart hosting the Institute of Solar Research. Other relevant actors in this early phase included the Fraunhofer Institute for Solar Energy Systems ISE in Freiburg, which entered the scene in 1981, and the Institute for Energy Supply Technologies ISET in Kassel, which was established in 1988 and later integrated into the Fraunhofer-Gesellschaft. The Fraunhofer Institute for Systems and Innovation Research ISI was established in 1972, with the aim of taking a more interdisciplinary approach, combining natural and social sciences as well as economic perspectives. Five years later, in 1977, the Oeko-Institut was founded as an independent research and consulting institute. Towards the end of the phase in 1988, the Centre for Solar Energy and Hydrogen Research Baden-Württemberg ZSW was founded, based in Stuttgart and Ulm. Not much later, in 1990, the Fraunhofer Institute for Environmental, Safety, and Energy Technology UMSICHT was set up with its main seat in Oberhausen. As Stadermann (2021) points out in his comprehensive compendium on the "solar turn", the renewable energy research community evolved gradually and was first and foremost also marked by individuals—scientists, engineers, technicians, economists—who broke away from "mainstream research" and delved into the topic of generating electricity from renewable sources, often motivated by ecological reasons.

2.2 *Renewables in the Early Energy Policy Debate*

Despite the growing interest in renewable energy research, there were also substantial doubts and even severe objections from industry, energy policy, and research organisations in the 1970s and 1980s. These concerned whether the use of renewables would have smaller environmental impacts than using fossil fuels or nuclear energy, and whether renewables could entirely substitute fossil fuels and nuclear energy. This opposition was also reflected in the considerable reduction of government funding spent on renewable technology R&D in the 1980s under the Kohl administration (Hake et al. 2015). The power industry questioned the future importance of renewables by labelling them "additive energies". For instance, the president of the German Power Industry purported in 1987 that the "electricity economy is prepared to undertake massive efforts in order to harness *additive* renewable energies. [...] It rejects any discrimination against electrical power. Power saving and the use of renewables will be inadequate in the foreseeable future to replace nuclear power. Treading the responsible path of power supply means accepting the realities, which safely lead to the objectives and not being lured by romantic visions" (Heidinger 1987, p. 1). Heidinger (1987) concluded that both nuclear energy and coal would be needed on a global scale to ensure an economic, environmentally responsible long-term supply of electricity. Regarding the impacts on health and the environment, Inhaber (1978), a researcher in a government-funded research institution, published a widely circulated report with the influential conclusion that the health hazards of deriving energy from wood, wind, and sunlight were comparable to those of using coal and oil and much greater than those of using nuclear power. These findings were severely criticised by Holdren et al. (1979) in a long report, which featured a harsh summary in the abstract: Inhaber's "conclusion is in no sense derived from the actual characteristics of the technologies involved. It is based entirely on mistakes of all varieties: conceptual confusions, inappropriate selection of systems and data, misreadings, and misrepresentations of literature, improper calculational procedures, and untenable assumptions and contentions" (Holdren et al. 1979, p. i).

Holdren et al. (1979) argued several major aspects in detail, which reflected the dissent and lack of knowledge in the late 1970s: "Inhaber offers no estimates of disease effects of oxides of nitrogen, hydrocarbons, or trace metals (mercury, lead, cadmium, nickel, etc.) emitted to the air by combustion of fossil fuels. He entirely neglects public disease from water pollution from nonnuclear energy activities, e.g., caused by hydrocarbons and trace metals released in extracting, processing, and transporting fossil fuels. And he ignores disease effects in generations beyond the present one, e.g., genetic illnesses caused by chemical mutagens from fossil fuels and radiation from nuclear power, and cancers produced in future generations by radiation from uranium-mill tailings and carbon-14 released in nuclear-fuel reprocessing". (Holdren et al. 1979, pp. 5–6).

In the spring of 1976, the Danish government published an energy plan for the period up to 1995. An essential part of this plan was the construction of five nuclear power plants. An alternative energy plan, which excluded nuclear power, was later published by a group of Danish scientists (Blegaa et al. 1977). This included an expansion of solar and wind energy and emphasised the use of decentralised fossil fuel plants with combined heat and power production and district heating. Their concluding remarks from four and a half decades ago are particularly interesting: "For a system with so much built-in inertia as the energy sector, there is a tendency to exclude qualitatively new solutions, especially if they require new types of institutional framework. In other words, new energy systems, such as those based on renewable energy sources, will suffer difficulties in receiving sufficient economic support to bring them through development into commercial large-scale production. One of the reasons that nuclear power may succeed in this respect lies in its military importance" (Blegaa et al. 1977, p. 93).

During the mid-1970s, newly founded energy systems analysis groups argued that increased oil prices would lead to higher energy costs and a more efficient use of energy; economic growth should be considered a linear annual per capita growth and its energy intensity would decline due to above-average growth in services and low-energy branches of industry. They expected a substantial slowdown in the growth of energy-intensive industries until saturation was reached in future decades. This "should result in projections of demand being considerably lower than currently available estimates" (Bossel and Denton 1977, p. 35). These low-energy demand scenarios also questioned the need for nuclear energy and pointed to renewables as a supply option (Goy et al. 1984). In 1980, the Enquete Commission published an interim report via its select committee on "Future Nuclear Energy Policy", concluding that the use of nuclear energy might not be mandatory in the future if (West) Germany could reduce its demand for energy and alternative energy sources could be developed. For the first time in Germany, this opened a window of opportunity for a debate about moving away from nuclear power and highlighting the potential role of renewable sources of energy as well (Hake et al. 2015). In general, the narrative around renewable energies started to broaden slowly, shifting away from the initial narrow focus on affordability, efficiency, and availability of energy sources. In the same year, the Oeko-Institut published its analysis "Energy transition. Growth and prosperity without oil and uranium" (Krause et al. 1980), which also coined the term "energy transition". A comprehensive critique of the developed theses was published in the same year by Schmitz and Voß (1980), who argued that the potential contribution of "regenerative energy sources" was substantially overestimated in the study. Despite these developments, research on the potential of renewables to reshape the energy system was slow to progress and the vision of a sustainable energy supply was yet to emerge. The study "Rational power utilisation and generation without nuclear energy: Potentials and assessment of effects on the power industry, ecology, and economy" (Masuhr et al. 1987), which was commissioned by the German Ministry of the Economy, focused on substituting the need for renewables by energy saving measures, investing in local and district heating systems and by co-generation in industry. The dissent concerning the future role of nuclear energy, energy efficiency, structural changes, and renewable energies eventually led to the foundation of the Forum of Future Energies in Germany in 1989.

While the significance of renewables for the power sector was not thoroughly investigated in the 1970s and 1980s, more interest was paid by research, policy, and media to their future role in heating. Again, the focus was more on technology solutions and other aspects were not addressed, especially socio-technological ones or the systemic importance of renewables in the heating sector. While the use of geothermal energy has a longer history in California or Iceland (Miethling 2011), Germany focused on other technologies including solar thermal energy, which was looked at for many applications including low-energy houses and passive buildings and houses, as well as greenhouses (Erhorn 1990). Solar thermal collectors were also considered for warm water generation and ancillary heating, particularly in one- and two-family houses as was heat storage in various media for short-term and long-term purposes (Reichert et al. (1980), Buchner (1980), Bakken (1981), Sørensen (1984), Jensen and Sorensen (1984)). Solar thermal systems were also examined for electricity generation and high-temperature applications in tower systems. Finally, biogas was originally considered as a substitute for natural gas in heat generation, but this was only efficient under very specific conditions (Kloss 1982). The use of biomass (such as wood, bio-based organic wastes, including the related biogas) was also investigated from various perspectives. Even the generation of green hydrogen by renewables was already studied in the mid-1970s (Nitsch 1976), although the objective here was to use it in rocket propulsion systems.

Increased attention was paid to renewable electricity generation from the early 1980s, when the first 30 kW wind turbines were successfully generating power in Denmark and the USA, particularly California. In Germany, on the other hand, the first large wind turbine (Growian) failed due to various unforeseen technical problems. This strengthened the claim of those backing nuclear power and fossil fuels that there were no suitable alternatives (Hake et al. 2015). Photovoltaics research started even earlier in the 1980s and was associated with high positive expectations due to its modular construction and no moving parts (Goetzberger 1982). Large PV fields were even planned in the upper atmosphere.

2.3 *Obstacles Hindering Fast Market Diffusion*

At the same time as obstacles to efficient energy use were observed (Jochem et al. (2024) in this anthology), similar barriers were identified by those analysing the market diffusion of renewable energies. Jarach (1989) surveyed the barriers reported in the literature to the diffusion of renewable energy sources at international level. These included financial (high capital cost vs. low operating cost), commercial, operational, social, and institutional factors and concerned various technologies such as solar thermal energy, wind power, and biomass-based energy production (thermochemical and biochemical processes). The evaluation carried out by Jarach (1989) also demonstrated the importance of economic barriers in terms of competitiveness with conventional energy sources that are often substantially subsidised. Operational barriers such as the lack of knowledge of planners, installers, or maintenance companies were also shown to be relevant. Consequently, Jarach recommended easy-to-run, automatic, and simple plants as well as intensifying research in the field to develop suitable renewable energy applications and overcome the barriers reported in the literature. Some of the first publications in Germany that tried to consider the efficient use of energy *and* maximise the use of the emerging renewables appeared in the early and late 1980s, e.g., Nitsch et al. (1981), and Luther (1989). Towards the end of the 1980s, and thus around the end of the first development phase, the general debate embraced the notion that renewables would have an important role to play in the future energy system. This slowly shifted the general narrative around renewables towards mitigating climate change. From 1987 onwards, another Enquete Commission worked on "preventive measures to protect the earth's atmosphere". Their final report, published in 1990, formed the early basis for German measures to protect the climate (Hake et al. 2015). Within the extensive study programme of this Enquete Commission, 13 studies were devoted exclusively to the potential of renewable energy in Germany. Their economic performance was seen as the major obstacle to the diffusion of renewables (Bölkow et al. 1990). Based on their study results, the Enquete Commission concluded in 1990 that renewables would make a limited contribution to reducing CO_2 emissions until 2005, but still called for immediate measures to foster their diffusion to bring down costs (German Bundestag 1991). This provided crucial support for the introduction of the Electricity Feed-in Law in 1991. During the 1980s, the first research was published that also considered social perspectives on renewable energies. Apart from the highly theoretical work of Meyer-Abich and Schefold (1981 and 1986) on future energy systems, the most notable project drawing on case study-based empirical evidence in this regard was the Landstuhl demonstration project, which combined energy-efficient construction and building design with the application of solar energy at the end of the 1980s. The inhabitants of the building were interviewed regarding their perceived quality of living and behavioural aspects (Gruber et al. 1988). This project, however, remained unique in its approach and non-technological research perspectives were the exception rather than the norm during this phase of development.

In 1990, the Renewable Energy Research Association (FVEE, formerly Solar Energy Research Association) was founded. This represented a crucial step towards forming and consolidating the research community as it brought together the relevant knowledge institutions in a nation-wide approach.

3 Market Expansion of Renewables in the Electricity Sector (1990–2009/2011)

3.1 *Implemented Policies and Regulations*

As pointed out in the previous section, the interest in renewable energy technologies of several industrialised countries was first sparked by the oil crises in the 1970s, which was followed by some efforts to develop renewable energy technologies including research and development programmes. Despite the perception of the heat sector as an important field of application for renewables, most research efforts concentrated on electricity. This focus on electricity can be at least partially explained by the major influence of the Chernobyl disaster and the resulting debate about a nuclear phase-out combined with the increasing importance of climate protection. However, decreasing oil prices in the 1980s led to a diminishing interest in renewables and their deployment could not be fostered on a larger scale. This changed in the 1990s with the start of the second development phase which was characterised by a growth in the deployment of renewables. The merit of renewable energy sources was rediscovered in light of the discussion about climate change and picked up by the EU in the 1990s. Relevant milestones included the Rio Summit in 1992, the first major international conference to discuss climate and environmental issues on a global scale and the Kyoto Protocol in 1997, which committed industrialised countries and economies in transition to reducing greenhouse gas emissions in combination with country-level targets.

Lively discussions about how to best support the deployment of renewable energy sources ensued in the research, science, and policy communities in the EU as Member States were free to choose their own approach. Feed-in tariffs were introduced as a dedicated support instrument in Germany and several other countries. In general, the targets set at both national and EU level strengthened the analysis of renewable energy potential as a separate field of research. Following the adoption of the Directive of the European Parliament and of the Council on the Promotion of Electricity Produced from Renewable Energy Sources in the Internal Electricity Market, referred to as the Renewable Energy Directive or RED (European Commission and European Parliament (2001)), an extensive descriptive literature emerged concerning policy differences and trends of renewable support in the EU, e.g. Meyer (2003), Reiche and Bechberger (2004), Johansson and Turkenburg (2004). The focus here was clearly on techno-economic analysis, with other factors discussed only on the sidelines, if at all. At the same time, this phase also saw the rise of energy system models incorporating increasingly high shares of intermittent renewables. According to Pfenninger et al. (2014), energy system models can be distinguished into four types: energy system optimisation models, energy system simulation models, power systems and electricity market models, and qualitative and mixed-methods scenarios. There are many different models available, which were designed for different purposes and cover different scopes. Some modelling approaches were also extended and combined with other perspectives during the subsequent development phases.

Figure 2 shows the evolution of renewable energy support policies over time in Germany and provides an interesting showcase for the development of a support regime. The figure clearly shows that a system of guaranteed fixed tariffs, which were linked with a high level of uncertainty for investors, has over time incorporated incentives to strengthen the market integration of renewables. Germany began its systematic support for renewables in the electricity sector in the 1990s with its Electricity Feed-in Law ("Stromeinspeisungsgesetz"). The basic principle of this law was to ensure grid access for electricity generated from renewable energy sources by obliging utilities to purchase electricity from RES at predefined fixed feed-in tariffs. The Stromeinspeisungsgesetz initiated a market diffusion process especially of onshore wind power plants in the 1990s, despite opposition from the incumbent electricity suppliers at the time. The evolution of the German support scheme can be seen as a consequence of political necessities and requirements, e.g. at EU level, a changing market environment and a continuous, scientifically-supported monitoring process (EEG-Erfahrungsberichte—Progress Reports for the Renewable Energy Sources Act).

Based on strong efforts by the EU to establish a liberalised internal EU market for energy and to unbundle generation and transmission, the German energy market opened up to third party generators. Extensive discussions on the efficient set-up of the energy system took place between policymakers and large energy companies, including the role of integrated energy companies and the need for regulators in the power system. Newcomers and private investors had been the drivers behind the initial market diffusion of renewable energies. Ownership-unbundling of the transmission grid assets of large utilities at the end of the 2000s was accompanied by strategic shifts in these utilities to invest more in renewables and play a major role in their development. Overall, the Stromeinspeisungsgesetz led not only to an

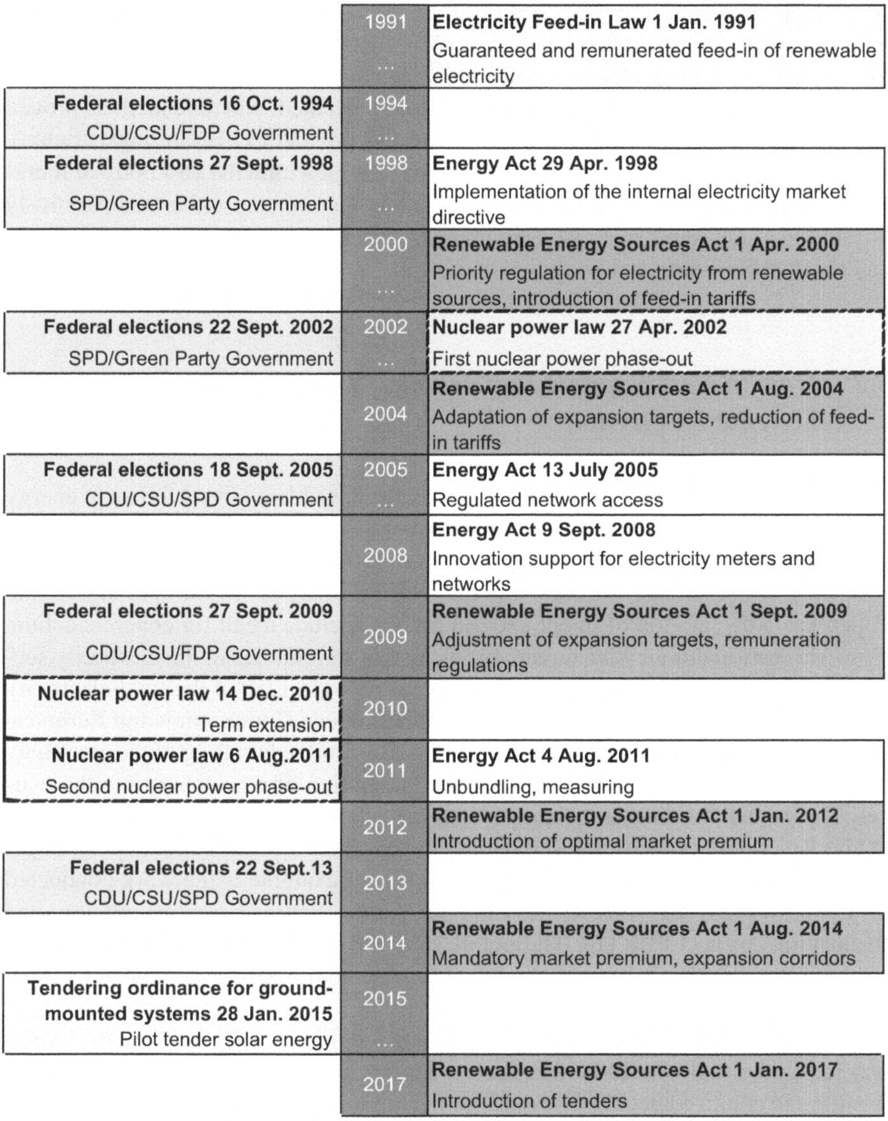

Fig. 2 Evolution of renewable energy support policies in Germany. Source: Own compilation

unprecedented expansion in installed capacity, but also to the creation of "learning networks" between suppliers of wind turbines and local component suppliers (Jacobsson and Lauber 2006).

As can be seen from the timeline, the Erneuerbare Energien Gesetz (Renewable Energies Sources Act, (EEG), which replaced the Stromeinspeisungsgesetz in 2000, has become the core policy instrument supporting renewables. This act can be seen as providing the basis for the successful market development of onshore wind and solar photovoltaics in Germany. In a Europe-wide comparison, Germany, together with Denmark and Spain were the forerunners in terms of renewable energy market development, especially for onshore wind.

The EEG has been continuously monitored, evaluated, and amended, as proven by the regularly commissioned EEG Progress Reports (EEG-Erfahrungsberichte). Thus, the EEG has proven a fruitful topic for research within the vast body of literature on the performance and monitoring of policy instruments, either comparatively or looking at measures in an isolated manner, e.g. Krewitt and Nitsch (2003), Bode and Groscurth (2006), or Langniß et al. (2009).

3.2 How to Best Support Renewables: The Support Scheme Discussion

The EU first stated that increasing the share of renewable energy sources in energy supply was a core objective in the White Paper "Energy for the future: Renewable sources of energy" in 1997 (European Commission 1997) due to their potential contribution to climate protection and the security of supply in Europe. This White Paper was a declaration of intent and did not yet include a call for concrete action. However, national indicative targets for the use of renewables in the electricity sector were stipulated in Directive 2001/77/EC to provide 12% of the total electricity consumption in the EU-25 by the year 2010 (European Commission and European Parliament 2001). In this context, the liberalisation and unbundling of the electricity market were the pre-conditions for a better integration of renewables into the market. The research community quickly embraced the new challenge posed by the White Paper, with a focus on establishing the baseline for future monitoring, target setting, and analysing instruments. One prominent example is the work conducted in the "*P*rogress of *R*enewable *E*nergy: *T*arget Setting, *I*mplementation and *R*ealisation" project PRETIR (Harmelink et al. 2001). This agenda was also driven by the European Commission, which had to put a monitoring and indicator framework in place to pave the way for implementation.

The decision about how to design the policy measures used to achieve the targets was left to the individual Member States, which employed a variety of instruments resulting in country-specific renewable policy mixes. This has been accompanied by a lively debate reflected in the large body of both theoretical and empirical scientific literature as well as policy documents on how to support renewable electricity, which focus on effectiveness while ensuring support costs remain at acceptable levels to society. Different policy mixes and the interactions of individual instruments increasingly became the focus of analysis.

A common taxonomy in the literature is to split the applied instruments into price- and quantity-based ones, which can then be further grouped according to different characteristics as shown in Table 1. It is also possible to distinguish direct and indirect instruments. While direct measures aim to stimulate the deployment of

Table 1 Characterisation of policy instruments for renewable energies

		Direct	Indirect	
		Price-driven	Quantity-driven	
Regulatory	Investment-focused	Investment incentives	Auctioning of investment grant	Environmental taxes Simplification of authorisation procedures (connection charges, balancing costs, ...)
		Tax credits Low interest/soft loans		
	Generation-based	Administratively set (fixed) feed-in tariffs Administratively set (fixed and sliding) premium system Production tax incentives	Auctioning of feed-in tariffs and premiums Quota obligation with a tradable Green Certificate (TGC) system	
Voluntary	Investment-focused	Shareholder programmes		Voluntary agreements
		Contribution programmes		
	Generation-based	Green tariffs		

Source: Based on Held (2010) and Haas et al. (2004)

renewable energy sources directly, indirect measures aim to foster a conducive framework. Over the years, a variety of methods, including case studies, simulations, and econometric modelling have been applied to study the effects of these different instruments, which has become a prolific field of research (Del Rio et al. 2012).

The two predominant and most controversially discussed support schemes in the 2000s were price-driven, fixed feed-in tariffs, and quantity-driven quota obligations with tradable green certificates. Feed-in Tariffs (FIT) represent a generation-based, price-driven approach. This means that a price per unit of electricity is predetermined by the government and has to be paid by the obliged actor, usually represented by a utility or the grid operator. This FIT can either be a fixed global tariff substituting the market price or a premium paid on top of the market price. In some cases, the time horizon for a tariff is fixed and provides additional planning security for potential investors. Generally, FITs allow technology-specific promotion of renewable energy technologies and can stimulate future cost reductions by considering certain criteria within the specific design of an FIT. In contrast, quota obligations based on Tradable Green Certificates (TGC) follow a generation-based but quantity-driven approach. Instead of predefining a price, a quota is established by the government. This quota then has to be fulfilled by one particular actor of the electricity supply chain, e.g. generators, suppliers, or consumers. Subsequently, the certificate price results from matching supply and demand in a market for TGC. The certificate price formed in this way serves as one revenue component in addition to the electricity market price. A penalty level may be defined, which must be paid if the obligated parties cannot prove quota fulfilment. In theory, there are different options for implementing technology diversification within TGC systems. However, these options are associated with several problems, e.g. loss of liquidity if markets are split up. Weighting certificates according to the respective technology option and its financial requirements may impede target setting and complicate the monitoring process of target fulfilment.

As more and more real-life evidence became available in the 2000s, research increasingly turned to monitoring experiences, shifting away from purely conceptual analysis. Most EU countries already applied feed-in-based support schemes, with only a few Member States using quota obligations in combination with tradable green certificates. Based on empirical evidence, some studies, e.g. Lauber and Toke (2005), Lehmann and Peter (2005), or Mitchell et al. (2006), concluded that feed-in tariffs and premiums outperform quota obligations, especially with regard to effectiveness while keeping support costs at an acceptable level. Backed by these analyses, more and more countries turned away from quota obligations. The majority of the studies focused on the effectiveness and efficiency of the employed instruments. Other investigated criteria included social acceptance, legal and/or political feasibility, and macroeconomic effects, e.g. impacts on the labour market (Ragwitz and Steinhilber 2014). A debate was triggered on the importance of different design elements with many studies concluding that there is no "single-right choice", but that design matters.

Another stream of research examined the evidence for either (top-down) harmonisation or (bottom-up) convergence of support schemes across Europe, e.g. Muñoz et al. (2007), Kitzing et al. (2012). Some of these analyses also took place in the context of EU-funded research projects, e.g. Huber et al. (2004), Bergmann et al. (2008), and were subsequently reflected in EU policy documents, including the 2005 Communication from the Commission on the support of electricity from renewable energy sources (European Commission 2005) and the Commission staff working document on the support of electricity from renewable energy sources accompanying the proposal for the subsequent Directive (European Commission 2008). In parallel, political developments in the overall energy sector moved on to a discussion about longer term targets and the target of 20% RES in gross final energy consumption was defined for 2020. In contrast to the Directive 2001/77/EC, this new proposal set targets for the whole energy sector, not just electricity. In addition, the targets were binding and no longer only indicative as in Directive 2001/77/EC. The Directive 2009/28/EC translated the required increase in the share of energy from renewable resources from 8.5% in 2005 to 20% in 2020 into individual targets for MS (The European Parliament and the Council of the European Union 2009). The existence of binding targets led to an emerging literature on depicting and analysing the trends and progress towards achieving these targets. In addition to an emerging body of scientific literature, many other institutions contributed to this endeavour in various projects for the European Commission and other clients, including the Progress project, for example, which assessed the progress in renewable energy and sustainability of biofuels, and Eur'Observer.[2] Approaches to monitoring and measuring the performance of policy instruments in practice were developed within the context of various research projects, such as OPTRES, RE-Shaping, or Towards2030. This approach was then applied by the European Commission in their official communications (COM(2005)627) as well as by the International Energy Agency to monitor the effectiveness and efficiency of policy instruments at international level (European Commission 2005). Similar to the findings of COM(2005)627, an updated evaluation supported by Fraunhofer ISI researchers concluded that well-designed FITs were generally the most effective and efficient policy measure for supporting RES-E (European Commission 2008).

3.3 *Assessing the Effects of the Expansion of Renewable Energy Sources*

Growing shares of renewables in electricity generation meant that the impacts of renewable energies on the electricity sector and the economy became increasingly important. This led to three domains of research, which differentiated positive and negative impacts at the micro, system, and macro levels and analysed them in detail (Breitschopf and Diekmann 2015; Breitschopf et al. 2016).

At the micro level, studies included the impacts of renewable electricity generation on the market electricity prices, the so-called merit-order effect (Sensfuß 2015), distributional aspects of the EEG levy on households (Diekmann et al. 2016a) and industries (Grave et al. 2015), and analyses of network expansion costs and their distributional effects (Diekmann et al. 2016b). Additional distributional aspects in the electricity, heat, and mobility sectors of energy efficiency and renewable energy expansion were systematically outlined and analysed (Lutz and Breitschopf 2016). Besides costs, some studies explored the benefits of the feed-in tariffs paid to photovoltaic and wind-based electricity generators (Breitschopf et al. 2014). At the system level, studies focused on the impact of renewable energy deployment on innovations (Groba and Breitschopf 2013), the contribution of renewables to energy supply security (Diekmann et al. 2016b), network expansion and enforcement costs as well as the contribution of variable renewables to providing capacity and the necessary back-up capacities. Forecasting tools became an important new area of research. These provide short-term information on current renewable generation levels, substantially improve the value of renewables, and are applied around the world.

Impacts at the macro level have received a lot of attention and have been compared to the costs of expanding renewable energies. A variety of studies analysed the impacts of expanding renewable energies on employment and growth. Over

[2] See EurObserv'ER | Measures the progress made by renewable energies in the European Union (eurobserv-er.org).

time, different approaches developed to assess these and other effects of renewables, which ranged from structural top-down models, such as Hillebrand et al. (2006), to computational general equilibrium (CGE) models, Bohringer and Loschel (2006), Schumacher and Sands (2006), or Abrell and Weigt (2008). These are applied at different levels, including the EU, national, and regional levels, and for different focus areas. Studies using macroeconomic models to analyse employment effects display different results, ranging from positive net effects, Fragkos and Paroussos (2018), to net losses, e.g. Frondel et al. (2010). The magnitude of effects is characterised by and depends on the underlying assumptions. This debate also sparked new approaches, such as the sectoral energy-economic econometric model (SEEEM) (Blazejczak et al. 2014).

Fraunhofer ISI elaborated a systematics of these different assessment approaches and outlined their underlying assumptions and input parameters that affect their results (Breitschopf et al. 2013; Winkler et al. 2018). A key focus was on the different types of effects and the resulting gross and net impacts. The latter were based on a scenario comparison and the results were incorporated in publications of IRENA and CEM (2014) and the European Parliament (Winkler et al. 2018).

A seminal work for assessing the impact of renewable energy policy on economic growth and employment in the European Union was the EmployRES study (Ragwitz et al. 2009), which combined top-down and bottom-up approaches for the first time when analysing renewable energies. This study followed the conceptual work of Walz (2006) and included first-mover advantages in its empirical analysis regarding the export of renewable energies. Several years later, Fraunhofer ISI updated its 2009 study for the EU (Duscha et al. 2016) and conducted macroeconomic impact assessments for Germany (Sievers et al. 2019). Both studies assessed the impact of renewable energy deployment on GDP and net employment, i.e. comparing a deployment scenario to a counterfactual scenario and thus accounting for all additional positive and negative aspects. Overall, research in this area became more and more interdisciplinary, increasingly considering additional effects.

Over the course of the market deployment of renewable energies, the focus of discussion slowly shifted from net or gross job creations associated with the expansion of renewable energies to the jobs and qualifications needed to power the energy transition. As the technology costs of wind power and photovoltaics decreased, another increasingly important issue was the cost of capital, determined by risks at project and country level (Breitschopf and Pudlik 2013).

Overall, there are numerous different approaches and focus areas in the literature during this phase of the market expansion of renewables, some of which covered different paradigms and mechanisms. The growing body of data regarding the deployment and effects of renewables triggered a wealth of empirical studies. Slowly, an increasingly systemic perspective was being embraced and interdisciplinary approaches started to become more widespread, combining sociology, energy economics, and agent-based modelling, etc.

4 Market Integration of Renewables (2011–2019)

With increasing shares of renewables in the energy systems in Germany and the EU, the related challenges continued to develop. First, the market and system integration of renewables became more important. Second, non-financial barriers were assessed in more detail and there was a focus on dismantling them to enable even faster expansion. The development of renewable technologies and their increasing deployment in the 2000s also led to research on the innovation process of renewable energy technologies. This research built on the innovation system approach, which had been developed for national, sectoral, and technological systems (see, e.g., Lundvall et al. (2002), Malerba (2005), Carlsson et al. (2002)). In addition, the multi-level perspective framework became more important, as the focus shifted to an increasingly systemic perspective.

New areas of research that emerged during the market integration phase included the impacts of renewable energy self-consumption as well as the global expansion of renewables. This led to the research perspective broadening considerably in the geographical sense and the global perspective becoming more and more important for renewables, along with the growing awareness that a country's energy problems cannot be viewed in isolation, but that global interdependencies exist. Furthermore, with the Paris Agreement and the UNFCCC process, more and more countries were having to embrace carbon-neutral development pathways and expand their renewable energy capacities. In this context, many researchers started to look at the specific conditions for expanding renewables in certain countries or regions. Equally, several projects emerged that focused on different geographical areas. Such projects needed to consider certain aspects that are not as relevant in the EU context, such as existing direct subsidies for fossil-based electricity, high grid losses, or a greater focus on supporting domestic industries.

4.1 Renewable Energy Innovation System Research

Studies investigating the innovation process related to renewable energies mainly used the technological innovation system (TIS) approach. With Germany becoming one of the key locations for renewable energy development, research on the renewable energy innovation system in Germany started with applications for wind energy, photovoltaics, and biomass (e.g. Jacobsson et al. (2004), Jacobsson and Lauber (2006), Walz (2007), Negro and Hekkert (2008)). In parallel, innovation researchers put forward the concept of a functional approach to account for the systemic nature of innovation (Smits and Kuhlmann (2004); Hekkert et al. (2007); Bergek et al. (2008), Hekkert and Negro (2009), Markard et al. (2015)). This type of TIS approach, linked to seven innovation functions, has been widely used since the late 2000s for the field of renewable energy technologies and beyond.

The TIS approach is so important for energy policy as it improved our understanding of the interplay between market formation policies and innovation. In particular, the approach was able to explain the role of energy policies and also provided evidence for the vital role of feed-in tariffs and policy mixes for promoting innovation in renewable energy technologies (Jacobsson and Lauber (2006), Walz (2007), Negro and Hekkert (2008), Dewald and Truffer (2011), Negro et al. (2012), Bergek et al. (2015), Reichardt et al. (2016), Reichardt et al. (2017)). This primarily case study-oriented research was supplemented by econometric research on the determinants of renewable energy technology innovation in the 2010s, which used patent counts as a proxy for innovation. Both Johnstone et al. (2010) and Costantini et al. (2015) focused on the effects of the different types of support mechanisms, particularly on FITs. Schleich et al. (2017) and Böhringer et al. (2017) found that policies increasing the demand for renewable energy technologies have a positive effect on innovation, but also that factors other than the instrument type are important.

To sum up, both qualitative case studies and quantitative statistical analyses underlined the importance of early market formation for further innovation in renewable energy technology. The fact that key energy policies increasing the deployment of renewable energy technologies were also demand-side innovation policies highlighted the fact that innovation research and energy policy research were increasingly becoming two sides of the same coin. This also involved the perspectives of research on sectoral energy policy and research on innovation policy moving closer together.

In addition to the paradigm of the technological innovation system, energy-related innovation research also began to embrace the multi-level perspective (MLP) approach, initially applied to analyse the energy transition. Fraunhofer ISI made an important contribution here by combining MLP and modelling, e.g. Rogge et al. (2020). Several studies examined the policy mixes facilitating renewable energy deployment, combining the lenses of technology push and demand pull. Using the German *Energiewende* as an example, Rogge and Reichardt (2016) developed a concept and framework analysis for policy mixes for sustainability transitions. This was picked up by several other works, including the GRETCHEN project, which explored the influence of policy mixes on technological and structural change in renewable power generation technologies in Germany.

4.2 From Market Integration to the System and Grid Integration of Renewable Energies

As renewable energy generation continued to grow across the EU, its market and system integration was probably the most important discussion point in policy and research during the period 2011–2019. Besides theoretical and empirical studies and analyses of integrating renewables into the market and electricity system as well as the diffusion of technologies, testing laboratories were established to try out innovative technological solutions in practice under the existing infrastructure and external framework conditions. One example is the SINTEG programme, which analysed the role of digital technologies in the energy transition based on experiments in several model regions. One core element was the regulatory sandbox approach to gather experience in adapting and further developing the legal framework to the innovative technologies (Brunekreeft et al. 2022).

While generation from the most relevant renewable energies of onshore and offshore wind and solar PV is variable and dependent on weather patterns, electricity demand needs to be met at all times and does not necessarily match the production output of renewables. If supply is no longer fully dispatchable, coordinating supply and demand and integrating renewables become even more important. Other reasons for including renewable energies in regular electricity markets included their growing share of generation and the liberalisation of electricity markets across Europe. It was simply no longer a viable option to exclude large shares of renewable electricity generation from the market.

Figure 3 gives an overview of the main aspects regarding the market integration of renewables over time that are analysed in the literature. It shows that the market integration of renewable energies has two different dimensions. The first one concerns measures to enable renewable energies to react to the price signals of regular electricity markets. The second concerns the use of competitive or market processes to determine support levels. The figure also shows the development of renewable

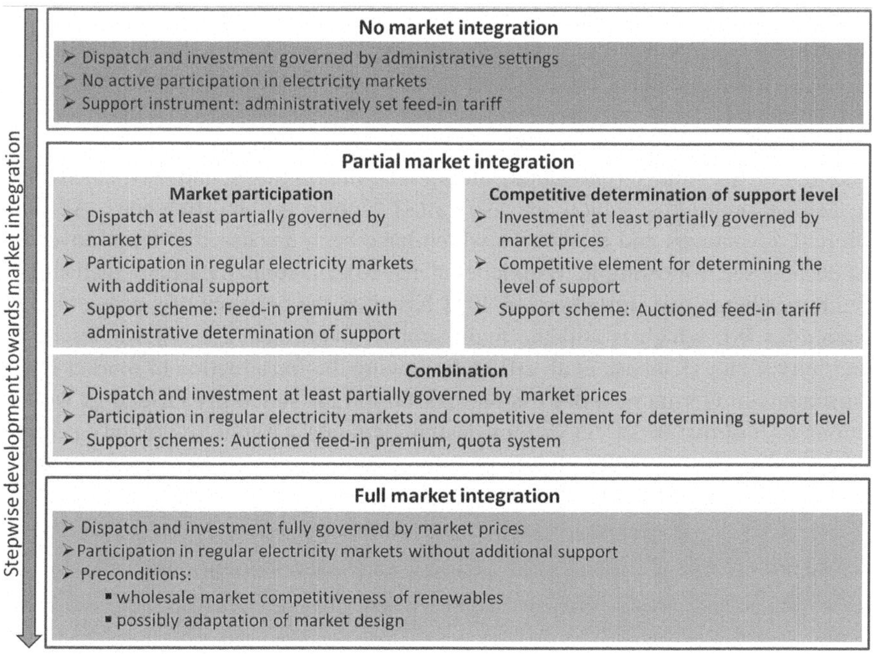

Fig. 3 Different aspects of the market integration of renewable energies. Source: Winkler (2017)

energy support over the last decade in Germany. Support in 2010 was still based on fixed feed-in tariffs that were administratively set. The market-premium system was introduced for larger plants, but still based on administratively set tariffs in 2011. Setting tariffs was changed to an auction system from 2015. The zero bids in onshore wind auctions in 2017 and the development of big PV plants without support indicate that we are gradually approaching full market integration.

Market-Premium Schemes

As described above, prior to the introduction of the Renewable Energy Directive (RED) at EU level in 2009, there had been a heated debate about how to support renewables in the electricity sector between proponents of the market (who preferred quota schemes to support renewables) and those of the state (who opted for fixed feed-in tariffs). In the years following this debate, there was a gradual shift among those supporting feed-in tariffs towards a more market-based integration of renewables.

Research was conducted on different market-premium systems, which were increasingly implemented in policies. The common factor shared by these premium schemes was that the renewable energy plants were obliged to sell the electricity produced on the regular electricity market. In addition, they received a premium payment to cover generation costs, which (at least throughout the previous decade) were typically still higher than those of fossil-based electricity generation. The most important change for the renewable energy plants in this context was that they were responsible for sticking to their generation forecast. As a result, the introduction of feed-in premium schemes led to a dramatic improvement in the generation forecasts for renewables.

As shown in Fig. 4, there are different types of feed-in premium schemes, namely fixed feed-in premiums, (one-sided) sliding feed-in premiums, and (two-sided) sliding feed-in premiums, which are also called "contracts for difference". All have different advantages and drawbacks which have been discussed at length over the last decade, see, for example, Winkler et al. (2020). In Germany, a one-sided sliding premium scheme was introduced in 2011 based on the proposal of a project led by Fraunhofer ISI, which is still the main support instrument for renewables in the electricity sector (Klobasa et al. 2013). Following the introduction of market-based instruments in Germany, a discussion regarding their suitability ensued as well as options to improve them via various design elements. Despite the mainly positive assessment of the one-sided sliding premium, there were also some critical voices, e.g. Bardt (2014) or EEX (2014).

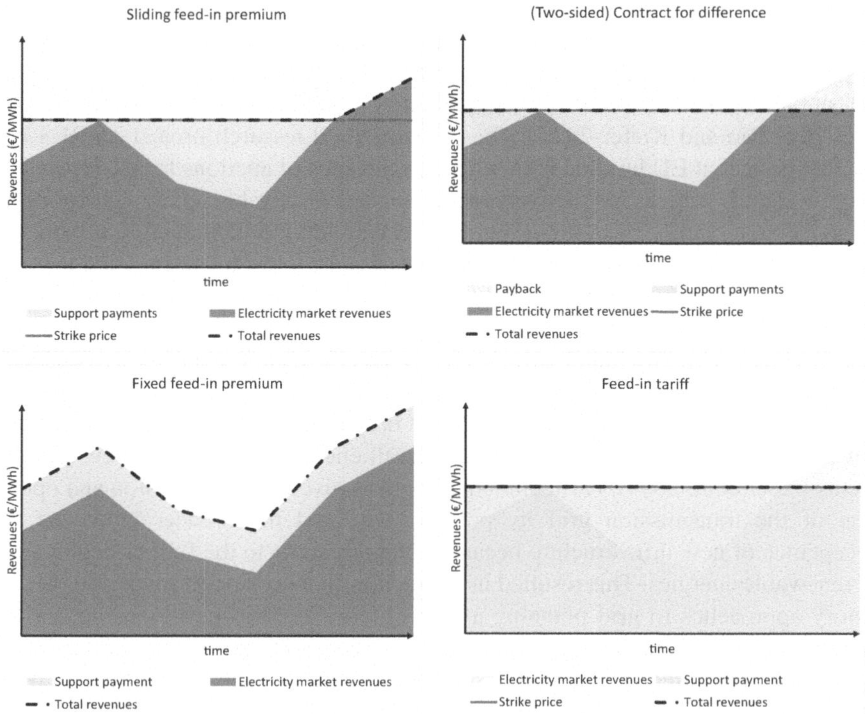

Fig. 4 Different types of feed-in premium schemes compared to feed-in tariffs. Source: Own compilation (DiaCore/ReShaping)

Auctions

The other aspect of the market integration of renewables is the market-based determination of the level of support. This can either be realised via a market for renewable energy certificates (as is the case in quota systems) or based on auctions or competitive bidding.

Auctions for allocating support to renewable energies were introduced based on the EU's new State Aid Guidelines, which were introduced in 2014 by the European Commission's Directorate for Competition (European Commission 2014). Contrary to the 2009 Renewable Energy Directive (RED), new support schemes were increasingly steered towards using auctioning.

Many other institutions and researchers have been heavily involved in research and policy advice projects for auctioning renewable energy support and developed the auction design for offshore wind in Germany. There is a vast and continually growing literature on auctions' effectiveness and efficiency under various circumstances as well as their optimal design due to their increasing importance on a global scale. In addition to country-level studies, e.g. Anatolitis and Welisch (2017) for Germany, Kitzing et al. (2022) for South Africa, and Mora et al. (2017) for several

EU countries, other studies focus on auctions' optimal design and the significance and role of different design elements. Methodologically speaking, auction theory approaches co-exist with other methodologies, but there is a dominance of empirical studies here, and within those, qualitative approaches dominate quantitative ones (Del Río and Kiefer 2023). The Horizon 2020 research project AURES is a notable project at EU level dealing with the relevance of auctions for the renewable energy sector, which has actively accompanied renewable energy auctions at Member State and EU level. Designing effective and efficient auction schemes is quite challenging, as many different parameters need to be considered and the specific market and framework conditions as well as the policy objectives need to be respected. The AURES project website provides detailed information on past and current auctions in EU Member States as well as an auction database and analyses of many specific aspects of auction design.

The expansion of renewable energies not only required their market integration, but also the adaptation of grids and the overall energy system to the new requirements (Auer et al. 2004). Particular attention was given to the extension and operation of the transmission grid by policymakers and the broader public as the acceptance of new infrastructure became a major barrier to the further deployment of renewable energies. This resulted in interactions between new groups and participatory approaches to grid planning and long-term scenario developments for the energy system. Substantial research efforts were needed to identify the best-practice approaches and integrate numerous stakeholders and research disciplines to draw up grid development plans. Grid planning in the past had been managed by a limited number of electrical engineering experts but has since become an issue of broad stakeholder participation and involves the consultation of different societal groups.

Many institutions supported this engagement and stakeholder participation process with advisory studies on techno-economic solutions for different grid operators and policymakers among other formats. Increased efforts have been made by policymakers on national and EU levels to optimise grid operation and increase the security of supply. An intensive discussion at EU level concerned whether capacity markets were needed for the security of supply. Other key topics for the grid integration of renewables included interconnection capacity with neighbouring countries as well as across the EU, and the establishment of ancillary service markets for grid operation.

In addition to their other unique characteristics, renewable plants are often not connected to the transmission grid but to distribution grids. This requires grid reinforcements but there are also other options to enhance flexibility at the local level. In the past, almost no generation units were connected to the distribution grid and no active operation of this level was foreseen. With the increasing number of renewable generators connected to distribution grids, however, the need for more active operation emerged. This gave rise to questions about what impacts could be expected at system level and how active distribution grids could contribute to an efficient power system. These questions have not yet been comprehensively and extensively addressed by the research community. They are the subject of on-going debate, and future research should contribute to closing these research gaps.

4.3 *Self-Consumption of Renewable Electricity and Non-financial Barriers*

Self-consumption refers to the on-site production and consumption of renewable power. Self-consumers are also called "prosumers" in the debate on this topic. As the costs of renewable energies and specifically rooftop PV have decreased, self-consumption or prosumerism has become increasingly attractive over the last decade. Although prosumerism helps to get more people actively involved in the energy system, it can lead to adverse distributional effects, as self-consumers are often exempt from paying grid fees or other energy taxes and levies. As these costs then have to be redistributed among the other consumers, prosumerism can result in a higher burden on socially disadvantaged households who are less likely to own a house and therefore have no possibility to install rooftop PV or who do not have the financial means to invest in self-consumption. The self-consumption of renewable power has also become a prolific research area, often associated with interdisciplinary approaches and the combination of multiple methods, including case studies from various European and non-European countries and involving different types of actors, modelling or interviews, see, for example, the EU-funded projects COMPILE or FlexCoop. Systematic literature reviews and comparative studies from different angles can form a good starting point for further analyses, e.g. Bauwens et al. (2016), Capper et al. (2022), or Lode et al. (2022).

4.4 *Refinancing the Expansion of Renewable Energies and Non-financial Barriers*

In Germany, support for renewable energies has long been financed via a levy on electricity prices. The increasing cost of support, especially due to the increase in solar rooftop capacity between 2009 and 2011 based on very high tariffs, meant the high renewable support levy in combination with other taxes and levies led to a high cost of electricity, also when compared to fossil energy carriers, especially natural gas and coal for heating. High electricity prices are problematic in the context of fostering the electrification of other sectors using sector coupling. From summer 2022, therefore, taxes have been used to finance the renewable energy levy. In addition, the introduction of a national CO_2 emissions trading system for transport and heating has helped to boost the competitiveness of heat pumps and electric cars, in particular.

Apart from financial support for renewable energies, reducing the non-financial barriers to their deployment is key for their expansion at low cost. The recent auction results for onshore wind in Germany are a clear example of this. Due to the low number of wind energy permits, there was not enough competition in most of the auctions. This led to auction results that were the same as or very close to the administratively set ceiling price and thus to very high support levels and profits for wind installations. Numerous research projects were conducted assessing the non-financial barriers to renewables, starting with a project about the relevant obstacles to wind energy between 2008 and 2010.[3] Based on this early experience, a comprehensive methodology for analysing barriers was developed (Boie 2016), which has already been applied to EU Member States in the RE-Frame project[4] and to non-EU countries as well, such as the six Western Balkan countries or Indonesia. Social acceptance issues regarding renewable energies are also moving into the focus of research, but these will become even more pronounced during the next development phase.

5 Speeding up the Transition to a Fully Renewable Electricity System (2019-Present)

The future energy system will have to rely fully on renewable energy, which will have major impacts on all its areas. In electricity, the share of renewables needs to increase to 100%. In order to integrate this renewable power and adapt to the fluctuating generation, the rest of the energy system must be very flexible. Direct electrification via heat pumps is necessary in the heating sector, including decentralised systems in buildings as well as centralised ones in district heating systems. In addition, the direct use of renewables must be expanded, e.g. solar thermal and geothermal as well as the use of waste heat. Direct electrification will also play a vital role in the transport sector. In addition, synthetic or renewable fuels will be necessary for long-distance aviation and shipping. It is not yet clear what the solutions will be for heavy-duty road transport. In industry, materials and processes need to be adapted. Apart from recycling and circular economy approaches, direct electrification but also hydrogen will contribute to an increasingly climate-neutral system. One approach is to develop future scenarios based on detailed modelling tools, which were already employed during the previous phase of research but have recently been expanded and become much more sophisticated. One example is the official modelling for the German Ministry for Economic Affairs and Climate Action, the Long-term Scenarios (Fraunhofer ISI et al. 2021). This will also be used as a basis for grid expansion plans in the future.[5] Sector coupling is another highly relevant topic, which is predominantly investigated using modelling approaches, e.g. Bernath et al. (2021). This attempts to integrate all the energy-consuming sectors of the economy into one system, including electricity, buildings/heating and cooling, transport as well as industry.

---

[3] See https://windeurope.org/policy/eu-funded-projects/windbarriers/.

[4] See http://www.re-frame.eu/footer/about/.

[5] Please see the Langfristenszenarien website for the most recent presentations, reports and updates: https://www.langfristszenarien.de/enertile-explorer-de/.

In addition to modelling approaches becoming more and more sophisticated, this phase is also characterised by a plurality of different research paradigms and approaches existing in parallel and cross-pollinating one another.

5.1 Ambitious Targets and EU Governance Structure for Renewables

Renewable energy is strongly embedded in the EU regulatory context. After a sector-specific target of 10% renewables in gross final electricity consumption had been set for the year 2010, the Renewable Energy Directive (2009) set a general target of 20% renewable energy sources in final energy consumption. Both the 2010 electricity target and the 2020 RES targets were broken down to Member State level.

The European Union launched its "Energy Union" strategy in February 2015 in order to align its energy and climate policies. This strategy aims at making the use of energy more secure, affordable, and sustainable. The Energy Union Strategy builds on existing legislation including the 2020 energy and climate policy framework, which set the targets of reducing greenhouse gas emissions by at least 20% by 2020 (compared to 1990), increasing the share of renewable energy sources in final energy consumption to 20% by 2020, and reducing energy demand by 20% (also compared to 1990).

The EU legislative package "Clean energy for all Europeans" was proposed and partially adopted to regulate EU climate and energy policy from 2021–2030. It covers energy performance in buildings, energy efficiency, renewable energy, electricity market design, and the governance structure of the Energy Union. The target for the share of renewable energy sources in gross final energy consumption is 32% by 2030, but this has not been broken down to Member State level. In its Fit-for-55 package, published in summer 2021, the European Commission proposed increasing this target to 40% by 2030.

In addition, the current target architecture laid out in the Governance Regulation requires Member States to prepare "Integrated National Energy and Climate Plans" (NECPs), an integrated planning and monitoring process for the five dimensions of the Energy Union, which includes plans for renewable energies. NECPs should include the Member States' proposed contribution in terms of renewable energy to the EU target of at least 32% (Governance Regulation, Article 3) as well as an indicative trajectory for several reference points, namely 2022, 2025, and 2027. A Member State shall not fall below a defined percentage of the total increase in the renewable energy share between that Member State's binding 2020 national target and its contribution to the 2030 target (Governance Regulation, Article 4). The indicative trajectory is slightly below a linear development between 2020 and 2030. The European Commission will assess and monitor the collective ambition and progress against the EU target and propose measures if required. This has also given rise to a prolific area of scientific activity concerned with the assessment of the EU's

2030 Climate and Energy Policy Framework as well as individual Member States' contributions to it, e.g. de Paoli and Geoffron (2019), Oberthür (2019), Mišík and Oravcová (2022).

5.2 *How to Achieve Higher Expansion Rates: Acceptance Issues and Limited Land Availability*

Research is now increasingly starting to tackle acceptance issues, a prominent non-financial barrier to the implementation of renewable energy projects. Various studies show that, in principle, the energy transition in the sense of a predominantly renewable power supply is viewed positively by the population, with an approval rating ranging from 60 to 80%, see BMU and UBA (2019) and Wolf et al. (2021). The highest level of support is for the expansion of rooftop solar power. There is less support for the construction of new onshore wind power plants compared to offshore expansion (Wolf et al. 2021). However, despite this general approval, the expansion of renewables often encounters resistance at local level. This is a major challenge of the energy transition and has been further investigated by many researchers, e.g. Kühne and Weber (2016) or Wüstenhagen et al. (2007). Many of these analyses are both qualitative and quantitative case studies and cover various groups, including residents, experts, and other stakeholders (Segreto et al. 2020). This topic was also explored within the research project Akzept, which shows that resistance is lower if citizens participate financially in the energy transition (Breitschopf et al. 2024). An important area of research is concerned with how to incorporate these issues into the planning processes of renewable energy projects and how to reduce local community opposition to wind energy. One prominent example is the WISE - POWER project (2014–2016, funded by the European Commission), which resulted in the development of the so-called "Social Acceptance Pathways" (SAP).

One way to increase the social acceptance of renewable energy projects (on site) is public participation (formal or informal, different intensities: information, consultation, cooperation, financial and non-financial) (Wolf et al. 2021). In the Akzept project (2020–2022, funded by the Federal Ministry for Economic Affairs and Climate Action), several research institutions investigated whether citizens who participate financially in the energy transition tend to have higher levels of acceptance than those who do not. The findings indicate a correlation between financial participation and acceptance. Those who benefit economically from the energy transition not only advocate it more often but are also more often willing to pay higher electricity prices, accept photovoltaic systems even if they do not hold a stake in them, and accept wind farms if they do hold a stake in them. In addition, the project found that financially involved citizens actively support the energy transition, for example by getting involved in initiatives and purchasing green electricity. However, it is important to point out that the direction of impact could not be investigated, i.e.

it is not clear how financial participation and acceptance are related (Breitschopf et al. 2024).

5.3 Future Support Requirements for Renewables

One very important question concerning the future of renewables is whether there will be a need to support them in the medium to long term, if their competitiveness improves considerably over time. The German Renewable Energy Act aims to end the support for new plants in 2030. Support can be phased out if plants refinance themselves from the market and market prices cover the increased risk premiums implied by fluctuating electricity prices. The following sections look at the research on identifying the important factors that determine whether to end support. These are divided into factors driving the costs of renewables and those driving the revenues of renewables. Both sides are somehow linked through the financing costs, which depend on the risk profile of plants. Higher revenue risks imply higher financing costs (if financing is still possible) and therefore increase the overall costs, which in turn require higher revenues.

There is an on-going scientific debate about how to design future support systems, led also by several projects at national and EU level. As an example, Held et al. (2019) provide an overview of the challenges to be addressed when phasing out support for renewables. The paper finds arguments for continuing the dedicated support for RES in order to create a predictable and secure investment framework.

The Future Costs of Renewables

In the past, costs for renewable energy generation, especially wind and solar, have decreased continuously. Most recently, however, this trend has stalled and sometimes even been reversed. The main cost driver of electricity generation costs is the resource quality at a specific site. The costs of raw materials, labour, etc. also play an important role as do risks and financing costs. This has opened up a new research area dedicated to facilitating the full market integration of renewables in the future.

A support system, especially one offered by a low-risk state like Germany and other EU countries, substantially reduces financing risks and therefore also the overall costs of electricity generation. Projects like AURES and DIA-Core found that financing costs have fallen across countries over the last decade. In terms of renewable support, the main factor driving down investment risks is a steady and reliable support system without retroactive changes. In terms of designing the premium, systems that smooth out electricity price fluctuations and lead to constant and predictable revenue flows imply the lowest risks and financing costs (Breitschopf and Alexander-Haw 2022).

Revenues of Renewable Energy Plants: The Market Side

Renewable energy sources, especially wind and solar energy, have very low or even close to zero variable costs. On the regular electricity market, prices are typically constructed based on these variable costs. Consequently, prices are typically low in hours when renewables set the price (i.e. renewables produce sufficient energy to cover the total electricity demand). This effect is called the "merit-order effect" or "cannibalisation effect" of renewables. Sensfuß et al. (2008) quantified this "Merit-Order Effect" for the German power market based on the agent-based simulation platform PowerACE (now Enertile). The merit-order or cannibalisation effect leads to decreasing revenues for renewables at higher expansion rates due to the simultaneity of their electricity generation combined with their low costs. The extent of the effect depends heavily on the elasticity of electricity demand and thus on the flexibility of the energy system as a whole. Bernath et al. (2021), for example, found that the existence of flexible heating grids considerably increases the revenues of renewable energies in electricity. In the short to medium term, the prices for gas and CO_2 also play a role (Winkler et al. 2016).

Apart from the level of future market revenues, the predictability of these revenues is important for investments in renewables. Again, the cost structure of renewables with their high investment expenditures and low variable costs makes them especially vulnerable to systemic changes in electricity prices, e.g. based on a re-arrangement of market zones or other political decisions. In this context, long-term contracts with private companies for trading electricity (power purchase agreements or PPAs) are often mentioned as an alternative solution to state-based support programmes. The potential for PPAs is, however, restricted on the part of industry and its interest in long-term contracts (at least 10 years) is limited, among others, because of the required company size in terms of electricity consumption and the creditworthiness of the offtaker. Furthermore, in the case of large purchase quantities, the PPA contract can adversely affect an offtaker's credit rating. Relevant research topics include the availability of and barriers to PPA contracts.

Discussions with regard to the longer term also address the market design for electricity in general. One topic is whether renewables and flexibility options will be able to recover their costs based on the current market design (energy-only market) or whether there is a need for specific revenue mechanisms called "capacity mechanisms". In Germany, there has been a very lively debate surrounding the market design for electricity over the last decade. Based on political discussions and research-based assessments, the energy-only market has been retained, but complemented by a strategic reserve. Recently, more general debates about capacity mechanisms are regaining momentum, also due to the increasing number of capacity mechanisms across EU Member States.

5.4 New Developments with Regard to Support Scheme Design

Apart from the question about phasing out support for renewables in the electricity sector, there are also new developments in the research field with respect to designing the allocation of support or de-risking tools. The most important discussions here concern different types of zero support auctions, on the one hand, and new support tools at EU level, on the other hand. Both are briefly outlined in the following.

Since 2017, especially offshore wind auctions but also solar PV and onshore wind auctions have sometimes resulted in zero bids by the auction participants. These plant operators no longer required support payments, but they participated in these auctions just to be able to secure a specific site or gain access to an existing grid connection. At the same time, merchant investments in solar PV and wind onshore plants were realised, mainly based on PPA contracts with private companies.

These results sparked a debate about how to design zero support auctions, e.g. Anatolitis et al. (2022) or Jansen et al. (2020). In Germany and many other EU countries, this especially concerned offshore wind. In the Netherlands, a tender design for zero support was already introduced in 2017. There, a "beauty contest" (de Rijke et al. 2017) selection process is used involving several criteria that are adapted for each tender. The most recent tenders also include a financial bid component, where bidders can offer what they are willing to pay for the opportunity to build and operate their wind park at the specific site. The UK has chosen another way to deal with the fact that market revenues are expected to fully cover the costs of offshore wind parks, at least in the medium term. The UK uses a Contract for Difference (CfD) system, where plant operators have to pay back money to the state whenever the electricity market price is higher than the auction strike price (Welisch and Poudineh 2020). Even though the grid connection costs are financed by the plant operators in the UK, the last auction resulted in a very low strike price, which, under the assumption of rising electricity prices, will lead to negative support payments due to the nature of the CfD system. Denmark also used a CfD design in the most recent tender for offshore wind in order to avoid zero offers. However, because payback was limited to a certain volume as part of the tender conditions and all auction participants expected higher profits than this amount, the auction still resulted in zero bids and a lottery was used to select the winner. In Germany, the first proposal of the government to include a negative bid component to select the winner in the case of more than one zero bid was not successful. In a new proposal, the main support instrument for offshore wind will change to a CfD system. A "beauty contest" combined with a negative bid component is proposed for additional areas without central predevelopment.

In the past, support systems for renewable energies were mainly organised on a national level. Although the Renewable Energy Directive has included cross-border cooperation to support renewables since 2009, this opportunity has only been used very rarely so far. As the new system only plans a binding target at EU level, and not at Member State level, the EU has proposed additional support programmes at EU level. The most important of these are the renewable energy financing mechanism and the new funding instrument for cross-border renewable energy projects under the Connecting Europe Facility. The financing mechanism is an auction for supporting renewables at EU level. Member States can participate by either offering renewable energy sites or buying renewables from the auctions. The new funding instrument supports cross-border renewable energy projects that fulfil certain criteria and participate in an application process.

6 Summary and Outlook

The research on renewable energy sources over the last 50 years has developed very dynamically. While the 1970s and 80s were characterised by finding technological solutions for renewables in a predominantly hostile environment dominated by incumbents, the market diffusion of renewable energy sources really took off in the electricity sector in the 1990s, stimulated by different support policies. In the early 1990s and 2000s, research was predominantly concerned with investigating, comparing, and analysing different diffusion policies and instruments, underpinned by a growing number of real-life experiences and data. As renewables expanded and the focus shifted to market and system integration, the range of research topics and approaches also expanded considerably. While research was heavily technology-driven in the first decades and non-technological perspectives were the exception rather than the rule, over time, inter- and multidisciplinary approaches started to become more important. Of course, technology-driven research still exists and is essential, but is not the focus of this article.

In light of policy developments at both the EU and national level in the 2000s, a large descriptive scientific literature has started to emerge on the existing support instruments for renewables and trends. A much-discussed subject has been the choice of recommended support regime, with feed-in tariffs and quota obligations the most prevalent instruments chosen by Member States. As numerous studies have concluded that feed-in tariffs and premiums tend to be more effective and efficient, many countries have abandoned quota-based mechanisms. Overall, however, the general conclusion is that there is no "one-fits-all" approach when it comes to supporting renewables and that it is vital to select appropriate design elements that are adapted to the unique conditions and system boundaries of a given country. In this context, a new research strand has emerged, which focuses on finding evidence of either harmonisation or the convergence of support schemes across the EU. At the same time, complex energy system models have become more widespread, which incorporate the growing contribution of intermittent renewables. Approaches to monitoring and measuring the performance of policy instruments in practice have also developed in the context of various research projects and investigate the effects of renewables expansion on the electricity sector and the economy. Positive and negative impacts are differentiated into effects at the micro, system, and macro levels and analysed in detail. At the system level, studies include the impacts of renewables in electricity generation on market electricity prices, the so-called merit-order effect. Further distributional aspects of energy efficiency and renewable energy expansion in the electricity, heat and mobility sectors have been systematically outlined and analysed. At the macro level, a variety of studies have analysed the impact of renewables expansion on employment and growth. Over time, different approaches have developed to assess these and other effects of renewables, ranging from structural top-down models to computational general equilibrium (CGE) models. Over the course of the market deployment of renewable energies, the focus of discussion has slowly shifted from net or gross job creations to the jobs and qualifications required to power the energy transition. Overall, there are many different approaches and focus areas in the literature on renewables during the phase of market deployment of renewables, some of which cover different paradigms and mechanisms.

Later on, in the third phase of development, a stronger market integration of renewable energies has been promoted, e.g. by introducing auction procedures for renewable energy support. This illustrates how renewables have developed from niche technological applications to a mass market, their dominant role in a future decarbonised energy system, and the challenges this poses for designing and continuously adapting policy. It has been possible to reduce electricity generation costs significantly, especially regarding solar PV, with Germany playing a major role in stimulating these cost reductions through large capacity additions. This phase is also marked by a broader research perspective in geographical terms and the global perspective of renewables becoming more important.

With increasing shares of renewables in the energy system in Germany and the EU, the market and system integration of renewables have become more important. Non-financial barriers have been assessed in more detail and priority given to dismantling them to encourage even faster expansion. In addition to market integration, the need for adjustments to the overall energy system and the grid also are increasingly important topics. New topics have emerged, such as prosumerism and self-consumption of renewable electricity, and are being addressed with interdisciplinary approaches and by combining multiple methods. A prolific area of research assesses the EU's 2030 Climate and Energy Policy framework as well as individual Member States' contributions to this. Acceptance issues, which had developed into a major barrier to the further expansion of renewable energies, have also become more relevant. Many acceptance analyses are qualitative and quantitative case studies and cover different groups, including residents, experts, and other stakeholders. A new and still on-going debate is on how to design future support systems for renewables, as many technologies are now more mature. Another current and highly relevant topic is sector coupling, i.e. shaping the energy system by integrating all the energy-consuming sectors of the economy, including electricity, buildings/heating and cooling, transport as well as industry.

To sum up, the first development phase focused on technology research. This was followed by the 1990s research into different policies supporting the diffusion of renewable technologies. The subsequent shift in research focus attached greater importance to exploring questions of systemic change using interdisciplinary approaches. Now more than ever before, research is called upon to take a systemic perspective and integrate different approaches, including systems and transformation research to monitor complex socio-technical changes and provide relevant impulses for holistic transformation of the energy sector. Interdisciplinary initiatives and associations that bundle different competencies in energy research, such as the Fraunhofer research cluster Integrated Energy Systems CINES, can play a major role in this context. In general, an increasing differentiation of the research questions can be observed over time, from technology-centric research to more interdisciplinary research. This is also valid for the methods applied.

The actor landscape has become more diversified over the years, with an increasing share of non-university research. Overall, it can be observed that the research field of renewables has developed from a niche to a mainstream topic, accompanied by increasing cross-pollination between research areas.

References

Abrell J, Weigt H (2008) The interaction of emissions trading and renewable energy promotion

Anatolitis V, Welisch M (2017) Putting renewable energy auctions into action—an agent-based model of onshore wind power auctions in Germany. Energy Policy 110:394–402. https://doi.org/10.1016/j.enpol.2017.08.024

Anatolitis V, Azanbayev A, Fleck A-K (2022) How to design efficient renewable energy auctions? Empirical Insights from Europe. Energy Policy 166:112982. https://doi.org/10.1016/j.enpol.2022.112982

Auer H, Stadler M, Resch G, Huber C, Schuster T, Nielsen LH, Twidell J, Swider DJ (2004) Pushing a least cost integration of green electricity into the European grid, GreenNet: Cost and technical constraints of RES-E Grid integration

Bakken K (1981) System Tepidus, high capacity thermochemical storage/heat pump. Int. Conf. Energy Storage 8. Brighton, Br. Hy-drodyn. Res. Assoc. Fluid Eng

Bardt H (2014) EEG 2.0: Ein zweiter Schritt muss folgen. Köln: Institut der deutschen Wirtschaft (IW) (IW Policy Paper, 5/2014). https://www.econstor.eu/handle/10419/95963

Bauwens T, Gotchev B, Holstenkamp L (2016) What drives the development of community energy in Europe? The case of wind power cooperatives. Energy Res Soc Sci 13:136–147. https://doi.org/10.1016/j.erss.2015.12.016

Bergek A, Jacobsson S, Carlsson B, Lindmark S, Rickne A (2008) Analyzing the functional dynamics of technological innovation systems: a scheme of analysis. Res Policy 37:407–429. https://doi.org/10.1016/j.respol.2007.12.003

Bergek A, Hekkert M, Jacobsson S, Markard J, Sandén B, Truffer B (2015) Technological innovation systems in contexts: conceptualizing contextual structures and interaction dynamics. Environ Innov Soc Trans 16:51–64. https://doi.org/10.1016/j.eist.2015.07.003

Bergmann, B., Bitsch, C., Behlau, V., Jensen, S., Held, A., Pfluger, B. Ragwitz, M., Resch, G. (2008) Harmonisation of support schemes–a European harmonised policy to promote RES-electricity–sharing costs & benefits. A report compiled within the European research project futures

Bernath C, Deac G, Sensfuß F (2021) Impact of sector coupling on the market value of renewable energies—a model-based scenario analysis. Appl Energy 281:115985. https://doi.org/10.1016/j.apenergy.2020.115985

Blazejczak J, Braun FG, Edler D, Schill W-P (2014) Economic effects of renewable energy expansion: a model-based analysis for Germany. Renew Sust Energ Rev 40:1070–1080. https://doi.org/10.1016/j.rser.2014.07.134

Blegaa S, Josephsen L, Meyer NI, Sørensen B (1977) Alternative Danish energy planning. Energy Policy 5:87–94. https://doi.org/10.1016/0301-4215(77)90002-7

BMU, UBA (2019) Umweltbewusstsein in Deutschland 2018: Ergebnisse einer repräsentativen Bevölke-rungsumfrage. https://www.umweltbundesamt.de/publikationen/umweltbewusstsein-in-deutschland-2018. Accessed 18 Oct 2022

Bode S, Groscurth H-M (2006) The effect of the German renewable energy act (EEG) on "the electricity Price". Hamburg Institute of International Economics (HWWA), Hamburg. (HWWA Discussion Paper, 358). https://www.econstor.eu/handle/10419/19388

Bohringer C, Loschel A (2006) Promoting renewable energy in Europe: a hybrid computable general equilibrium approach. Energy J 27:SI2006. https://doi.org/10.5547/ISSN0195-6574-EJ-VolSI2006-NoSI2-7

Böhringer C, Cuntz A, Harhoff D, Asane-Otoo E (2017) The impact of the German feed-in tariff scheme on innovation: evidence based on patent filings in renewable energy technologies. Energy Econ 67:545–553. https://doi.org/10.1016/j.eneco.2017.09.001

Boie I (2016) Determinants for the market diffusion of renewable energy technologies – an analysis of the framework conditions for non-residential photovoltaic and onshore wind energy deployment in Germany, Spain and the UK. University of Exeter; College of Life and Environmental Sciences

Bölkow L, Selzer H, Meliß M, Ziesing H-J (1990) Erneuerbare Energiequellen: Zusammenfassung der Ergebnisse des Studienkomplexes A.2. In: Enquete-Kommission "Vorsorge zum Schutz der Erdat-mosphäre" des deutschen Bundestags (ed) Energie und Klima: Band 3 Erneuerbare Energien. Economica Verlag, Bonn, pp 963–1100

Bossel H, Denton R (1977) Energy futures for the Federal Republic of Germany. Energy Policy 5:35–50. https://doi.org/10.1016/0301-4215(77)90058-1

Breitschopf B, Diekmann J (2015) Gesamtwirtschaftliche Auswirkungen des Ausbaus Erneuerbarer Energien. In: Gerhard M (ed) Finanzierung Erneuerbarer Energien. 2, überarb. edn. Frankfurt-School-Verlag; Frankfurt-School-Verl, Frankfurt am Main, pp 65–89

Breitschopf B, Pudlik M (2013) Basel iii and solvency ii: are the risk margins for investments in Pv and wind adequate? Energy Environ 24:171–194. https://doi.org/10.1260/0958-305X.24.1-2.171

Breitschopf B, Nathani C, Resch G (2013) Employment impact assessment studies – is there a best approach to assess employment impacts of RET deployment? Renew Energy Law Policy Rev:93–104

Breitschopf B, Alexander-Haw A (2022) Auctions, risk and the WACC. How auctions and other risk factors impact renewable electricity financing costs. In: Energy Strategy Reviews 44, S. 100983. https://doi.org/10.1016/j.esr.2022.100983

Breitschopf B, Bürer S, Lürich L (2014) Verteilungswirkungen der Marktförderung des EEG in den Bereichen Photovoltaik und Windenergie (onshore) [distributional effects of the RE levy in photovoltaics and wind onshore] ImpRES Projektpaper

Breitschopf B, Held A, Resch G (2016) A concept to assess the costs and benefits of renewable energy use and distributional effects among actors. The example of Germany. Energy Environ 27(1):55–81. https://doi.org/10.1177/0958305X16638572

Breitschopf B, Keil J, Burghard U, Scheller F (2024) Forms of financial participation and acceptance of the energy transition – are there any relations? ZfE 1:8–57. https://doi.org/10.1007/s12398-024-1256-6

Brunekreeft G, Buchmann M, Kusznir J (2022) Regulatorische Experimente und Anreizregulierung—Erfahrungen mit der SINTEG-V. Z Energiewirtsch 46:195–206. https://doi.org/10.1007/s12398-022-00327-2

Buchner H (1980) Thermal energy storage using metal hydrides. In: Energy storage. Pergamon, pp 327–334. https://doi.org/10.1016/B978-0-08-025471-5.50033-6

Capper T, Gorbatcheva A, Mustafa MA, Bahloul M, Schwidtal JM, Chitchyan R, Andoni M, Robu V, Montakhabi M, Scott IJ, Francis C, Mbavarira T, Espana JM, Kiesling L (2022) Peer-to-peer, commu-nity self-consumption, and transactive energy: a systematic literature review of local energy market models. Renew Sust Energ Rev 162:112403. https://doi.org/10.1016/j.rser.2022.112403

Carlsson B, Jacobsson S, Holmén M, Rickne A (2002) Innovation systems: analytical and methodological issues. Res Policy 31:233–245. https://doi.org/10.1016/S0048-7333(01)00138-X

Cook EF (1976) Man, energy, society. W.H. Freeman and Company, San Francisco. (Texas A and M University, College of Geosciences, College Station, TX). https://www.osti.gov/biblio/7357737

Costantini V, Crespi F, Martini C, Pennacchio L (2015) Demand-pull and technology-push public support for eco-innovation: the case of the biofuels sector. Res Policy 44:577–595. https://doi.org/10.1016/j.respol.2014.12.011

de Paoli L, Geoffron P (2019) Introduction. A critical overview of the European national energy and climate plans. Econ Policy Energy Environ 1:31–41. https://doi.org/10.3280/EFE2019-001002

de Rijke M, Dingenen S, Kyrberg L, Beetstra T (2017) What you need to know about the coming Dutch offshore wind tender. https://www.twobirds.com/en/insights/2017/netherlands/what-you-need-to-know-about-the-coming-dutch-offshore-wind-tender

Del Río P, Kiefer CP (2023) Academic research on renewable electricity auctions: taking stock and look-ing forward. Energy Policy 173:113305. https://doi.org/10.1016/j.enpol.2022.113305

Del Río P, Ragwitz M, Steinhilber S, Resch G, Busch S, Klessmann C, de Lovinfosse I, Nysten J, Fouquet, D, Johnston A (2012) Assessment criteria for identifying the main alternatives. Advantages and drawbacks, synergies and conflicts. A report compiled within the project beyond 2020, supported by the EACI of the European Commission within the "Intelligent Energy Europe" programme. CSIC, Madrid. https://res-policy-beyond2020.eu/. Accessed 27 Mar 2024

Dewald U, Truffer B (2011) Market formation in technological innovation systems—diffusion of pho-tovoltaic applications in Germany. Ind Innov 18:285–300. https://doi.org/10.1080/13662716.2011.561028

Diekmann J, Breitschopf B, Lehr U (2016a) Social impacts of renewable energy in Germany: size, history and alleviation (GWS Discussion Paper, 2016/07). https://www.econstor.eu/bitstream/10419/156294/1/861879155.pdf. Accessed 7 Sep 2018

Diekmann J, Schill W-P, Breitschopf B, Sievers L, Klobasa M, Lehr U, Horst J (2016b) Impacts of renewable energy deployment—summary and conclusions. Final report in the framework of the project "impacts of renewable energy sources (ImpRES) supported by the Federal Ministry of Economics and Energy. Deutsches Institut für Wirtschaftsforschung e V. (DIW); Fraunhofer Institut für System- und Innovationsforschung (Fraunhofer ISI); Gesellschaft für Wirtschaftliche Strukturforschung mbH (GWS); Institut für ZukunftsEnergieSysteme (IZES). http://www.impres-projekt.de/impres-en/content/abschlussworkshop.php

Directive 2009/28/EC of the European Parliament and of the Council of 23 April 2009 on the promotion of the use of energy from renewable sources and amending and subsequently repealing Directives 2001/77/EC and 2003/30/EC. Available online at Directive - 2009/28 - EN - Renewable Energy Directive - EUR-Lex (europa.eu)

Duscha V, Fougeyrollas A, Nathani C, Pfaff M, Ragwitz M, Resch G, Schade W, Breitschopf B, Walz R (2016) Renewable energy deployment in Europe up to 2030 and the aim of a triple dividend. Energy Policy 95:314–323. https://doi.org/10.1016/j.enpol.2016.05.011

EEX (2014) EEX-Stellungnahme zum EEG-Referentenentwurf

Erhorn H (1990) Low-energy houses. Environmental daydream or design of the future. Niedrigenergie-haeuser. Umwelttraeumerei oder bautechnische Zukunft. Clima Commerce International; (Germany, F.R.) 24:10

European Commission (1997) Energy for the future: renewable sources of energy. White Paper for a Community Strategy and Action Plan. COM (97) 599 final, 26 November 1997 [follow-up to the Green Paper]

European Commission (2005) Communication from the Commission: the support of electricity from renewable energy sources

European Commission (2008) European commission staff working document "The support of electricity from renewable energy sources", accompanying document to the proposal for a Directive of the Euro-pean Parliament and of the Council the promotion of the use of energy from renewable sources

European Commission (2014) Communication from the Commission — Guidelines on State aid for envi-ronmental protection and energy 2014–2020. https://eur-lex.europa.eu/legal-content/EN/TXT/?uri=CELEX%3A52014XC0628%2801%29

European Commission and European Parliament (2001) Directive 2001/77/EC. https://eur-lex.europa.eu/legal-content/EN/TXT/?uri=CELEX:32001L0077. Accessed 18 Oct 2022

Fragkos P, Paroussos L (2018) Employment creation in EU related to renewables expansion. Appl Energy 230:935–945. https://doi.org/10.1016/j.apenergy.2018.09.032

Fraunhofer ISI, Consentec, ifeu, TU Berlin (2021) Langfristszenarien für die Transformation des Energie-systems in Deutschland 3: Kurzbericht: 3 Hauptszenarien

Frondel M, Ritter N, Schmidt CM, Vance C (2010) Economic impacts from the promotion of renewable energy technologies: the German experience. Energy Policy 38:4048–4056. https://doi.org/10.1016/j.enpol.2010.03.029

German Bundestag (1991) Protecting the earth: a status report with recommendations for a new energy policy, vol I+II, Bonn

Goetzberger A (1982) Stand der Forschung auf dem Gebiet der Photovoltaik. In: Optoelektronik in der Technik / Optoelectronics in Engineering. Springer, Berlin, Heidelberg, pp 547–558

Goy GC, Wittke F, Ziesing HJ, Rehbock M, Jaeger F, Kunz P, Mannsbart W, Poppke H (1984) Assessment of the potential of renewable energy sources in the Federal Republic of Germany. Abschaetzung des potentials erneuerbarer Energiequellen in der ... technical report

Grave K, Breitschopf B, Lutz C, Mandana H, Boeve S, von Blücher F, Bourgault C, Friedrichsen N, Arens M, Aydemir A, Pudlik M, Duscha V, Ordonez JA, Großmann A, Flaute M (2015) Electricity costs of energy intensive industries – an international comparison. http://www.isi.fraunhofer.de/isi-wAssets/docs/x/de/projekte/Strompreiswirkung_330639/Industriestrompreise_englisch.pdf. Accessed 17 Feb 2016

Groba F, Breitschopf B (2013) Impact of renewable energy policy and use on innovation. Discussion Papers of DIW, Berlin

Gruber E, Jaeckel G, Meyer T (1988) The Landstuhl project: solar architecture and energy saving. In: Advances in solar energy technology. Elsevier, pp 3671–3675

Hake J-F, Fischer W, Venghaus S, Weckenbrock C (2015) The German Energiewende—history and status quo. Energy 92:532–546. https://doi.org/10.1016/j.energy.2015.04.027

Harmelink M, Voogt M, Joosen S, Jager D de, Palmers G, Shaw S, Cremer C (2001) PRETIR. Implementation of renewable energy in the European Union until 2010. Project executed within the framework of the ALTENER Programme of the European Union

Haas R, Eichhammer W, Huber C, Langniss O, Lorenzoni A, Madlener R, Menanteau P, Morthorst P-E, Martins A, Oniszk A, Schleich J, Smith A, Vass Z, Verbruggen A (2004) How to promote renewable energy systems successfully and effectively. Energy Policy 32(6):833–839

Heidinger PF (1987) Electricity for the future – our responsibility between risk and chance. Bulletin des Schweizerischen Elektrotechnischen Vereins 78:1252–1267

Hekkert MP, Negro SO (2009) Functions of innovation systems as a framework to understand sustainable technological change: empirical evidence for earlier claims. Technol Forecast Soc Chang 76:584–594. https://doi.org/10.1016/j.techfore.2008.04.013

Hekkert MP, Suurs R, Negro SO, Kuhlmann S, Smits R (2007) Functions of innovation systems: a new approach for analysing technological change. Technol Forecast Soc Chang 74:413–432. https://doi.org/10.1016/j.techfore.2006.03.002

Held AM (2010) Modelling the future development of renewable energy technologies in the European electricity sector using agent-based simulation. Available online at https://publikationen.bibliothek.kit.edu/1000021841/1809689

Held A, Ragwitz M, del Río P, Resch G, Klessmann C et al (2019) Do almost mature renewable energy technologies still need dedicated support towards 2030? – ProQuest. Econ Energy Environ Policy 8(2):81–98

Hillebrand B, Buttermann HG, Behringer JM, Bleuel M (2006) The expansion of renewable energies and employment effects in Germany. Energy Policy 34:3484–3494. https://doi.org/10.1016/j.enpol.2005.06.017

Holdren J, Anderson K, Gleick P, Mintzer I, Morris G, Smith K (1979) Risk of renewable energy sources: a critique of the Inhaber report. United States https://www.osti.gov/servlets/purl/5565788

Huber C, Faber T, Haas R, Resch G, Green J, Ölz S et al. (2004) Action plan for deriving dynamic RES-E policies: report of the project Green-X – a research project within the Fifth Framework Programme of the European Commission, Supported by DG Research. https://core.ac.uk/display/57008399

Inhaber H (1978) Is solar power more dangerous than nuclear. New Scientist 78:444–446

IRENA, CEM (2014) The socio-economic benefits of solar and wind energy. EconValue report Edited by IRENA, Clean Energy Ministerial (CEM). http://www.irena.org/DocumentDownloads/Publications/Socioeconomic_benefits_solar_wind.pdf. Accessed 19 Feb 2018

Jacobsson S, Lauber V (2006) The politics and policy of energy system transformation—explaining the German diffusion of renewable energy technology. Energy Policy 34:256–276. https://doi.org/10.1016/j.enpol.2004.08.029

Jacobsson S, Sandén B, Bångens L (2004) Transforming the energy system — the evolution of the Ger-man technological system for solar cells. Tech Anal Strat Manag 16:3–30. https://doi.org/10.1080/0953732032000199061

Jansen M, Staffell I, Kitzing L, Quoilin S, Wiggelinkhuizen E, Bulder B, Riepin I, Müsgens F (2020) Off-shore wind competitiveness in mature markets without subsidy. Nat Energy 5:614–622. https://doi.org/10.1038/s41560-020-0661-2

Jarach M (1989) An overview of the literature on barriers to the diffusion of renewable energy sources in agriculture. Appl Energy 32:117–131. https://doi.org/10.1016/0306-2619(89)90073-1

Jensen J, Sorensen B (1984) Fundamentals of energy storage. John Wiley and Sons, Inc, New York, NY, USA

Johansson TB, Turkenburg W (2004) Policies for renewable energy in the European Union and its member states: an overview. Energy Sustain Dev 8:5–24. https://doi.org/10.1016/S0973-0826(08)60387-7

Johnstone N, Haščič I, Popp D (2010) Renewable energy policies and technological innovation: Evi-dence based on patent counts. Environ Resour Econ 45:133–155. https://doi.org/10.1007/s10640-009-9309-1

Kitzing L, Mitchell C, Morthorst PE (2012) Renewable energy policies in Europe: converging or diverging? Energy Policy 51:192–201. https://doi.org/10.1016/j.enpol.2012.08.064

Kitzing L, Siddique MB, Nygaard I, Kruger W (2022) Worth the wait: how South Africa's renewable en-ergy auctions perform compared to Europe's leading countries. Energy Policy 166:112999. https://doi.org/10.1016/j.enpol.2022.112999

Klobasa M, Winkler J, Sensfuß F, Ragwitz M (2013) Market integration of renewable electricity generation â€ the German market premium model. Energy Environ 24:127–146

Kloss R (1982) Elektrizität oder Wärme aus Biogas? https://doi.org/10.24355/dbbs.084-201205091003-0

Krause F, Bossel H, Mueller-Reissmann KF (1980) Energie-Wende. Wachstum und Wohlstand ohne Erdöl und Uran

Krewitt W, Nitsch J (2003) The German renewable energy sources act—an investment into the future pays off already today. Renew Energy 28:533–542. https://doi.org/10.1016/S0960-1481(02)00064-2

Kühne O, Weber F (2016) Zur sozialen Akzeptanz der Energiewende. uwf 24:207–213. https://doi.org/10.1007/s00550-016-0415-6

Langniß O, Diekmann J, Lehr U (2009) Advanced mechanisms for the promotion of renewable energy—models for the future evolution of the German renewable energy act. Energy Policy 37:1289–1297. https://doi.org/10.1016/j.enpol.2008.11.007

Lauber V, Toke D (2005) Einspeisetarife sind billiger und effizienter als Quoten-/Zertifikatssysteme. Der Vergleich Deutschland-Großbritannien stellt frühere Erwartungen auf den Kopf. Zeitschrift für Neues Energierecht:132–139. Available online at https://uni-salzburg.elsevierpure.com/en/publications/einspeisetarifesind-billiger-und-effizienter-als-quoten-zertifik

Lehmann H, Peter S (2005) Analyse der Vor-und Nachteile verschiedener Modelle zur Förderung des Ausbaus von Offshore-Windenergie in Deutschland. Aachen (ISUSI Endbericht für das BMU)

Lode ML, te Boveldt G, Coosemans T, Ramirez Camargo L (2022) A transition perspective on energy communities: a systematic literature review and research agenda. Renew Sust Energ Rev 163:112479. https://doi.org/10.1016/j.rser.2022.112479

Lundvall B-Å, Johnson B, Andersen ES, Dalum B (2002) National systems of production, innovation and competence building. Res Policy 31:213–231. https://doi.org/10.1016/S0048-7333(01)00137-8

Lutz C, Breitschopf B (2016) Systematisierung der gesamtwirtschaftlichen Effekte und Verteilungswirkungen der Energiewende. [systemic overview of macroeconomic and distributional effects of the energy transition] (GWS Research Report, 2016/1). http://papers.gws-os.com/gws-researchreport16-1.pdf. Accessed 7 Sep 2018

Malerba F (2005) Sectoral systems of innovation: a framework for linking innovation to the knowledge base, structure and dynamics of sectors. Econ Innov New Technol 14:63–82. https://doi.org/10.1080/1043859042000228688

Markard J, Hekkert M, Jacobsson S (2015) The technological innovation systems framework: response to six criticisms. Environ Innov Soc Trans 16:76–86. https://doi.org/10.1016/j.eist.2015.07.006

Masuhr KP, Eckerle K, Hofer P, Meckel H, Wolff H, Frey B, Schilling K, Brand M, Jochem E (1987) Ra-tional power utilization and generation without nuclear energy: potentials and assessment of effects on the power industry, ecology, and economy. Prognos AG, Köln

Meadows DH, Meadows DL, Randers J, Behrens WW III (1972) The limits to growth – Club of Rome. Club of Rome

Meyer NI (2003) European schemes for promoting renewables in liberalised markets. Energy Policy 31:665–676. https://doi.org/10.1016/S0301-4215(02)00151-9

Meyer-Abich KM, Schefold B (1981) Wie möchten wir in Zukunft leben: der "harte" und der "sanfte" Weg. Eine Publikation der Vereinigung Deutscher Wissenschaftler (VDW). Beck Verlag, München

Meyer-Abich KM, Schefold B (1986) Die Grenzen der Atomwirtschaft. Beck Verlag, München

Miethling B (2011) Different but similar: geothermal energy and the role of politics in Germany, Iceland and the United States. Z Energiewirtsch 35:287–296. https://doi.org/10.1007/s12398-011-0063-z

Mišík M, Oravcová V (2022) Ex ante governance in the European Union: energy and climate policy as a 'test run' for the post-pandemic recovery. Energy Policy 167:113076. https://doi.org/10.1016/j.enpol.2022.113076

Mitchell C, Bauknecht D, Connor PM (2006) Effectiveness through risk reduction: a comparison of the renewable obligation in England and Wales and the feed-in system in Germany. Energy Policy 34(3):297–305. https://doi.org/10.1016/j.enpol.2004.08.004

Mittlefehldt S (2018) From appropriate technology to the clean energy economy: renewable energy and environmental politics since the 1970s. J Environ Stud Sci 8:212–219. https://doi.org/10.1007/s13412-018-0471-z

Mora D, Islam M, Rosenlund Soysal E, Kitzing L, Kelm T, Amazo, Blanco, Ana Lucia, Förster S, Tiedemann S, Wigand F (2017) Experiences with auctions for renewable energy support. Accessed 16 Dec 2022

Muñoz M, Oschmann V, David Tàbara J (2007) Harmonization of renewable electricity feed-in laws in the European Union. Energy Policy 35:3104–3114. https://doi.org/10.1016/j.enpol.2006.11.006

Negro SO, Hekkert MP (2008) Explaining the success of emerging technologies by innovation system functioning: the case of biomass digestion in Germany. Tech Anal Strat Manag 20:465–482. https://doi.org/10.1080/09537320802141437

Negro SO, Alkemade F, Hekkert MP (2012) Why does renewable energy diffuse so slowly? A review of innovation system problems. Renew Sust Energ Rev 16:3836–3846. https://doi.org/10.1016/j.rser.2012.03.043

Nitsch J (1976) Production of chemical energy carriers by non-expendable energy sources

Nitsch J, Luther J (1989) Energy supply of the future—efficient use of energy and renewable Energies. Springer, Berlin, Heidelberg

Nitsch J, Schott T, Klaiß H, Jochem E, Herz H, Hudetz W (1981) Developing secondary energy Systems in the Fed. Rep. of Germany until the year 2000. Köln, Karlsruhe (Programme applied systems analysis (ASA) of the Fed. Research Institutes, report ASA-ZE/18/80)

Oberthür S (2019) Hard or soft governance? The EU's climate and energy policy framework for 2030. PaG 7:17–27. https://doi.org/10.17645/pag.v7i1.1796

Pfenninger S, Hawkes A, Keirstead J (2014) Energy systems modeling for twenty-first century energy challenges. Renew Sust Energ Rev 33:74–86. https://doi.org/10.1016/j.rser.2014.02.003

Ragwitz M, Schade W, Breitschopf B, Walz R, Helfrich N, Rathmann M, Resch G, Panzer C, Faber T, Haas R, Nathani C, Holzhey M, Konstantinaviciute I, Zagame P, Fougeyrollas A, Le Hir B (2009) EmployRES. The impact of renewable energy policy on economic growth and employment in the Euro-pean Union. Final report

Ragwitz M, Steinhilber S (2014) Effectiveness and efficiency of support schemes for electricity from renewable energy sources. WENE 3(2):213–229. https://doi.org/10.1002/wene.85

Reichardt K, Negro SO, Rogge KS, Hekkert MP (2016) Analyzing interdependencies between policy mix-es and technological innovation systems: the case of offshore wind in Germany. Technol Forecast Soc Chang 106:11–21. https://doi.org/10.1016/j.techfore.2016.01.029

Reichardt K, Rogge KS, Negro SO (2017) Unpacking policy processes for addressing systemic problems in technological innovation systems: the case of offshore wind in Germany. Renew Sust Energ Rev 80:1217–1226. https://doi.org/10.1016/j.rser.2017.05.280

Reiche D, Bechberger M (2004) Policy differences in the promotion of renewable energies in the EU member states. Energy Policy 32:843–849. https://doi.org/10.1016/S0301-4215(02)00343-9

Reichert J, Jäger F, Herz H (1980) Überprüfung eines Erd-Wärmespeichers. Fraunhofer ISI. Karlsruhe (BMFT-FB-T 81-200)

Rogge KS, Reichardt K (2016) Policy mixes for sustainability transitions: An extended concept and framework for analysis. Res Policy 45:1620–1635. https://doi.org/10.1016/j.respol.2016.04.004

Rogge KS, Pfluger B, Geels FW (2020) Transformative policy mixes in socio-technical scenarios: the case of the low-carbon transition of the German electricity system (2010–2050). Technol Forecast Soc Chang 151:119259. https://doi.org/10.1016/j.techfore.2018.04.002

Schleich J, Walz R, Ragwitz M (2017) Effects of policies on patenting in wind-power technologies. Energy Policy 108:684–695. https://doi.org/10.1016/j.enpol.2017.06.043

Schmitz K, Voß A (1980) Energiewende?: Analysen, Fragen und Anmerkungen zu dem vom ÖKO-Institut vorgelegten "Alternativ-Bericht". Universität Stuttgart

Schumacher K, Sands RD (2006) Innovative energy technologies and climate policy in Germany. Energy Policy 34:3929–3941. https://doi.org/10.1016/j.enpol.2005.09.022

Segreto M, Principe L, Desormeaux A, Torre M, Tomassetti L, Tratzi P, Paolini V, Petracchini F (2020) Trends in social acceptance of renewable energy across Europe—a literature review. IJERPH 17:9161. https://doi.org/10.3390/ijerph17249161

Sensfuß F (2015) Analysen zum merit-order-Effekt Erneuerbarer Energien (analysis of the merit-order effect of RE use). Veröffentlichungen | ImpRES – impact of renewable energy sources. http://www.impres-projekt.de/impres-wAssets/docs/Sensfuss_Gutachten-Merit-Order-2013_2014.pdf. Accessed 18 Nov 2015

Sensfuß F, Ragwitz M, Genoese M (2008) The merit-order effect: a detailed analysis of the price effect of renewable electricity generation on spot market prices in Germany. Energy Policy 36:3086–3094. https://doi.org/10.1016/j.enpol.2008.03.035

Sievers L, Breitschopf B, Pfaff M, Schaffer A (2019) Macroeconomic impact of the German energy transi-tion and its distribution by sectors and regions. Ecol Econ 160:191–204. https://doi.org/10.1016/j.ecolecon.2019.02.017

Smits R, Kuhlmann S (2004) The rise of systemic instruments in innovation policy. IJFIP 1:4. https://doi.org/10.1504/ijfip.2004.004621

Sørensen B (1984) Energy storage. Annu Rev Energy 9:1–29

Stadermann G (2021) Das Notwendige möglich machen: Die solare Forschungswende in Deutschland. Springer

Walz R (2006) Increasing renewable energy in Europe—impacts on competitiveness and lead markets. Energy Environ 17:951–975

Walz R (2007) The role of regulation for sustainable infrastructure innovations: the case of wind energy. Int J Public Policy 2:57–88

Welisch M, Poudineh R (2020) Auctions for allocation of offshore wind contracts for difference in the UK. Renew Energy 147:1266–1274. https://doi.org/10.1016/j.renene.2019.09.085

Winkler J (2017) Market integration of renewables in the electricity sector. Impact on electricity markets and renewable support policy as well as interactions with system flexibility. Dissertation, University of Freiburg

Winkler J, Pudlik M, Ragwitz M, Pfluger B (2016) The market value of renewable electricity—which factors really matter? Appl Energy 184:464–481. https://doi.org/10.1016/j.apenergy.2016.09.112

Winkler J, Breitschopf B, Harmelink M, Cail S, Mistre M, Ragwitz M (2018) Renewable energy directive target. Study for the ITRE committee. Edited by directorate-general for internal policies, policy department, Econmic and scientific policy a. http://www.europarl.europa.eu/RegData/etudes/STUD/2018/614201/IPOL_STU(2018)614201_EN.pdf. Accessed 12 Mar 2018

Winkler J, Held A, Bangert L, Maurer C, Tersteegen B, Kahles M, Geipel J, Resch G, Schöniger F, Gephart M, Klessmann C, Gül S, Jachmann H, Kelm T, Metzger, Jochen (2020) Zukunftsszenarien für erneuerbare Energien – wie viel und welche Förderung wird zukünftig benötigt? https://www.bundesregierung.de/breg-de/service/publikationen/zukunftsszenarien-fuer-erneuerbare-energien-wie-viel-und-welche-foerderung-wird-zukuenftig-benoetigt%2D%2D1789526

Wolf I, Fischer A-K, Huttarsch J-H (2021) Soziales Nachhaltigkeitsbarometer der Energie- und Verkehrs-wende 2021: Kernaussagen und Zusammenfassung der wesentlichen Ergebnisse

Wüstenhagen R, Wolsink M, Bürer MJ (2007) Social acceptance of renewable energy innovation: an in-troduction to the concept. Energy Policy 35:2683–2691. https://doi.org/10.1016/j.enpol.2006.12.001

Julia Panny studied economics and Slavistics at the Vienna University of Economics and Business and the University of Vienna. She earned a Master of Advanced International Studies through a joint programme with the Diplomatic Academy Vienna. Her career includes roles at the UNFCCC, the German Commission for UNESCO and EIT Climate-KIC, where she led climate innovation programmes across Europe and developed a cleantech startup accelerator. Since June 2021, she is a research associate at Fraunhofer ISI's Competence Centre Energy Policy and Energy Markets, working on renewable energies.

Anne Held is coordinator of the Business Unit Renewable Energies in the Competence Centre Energy Policy and Energy Markets at the Fraunhofer Institute for Systems and Innovation Research ISI. Previously, she worked as a researcher at the European Commission's Joint Research Centre. Her research focusses on energy policy design and energy system modelling. Results of her work have contributed to discussions on the development of energy policy at EU and national level, including the 2030 framework for climate and energy policies. She has excellent knowledge on the economic, technological, and social dimensions of the energy transition.

Jenny Winkler holds a master's degree in Environmental Technology from Imperial College London and a Diploma in International Economics from the University of Tübingen. Between 2011 and 2023, she worked at Fraunhofer ISI on different topics with respect to renewable energy, support frameworks for the energy transition and market design. She was involved in numerous advisory projects on policy design and evaluation, mainly for German Ministries and the European Commission. From 2018 to 2023, she co-led the business unit Renewable Energy. Currently, Jenny is an analyst for the PPA department of wpd GmbH.

Barbara Breitschopf worked as a trainee and employee in a bank after graduating from the University of Hohenheim with a degree in Agricultural Science. After receiving an M.A. in Economics from Ohio State University, she worked as a consultant for various economic cooperation organisations before completing her PhD. Subsequently, she conducted research in the field of innovation economics at KIT and Fraunhofer ISI. Since 2008, she has been working in the field of energy policy and energy markets, with a special focus on socio-economic issues of energy and heat transitions. She leads several projects for the European Commission and national clients.

Eberhard Jochem is senior scientist of the Fraunhofer Institute of Systems and Innovation Research ISI. Internationally acknowledged expert in technical and socio-economic research, mainly in the field of energy, material efficiency, and climate change. Major contributions to the development and implementation of a sustainable energy and resources policy. Member of various national and international scientific organisations and advisory committees during the last five decades. Author of several books and numerous publications in the field of energy, resource efficiency, and environment with regard to R&D, innovation, obstacles and policy issues, and energy economics. Promoter of energy efficiency and climate protection networks in industry.

Rainer Walz is Deputy Director of the Institute and Head of the Competence Centre Sustainability and Infrastructure Systems at the Fraunhofer Institute for Systems and Innovation Research ISI in Karlsruhe. He obtained his PhD and "habilitation" in economics. In 1991, Prof. Dr. Walz joined the Fraunhofer ISI. Prior employment includes the University of Wisconsin and the Enquête Commission "Protecting the Earth's Atmosphere" of the German Bundestag. He is also an adjunct professor of economics at the Karlsruhe Institute of Technology (KIT) and member of the EU expert group "Economic and Social Impact of Research".

Energy Demand and Modelling of Energy Systems: Five Decades from Little Knowledge to Differentiated Know-How

Eberhard Jochem, Harald Bradke, Elisabeth Dütschke, Marian Klobasa, Martin Wietschel, Patrick Plötz, and Tobias Fleiter

Abstract Policy-oriented research of efficient energy use and energy demand during the last five decades developed from the scratch to a quite complex research field with many perspectives: new and improved energy-efficient buildings, vehicles, and production processes, structural changes in industry, income, rebound, and saturation effects. Although energy-efficient solutions were (and are) highly profitable, several obstacles prevent their full realisation. Energy policy "discovered" energy efficiency as the "fifth energy source" in the 1980s and labelled its policy priority after the increase of oil prices in the early 2010s by "efficiency first", although policy analysts may have doubts regarding the real energy policy and allocation of resources. The liberalisation of grid-based energy supply triggered a strong push for demand-side measures (flexible demand; energy services). Electricity demand models became much more dynamic in terms of time to match the increasingly fluctuating electricity supply and load shifting options. Climate policy since the 2010s induced a new wave of energy-efficient applications such as electric vehicles or heat pumps. Regarding the tough climate protection goals of a maximum temperature increase below 2.0 °C, more efficient energy use, conversion, and storage are likely to play a major role, particularly in using the large waste heat from useful energy applications.

1 Introduction

Analysing energy demand and energy-efficient solutions in final energy sectors or in the transformation sector is a difficult and cumbersome task. However, even more challenging are projections of future energy demand in these sectors and of

E. Jochem (✉) · H. Bradke · E. Dütschke · M. Klobasa · M. Wietschel · P. Plötz · T. Fleiter
Fraunhofer Institute for Systems and Innovation Research ISI, Karlsruhe, Germany
e-mail: Eberhard.Jochem@isi.fraunhofer.de; Harald.Bradke@isi.fraunhofer.de; Elisabeth.Duetschke@isi.fraunhofer.de; Marian.Klobasa@isi.fraunhofer.de; Martin.Wietschel@isi.fraunhofer.de; Patrick.Ploetz@isi.fraunhofer.de; Tobias.Fleiter@isi.fraunhofer.de

J. Edler, R. Walz (eds.), *Systems and Innovation Research in Transition*, Sustainability and Innovation, https://doi.org/10.1007/978-3-031-66100-6_9

thousands of possible more efficient applications, particularly in industry. Research on energy demand and its possible reduction by more efficient energy use starts in the early 1970s when the report of the Club of Rome was published in 1972 and, more importantly, the first oil embargo was initiated by the OPEC-countries in late 1973 and the second oil price increase in 1979–1980. Until this time, energy policy focused on sufficient energy supply: oil, coal and nuclear energy, the new abundant and inexpensive form of primary energy.

In the 1950s and 1960s, energy demand projections were a domain of energy economists who calculate the final energy demand at a high level of aggregation by the elasticities of energy prices and demand drivers such as the number of private households, of cars or the net production of industry (Morrison and Readling 1968). Knowledge of engineering was focused on the transformation sector (i.e., power plants, refineries, coal and coke production). Foreseeable structural changes in industry, in the economy, or saturation effects in private households were not yet considered in energy demand projections (Kraus 1988). The energy future of the decades until 2000 seemed to be quite clear and straightforward.

During the five decades starting in 1970 the attention given to energy demand substantially changed with regard to the political debate and policymaking as well as to research on energy using technologies. We portray how both new political challenges and related energy policy strategies led to new energy-efficient technologies and solutions. These developments induced the need for more technological content in the energy demand projections leading to advances in methodologies of energy systems research. The new energy demand models also captured the dynamics of technical innovation, structural change of the economy, saturation effects, and patterns of decision making. This chapter describes how energy systems research developed from an area which is little known into a differentiated field of research within five decades.

This chapter is organised by decades beginning with the 1970s. There are developments covering two decades; however, they may be more important for future energy demand in their initial phase or at a later stage. One development, which can be observed throughout the five decades is the increasing attention given to energy-efficient solutions. However, the present energy policy in many countries pays lip service to catchphrases such as "Efficiency First". This changing attention was reflected in the policy of the IEA (International Energy Agency n.d.) which focused on nuclear energy in the 1970s and on efficient energy use today (Lantzke 1980; Geller and Attali 2005; IEA 2022a).

2 The 1970s: The End of the Energy Growth Dream

The early 1970s were characterised by the two preceding decades of substantial economic growth in all OECD countries. Primary energy consumption increased until 1970 with a demand elasticity of around 1.0 in many OECD countries, and the elasticity of electricity demand often even surpassed that level (Lebert 1977; Berndt

1978; Bohnen 1982). Most energy economists of the 1970s were convinced that energy demand increases at the same rate as the gross national product. National energy policy focused predominantly on increasing energy supply, particularly in the fields of mineral oil and nuclear energy.

Energy demand projections of the 1970s until the target year 2000 expected further substantial increases in the demand for heating oil, gasoline, and diesel as well as coal and nuclear energy for electricity generation. However, three events dramatically changed the scene of energy economics research and energy policy in OECD countries:

- In 1972, the Club of Rome published its famous analysis "Limits to Growth", which applied newly developed system dynamics models to worldwide economic developments and their consumption of natural resources like fossil fuels and basic materials, but also to food production and environmental pollution (Meadows et al. 1972).
- In September 1973, the West German government released its First Energy Programme (BMWi 1973), 4 weeks before OPEC decided to reduce oil production by 25% and placed an embargo on the large oil importing harbour of Rotterdam due to the Suez crisis. The oil price (fob) quadrupled to around 12 US$/ barrel. The second oil crisis in 1979 accelerated the challenge to the paradigm of ever increasing energy demand.
- In the early 1970s, new computer systems became available that were able to calculate complex optimisation and simulation model runs which had not been possible in the 1960s.

There were numerous reactions of different governments and the research community:

- OECD countries established the International Energy Agency (IEA), responsible for planning oil storage and the supply of crude oil and oil products in the wake of the oil crisis, but also for strategic energy forecasts and for suggesting new energy-related policies (Lantzke 1980).
- Following the first oil crisis, energy efficiency options in buildings (Prömmel 1978), industry (Jochem et al. 1978), and transport were for the first time clearly addressed by technological-based energy research.
- In the 1970s and at the beginning of the 1980s, this led to considerable controversy and tension among energy researchers in many OECD countries, although this is hardly communicated internationally. Instead, the international scene was dominated by the International Association of Energy Economists (IAEE) and representatives of large publicly owned energy research organisations with their focus on nuclear energy. Attempts to report energy demand and the potentials of efficient energy use in energy journals or conferences often came to nothing.
- In West Germany, for instance, an intensive scientific discourse between three traditional energy economics institutes (EWI, RWI, ifo) and Prognos, and the "newcomers" of energy systems analysis (ASA der Großforschungseinrichtungen: Bohn et al. 1977; Nitsch et al. 1981) and Fraunhofer ISI (Bossel and Denton 1977)

began to address the future impacts of energy efficiency policies, the changes to a service-oriented economy, as well as the structural changes within industry and individual sectors.

The West German government published its first amendment to the Energy Programme in October 1973 (BMWi 1973). At national and federal state level, the ministries of economics were responsible for energy issues, but in the 1970s there was not a single organisational unit that dealt with energy demand or energy efficiency.

After the first oil crisis, the traditional energy economics institutes continued to project substantial increases in primary and final energy demand until 2000. For instance, DIW et al. (1978) projected a gross electricity demand in 1985 of between 1830 and 2030 PJ, implying demand elasticities between 1.4 and 1.55; electricity demand in 2000 was projected to grow to 3000 to 3200 PJ. Actually, however, the electricity consumption of the Federal Republic of Germany amounted to only 1230 PJ in 1985 (overestimated by about 50% for a period of only 7 years) and to 1780 PJ in 2000 for a reunified Germany (overestimated by around 70%).

During the mid-1970s, the new technology-oriented energy systems analysis groups argued that increased energy prices will encourage more efficient use of energy, that economic growth is likely to slow down to a linear per capita growth and will be induced more by growth in services and low-energy branches of industry like investment goods, durables, and consumer goods. They expected a substantial decrease in the growth of energy-intensive industries—and sometimes stagnation—in future decades. This "should result in projections of demand being considerable lower than currently available estimates" (Bossel and Denton 1977; Möller and Ströbele 1978; Neu 1978).

In the 1960s and early 1970s, there were only very few research institutes specialised in the efficient use of energy in OECD countries. Taking Germany as an example, there was only one research institute at the Technical University of Karlsruhe (Mueller 1957) which examined energy efficiency. In addition, some branch-specific institutes covered the topic of energy efficiency as a "by-product" while conducting research to improve branch-specific production processes (e.g., VDEH (Verein Deutscher Eisenhüttenleute, the Association of German Steel Manufacturers), the Brick Research or Textile Research Institute, etc.). Berg (1976) characterised the situation by his introductory first sentence in the first volume of the Journal "Annual Review of Energy": "In the winter of 1973–1974 energy conservation became a popular subject of discussion, so much so that the subject was often, and not entirely inaccurately, referred to as energy conversation".

The first studies on the potentials of energy efficiency in energy-intensive industries were published in the late 1970s (Jochem et al. 1978; Berndt 1978). The methodology applied included interviews with energy managers, process engineers, and applied researchers as well as analytical statistics. The results of these efficiency studies were used to simulate and project industrial energy demand under higher energy prices in OECD countries. The first macroeconomic studies on a reduced oil supply with the associated impacts on growth, employment, and emissions appeared in 1976 with models by Bossel et al. (1976a). Scenario methodology in energy systems analysis was applied systematically for the first time. This approach was developed by the Batelle Institute in 1976, but has its origins in 1969 and was originally designed to consider military strategies (see Cuhls et al. 2024 in this anthology).

These first approaches of energy system modelling as a basis for energy demand projections were—compared to today—still rudimentary, among other things because national energy balances were usually very rough; energy engineering-based models had not yet become available. To start with, systems dynamics methods are taken up by natural science/engineering groups (Mesarovic, Pestel, Kortzfleisch in Germany), and other computer-based modelling is initiated by many other research groups (Bossel et al. 1976b). However, these approaches were abandoned in the 1980s, also because the parameters required for science-based modelling could not be estimated with sufficient accuracy due to the lack of empirical data.

3 The 1980s: Starting Energy Efficiency Policy—Supported by a First Generation of Technically Based Energy Models

3.1 First Energy Efficiency Policies

The second oil crisis of 1979 induced first substantial policy measures on the efficient use of energy, in particular regulations and standards for new buildings (Schipper et al. 1979; Gruber et al. 1982). In the mid-1980s, low-energy buildings were already a legally required energy standard for new buildings in Sweden and Denmark. Even at that time, it was considered to further develop the principles of low-energy housing, such as first-rate insulation, prevention of thermal bridges, airtightness, insulated glazing, and controlled ventilation. Based on these considerations, the "Passive House" was launched by Bo Adamson and W. Feist in Sweden in 1988 (Levine and Adamson 1988).

The complex interaction of the factors driving energy demand addressed by Darmstadter et al. (1977) led to a method explaining how various drivers ex post contribute to the observed changes in energy demand and intensities by energy conservation indicators—for the EU (Morovic and Schön 1987, 1989). The individual influencing factors such as inter-industrial and intra-industrial structural change, short-term structural changes in industry by the business cycle (Garnreiter et al. 1986), structural change in passenger and freight transport (Schipper et al. 1997), the influence of the heating and summer period as well as the more efficient use of energy were analysed using time series and provided indications for scenarios and projections of energy demand in the future.

These studies were flanked by detailed technological examinations of energy-intensive industries at the sectoral level and process technologies, e.g. specific analyses for the basic chemical industry regarding energy efficiency and raw material substitution as well as demand reduction. Another example is the non-metallic minerals industry with its very energy-intensive production processes and energy efficiency potentials offered by innovative technologies for cement, brick, lime, and ready-mixed concrete as a service. The textile industry has almost completely been outsourced and only finishing partly still takes place in Germany and in most OECD countries. This is an example of intra-industrial structural change in the 1970s and early 1980s.

Based on these insights, energy efficiency policy measures were increasingly developed and proposed to national governments or the European Commission by administration, researchers, and trade associations. These include, e.g. regulatory policies for energy efficiency standards for new buildings, speed limits in road transportation that are implemented, eco-design guidelines for mass-produced products, energy labels for domestic appliances, EU energy efficiency directives, energy management systems such as ISO 50001, EMAS, and lists of energy efficiency consultants. Financial incentives for new energy-efficient solutions and consultation were developed for industry, as well as accelerated depreciation, which also applied to private consumers and based on scientific findings. These include energy taxes with exemptions for energy-intensive or internationally competitive industries to help them retain their competitiveness.

Rather late on January 1, 2005, the European Union Emissions Trading System (EU ETS) was implemented in order to limit the emissions of greenhouse gases of energy-intensive industries and electricity generation. And even three years later, the first scientific journal devoted to efficient use of energy, Energy Efficiency, started its first volume.

3.2 First Generation of Technically Based Energy Models

Early bottom-up models emerged in the 1980s and quickly became more and more complex. Based on how the models consider technology choice and adoption, they can be grouped into accounting models, simulation and optimisation models (Fleiter et al. 2011). Accounting models are simply based on exogenous assumptions of, e.g. energy efficiency progress or technological change, to calculate resulting energy demands and CO_2 emissions. They aim to answer "What if…?" questions. Examples include early models like MEDEE or MED-Pro (Chateau and Lapillonne 1978; Chateau and Lapillonne 1990; Lapillonne and Chateau 1981), PRIMES (Capros 1995), IKARUS in Germany (Hake et al. 1994), or the leap model framework. In the 1980s and 1990s, this led to optimisation models which were linked to input–output models and macroeconomic models, usually transferring the respective results and necessary data manually, with iterative computer runs between the different types of model (see also Figs. 1 and 2).

The optimisation models at the beginning of the 1980s, initially designed and used for optimal structures of electricity generation and other energy supply, were increasingly extended to include additional parts modelling the final energy demand of buildings, electrical appliances, road, rail, ship, and air transport, as well as

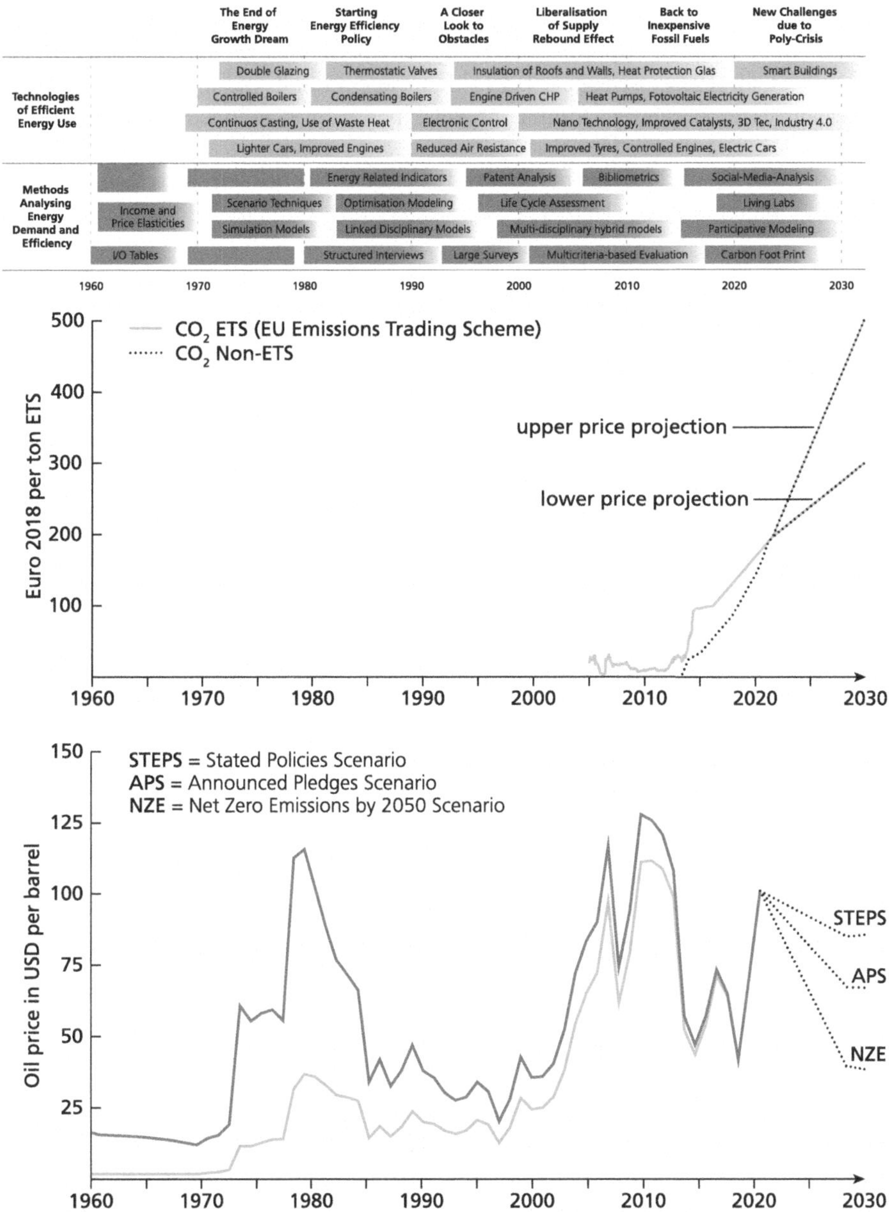

Fig. 1 Development of technologies of efficient energy use and of methods projecting energy demand in final energy sectors, 1960 to 2030. Projection: ISI's own estimate. Top: Development of the technologies for efficient energy use and of methods for projecting energy demand in final energy sectors, from 1960 to 2030; Middle: Development of historical annual average prices of crude oil from 1861 to 2021, source S&P Global Platts. Lower light grey line: historical nominal price; upper dark grey line: real price adjusted for purchasing power in 2020; Bottom: Development of the European CO2 emission allowance price (EU ETS) and the CO2 allowance price for residential buildings and transport in Germany, which is not yet subject to the EU ETS, as well as their lower and upper price projections. Source: Past: German Emissions Trading Authority DEHSt

energy-intensive sectors and production processes. Examples include the development of simulation and optimisation programmes such as MARKAL in the USA (Abilock et al. 1979; Sweeney 1981) that were initially adopted by the major European research centres. Similar model types were developed in the EU such as MEDEE.

Over time, these different models and energy fields became increasingly electronically interconnected (Herbst et al. 2012, see also Fig. 1).

4 The 1990s: Moderate Energy Efficiency Improvements—Obstacles, but the New Driver: Climate Protection

During this decade, globally important events occurred which influenced energy perspectives and related research: the re-organisation of the countries that had belonged to the former Eastern bloc, the Conference on Environment and Development in Rio de Janeiro in 1992, and a new legislation on the liberalisation of energy markets in many OECD countries. The Framework Convention on Climate Change (FCCC) was concluded and signed at the Rio Conference on June 4, 1992 (Adede 1995). The Kyoto Protocol to the United Nations Framework Convention on Climate Change (UNFCCC) of 1997 is the most important allegiance for the future of efficient energy solutions. It contains legally binding commitments of industrialised and emerging countries for reducing anthropogenic greenhouse gas emissions. This Protocol has led to many in-depth studies including efficient energy use and energy conservation since the 1990s (IPCC 2001a).

4.1 Examining the Details of Efficient Energy Use, Options Reducing Energy Demand, and Their Social Benefits

In the 1990s, the price of oil was cut in half and there was a corresponding reduction of the interest in energy efficiency and renewable energy sources. Some energy researchers slowly became aware that the market prices for fossil and nuclear energy do not cover the external costs of energy consumption, such as soil acidification ("forest dieback"), lung diseases caused by particulates and ground-level ozone, etc. (Hohmeyer 1988). This implies that energy prices cannot be left purely up to the market, but require government intervention such as taxes on the consumption of fossil fuels or subsidies for energy-saving measures and renewable energy sources. However, there was great opposition to this idea, from circles of conservative scientists and politics (Rennings 2000).

Simultaneously, the calls to protect the climate grew louder with the 1988 Conference on the Changing Atmosphere in Toronto, the Enquete Commission's recommendations (German Bundestag 1991), or the 1992 Earth Summit in Rio (sustainability). These events pushed energy topics higher up the agenda in science and politics. In practice, however, and due to low energy prices there was only minor progress in energy efficiency up to the end of the 1990s. Therefore, the 2nd and 3rd report of the Intergovernmental Panel on Climate Change (IPCC 2001) contain warnings about this trend, in particular about stranded investments and lock-in situations in OECD countries due to the adherence to old fossil-based energy sources.

Nevertheless, some practical progress was made. Wind power and (10 years later) photovoltaic systems spread rapidly due to the politically created favourable framework conditions in some countries. Houses with very low heating energy demand (so-called passive houses), thermal solar panels, small and more energy-efficient passenger cars, boilers with condensing technology and thermal solar support also found their way onto the market (see Panny et al. 2024 in this anthology).

In the 1990s, policy makers in the field of energy efficiency also wanted to know how effective individual policy measures were in order to learn from them for future measures. The main results for industry were (Diekmann et al. 1999): It makes sense to distinguish between electricity and the other final energy carriers. Real net production value is considered to be the suitable activity in energy intensity studies. Both inter-industrial and intra-industrial structural change (shifts in product structure, product-based services) significantly influence the development of energy intensity. Structural change caused by business cycles is also taken into account as an explanatory variable in the case varying energy intensities in industry (Morovic and Schön 1989). In this decade, it was hardly possible to specify or project the influence of (often new) energy efficiency policies, because results of measure-specific empirical evaluations were missing.

Although energy efficiency and energy intensity indicators were clearly defined by the end of the 1990s in terms of methodology and data and were recognised as very useful for understanding past energy consumption, today—20 years later—they are still hardly used and often only in a very undifferentiated form, despite the energy policy mantra of "Efficiency First".

Only sluggish progress was made in implementing energy efficiency potentials at two levels of energy use, i.e.

1. Cross-Cutting Technologies—improving their efficiency in energy and exergy terms, while converting final energy into useful energy
2. Reducing the useful energy demand of production technologies in industry and crafts by improving and substituting processes, or by reducing final energy demand in industry by intensive use of waste heat.

Final energy demand in industry could also be diminished by reducing demand of basic products due to increased resource efficiency—today this is labelled "Circular Economy" (Jochem 1991; Angerer 1995; Radgen and Tönsing 1996):

3. Increased recycling and improved material efficiency of energy-intensive materials.
4. Substituting materials with less energy-intensive ones.
5. Intensifying the use of durables by sharing and leasing.

Although numerous energy efficiency improvements became highly profitable following the second oil price increase in 1979, few efficiency improvements were observed in the early 1980s in most OECD countries. High potentials of the *"fifth energy resource"* were overlooked by many companies, administrations, and private households or judged to be "purely theoretical" or "unfeasible". The heterogeneity and diversity of energy consumers, the variety of energy-efficient solutions and of the related manufacturers of energy-efficient equipment contributed a low perception of the potentials offered by energy efficiency in the 1980s. Because of this variety and complexity, energy efficiency—in many cases quite profitable—had (and even still today at much higher energy prices) very little appeal for either the media or politicians (Jochem 1991). This sluggish progress in more efficient energy use despite high energy prices and despite new efficient solutions by research and development led to substantial socio-economic and psychological research on decisions of efficiency investment and behaviour of various target groups in the 1990s.

4.2 Obstacles and Market Imperfections

Obstacles and market imperfections of energy efficiency in end-use sectors have been observed and reported for more than 30 years. Although limited, the empirical quantitative research on these barriers highlighted the large diversity of individual investors (e.g., thousands of firms (capital-based, family-owned, small or large), hundreds of thousands of landlords or homeowners, and millions of consumers in a single country).

In theory, given all the benefits of energy efficiency at the business and macroeconomic levels (Hohmeyer 1991), a perfect market would optimally allocate the rewards from these energy-efficient solutions. In practice, however, researchers and consulting engineers in the 1980s and 1990s observed many obstacles and market imperfections that prevent profitable energy-efficient solutions from being fully realised. Energy policy researchers began publishing these observations in the early 1990s (Jochem and Gruber 1990; Hirst 1991; Jhirad and Mintzer 1992; Weber 1997). This led to a growing body of literature on the so-called barriers to energy efficiency. Barriers are described as "a mechanism that inhibits a decision or behaviour that appears to be both energy efficient and economically efficient" (Sorrell et al. 2004), a phenomenon that has also been described as the energy efficiency gap.

In fact, these findings in the area of energy efficiency are simply an example of market and system failure. This fact, however, has to be differentiated by several aspects, reasons, and target groups in order to set up an adequate policy design.

Although, in principle, the types of obstacles and market imperfections are universal, their importance differs among sectors, institutions, and world regions, depending on many factors including technical education and training, entrepreneurial and household traditions, the availability of capital, and existing legislation (see Fig. 2).

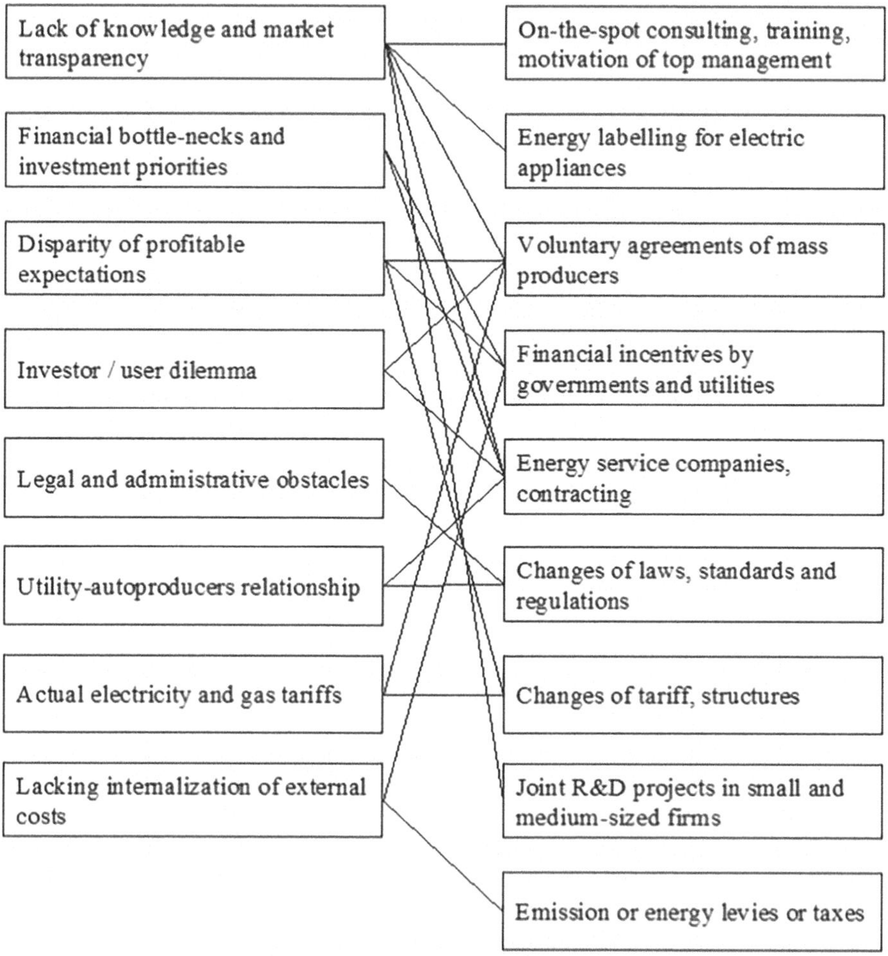

Fig. 2 Obstacles and market imperfections of energy efficiency and related policies—a scheme for policy options and integrated efficiency policies. Source: Jochem et al. 2000b

- Market imperfections include many forms of *subsidies* and traditional legislation and rules, but also the traditions and motivations of behaviour in private households, and of decision making in companies and administrations (Sanstad and Howarth 1994).
- The *"invisibility" of energy efficiency measures* (in contrast to photovoltaic or solar thermal collectors) and the problems with demonstrating and quantifying their impacts are also important factors for private households, companies, and car investments due to social prestige aspects (Sanstad and Howarth 1994).
- Psychological reasons comprise another problem affecting energy efficiency measures. These include *lack of attention, knowledge, know-how and technical skills, and unspecified transaction costs.* Improved energy efficiency is brought

about by new technologies or just incremental changes to a known product, process or vehicle, and by changed forms of organisation. This implies that investors and energy users are able to get to know and understand the benefits of technical efficiency improvements as well as to evaluate the possible risks against perceived benefits. This also implies that investors or users have to be prepared to realise improvements and to give themselves time to absorb new information and to evaluate the innovations (Levine et al. 1995; Sioshansi 1991; DeCanio 1998). Private households and car drivers, small and medium-sized companies, small public administrations or banks do not have enough knowledge about the possibilities and risks of energy savings or sufficient technical skills to implement them. Managers, preoccupied with daily routines and core business areas, are able to only engage in the most important and immediate tasks (Velthuijsen 1995; Ramesohl 1999). Energy efficiency, with its minor role in running a business or its potential to reduce only a small share of the energy costs of total production or household costs, was placed on the back burner.

- *Energy consumers may also face a lack of access to capital, or they may follow historically or socially formed investment patterns*. Even if they acquire the knowledge they need, they often face difficulties in raising funds for energy efficiency investments. Their own capital may be limited and additional credit may be considered as too expensive. Especially if interest rates are high, small firms and private households prefer to accept higher current costs and the risk of rising energy prices instead of a later energy credit.
- *Relying on investment risk decisions and neglecting the profitability of energy-efficient investments* was (and still is today) a major obstacle. Energy consumers demand payback periods of between 1 and 4 years, which are equivalent to an internal rate-of-return of about 25% to 50% (DeCanio 1998; Gruber and Brand 1991; Schröter et al. 2009). This rate-of-return expectation rules out highly profitable efficiency investments and favours investments in energy supply, resulting in an inter-sectoral disparity of profitability expectations of at least 10% to 30% distortion of energy-saving investments (Jochem and Gruber 1990).
- *Legal and administrative obstacles* are observed in almost all end-use sectors. These are mostly country-specific and often date back to before 1973, when there were low and in real terms declining energy prices, and there was no awareness of the threat of global warming.
- The *investor/user dilemma* describes the fact that, for rented dwellings or leased buildings, there are few incentives for tenants to invest in property they do not own. In the same way, there are also few incentives for landlords, builders, or owners because of the uncertainty of being able to recover the investment through a higher rent (Fisher and Rothkopf 1989).

For every obstacle and market imperfection discussed, there are *interrelated measures of energy efficiency policy* that could remove or reduce them, as illustrated by a few examples in Fig. 2. Yet, the choice of which policies to pursue has to be made with care as their effectiveness depends on many regional, cultural, political, and societal factors.

At the beginning of the 1990s, pilot tests and field test demonstrations became an additional component of research projects as demand-side management always includes the users of technologies. In addition to technical solutions, the role of economic incentives and prices became more important (Brand et al. 1988). Researchers supported and accompanied many of these first field tests, e.g. (Jochem and Gruber 1990; Hennicke et al. 1998). Economic efficiency was one of the main drivers of demand-side management approaches as was the research on demand elasticity and incentives for efficient energy use. Key outcomes identified and assessed the impacts of economic incentives and possible policy measures to improve the energy system (Zweifel et al. 1997).

The continued analyses of obstacles in industry led to further proposals for policy measures as more and more obstacles were identified that can only be overcome with bundles of policies. From the early 1990s, these were documented in the reports of the Enquete Commission for "The Protection of the Earth's Atmosphere" (German Bundestag 1991, pp. 378–395), and also in reports at EU level and by the IPCC Group III Report (IPCC 2001 and related publications such as Jochem et al. 2000a).

5 The Noughties—Liberalised Grid-Based Energy Markets and the Takeoff of Electricity Producing Renewables: New Boundary Conditions for Energy Demand and Challenges for Energy Modelling

One important obstacle for efficient energy use, the monopolistic structure of grid-based energy industries, was widely discussed in the 1990s (Walz 1994) and gradually diminished by new legislation in all industrialised countries between the mid-1990s and the beginning of the noughties (Finon and Midttun 2004). A new business field, energy services, was widely discussed, and it was particularly to be offered by gas and electricity distribution companies. This change, however, from maximising energy sales to optimising sales *and* savings, was perceived as a mayor cultural change in the traditional energy supply companies. Financial incentives by governments, strict control by the antitrust authority, and stiff competition by energy service companies from investment goods industries and consulting companies gradually forced functioning liberalised energy markets in the noughties.

Early in the noughties, only a few researchers estimated the liberalised energy markets as an absolute precondition for very high market shares of renewables in the future. It is a common understanding that "more than 30 per cent of fluctuating electricity from wind and photovoltaics will not be substantially surmounted" (Nitsch 2000).

More regulation, not less, was temporarily necessary, if effective competition was to be established in grid-based energy industries. Traditional optimisation models simulating the decisions of a monopolistic energy market had to be replaced at the branch level. The liberalised markets placed new requirements on computable models: they were to provide realistic descriptions of technologies (demand, production, transport, and distribution), but also of markets and institutions. Industrial economics and computation of economic equilibrium were to help achieve this dual requirement (Smeers 1997).

The scarcity of conventional crude oil was highlighted by the buzzword peak oil resulting from an increase in the oil price to more than 100 US$ per barrel (nominal) in the period 2008 to 2014. This caused a renaissance in politics, society, and research of energy efficiency and renewable energies. Their impacts on energy demand, however, were smaller than expected due to rebound, income, and price effects.

5.1 New Policy Instruments Implementing Energy Efficiency

Expectations of further increases in energy prices (including increasing prices for emissions of energy-related CO_2) and rising greenhouse gas emissions in the mid-noughties steer research away from simply analysing the obstacles to efficient energy use and more towards exploring which instruments can be used to overcome them—if possible several of them simultaneously.

A successful example of a new instrument is the *"Learning Energy Efficiency Network"*. This had its origins in a group of Zürich entrepreneurs in 1985 (Bürki 1999) and was further developed in the 1990s in the Swiss industry. Since 2002, the format has been adopted to the situation in Germany (Gruber and Jochem 2007). In this scheme, between 10 and 15 energy managers of companies in a region come together and agree on a joint energy efficiency target and CO_2 mitigation for the network, which should be achieved within about 4 years. Prior to this, specially trained consultant engineers have analysed the energy efficiency potentials in the participating companies, and each company has to set itself a respective target. At around four meetings per year, the measures already implemented by a participating company are inspected and the experience gained is shared. In addition, external experts are invited to talk about new and interesting energy efficiency technologies and know-how is exchanged among the participants.

This approach was so successful that more than 20 associations of German industry had pledged in a voluntary agreement with the German Government in 2014 to establish and operate 350 new energy efficiency networks between 2015 and 2020 (Dütschke et al. 2016). This approach was also modified for small enterprises and local authorities and spread to Austria, Sweden (Palm and Backman 2020), France, China and, in the late 2010s, to countries in South and Central America, Asia and Africa (Durand and Damian 2019). Principally, participating in those networks substantially speeds up the implementation of energy-efficient solutions. Energy efficiency and climate protection networks can be understood as "group energy management systems". Scope 3 emissions from upstream and downstream processes are an integral part of climate protection networks since the early 2020s (Eberle et al. 2022).

Competitive bidding for energy-efficient investments was also introduced in several countries as an incentive for industry and commerce. Energy service companies, consulting engineers, or manufacturers can compete with their efficiency investments that cannot (yet) be standardised. This instrument strengthens the competition among energy service companies and realises branch-specific efficiency potentials which are not covered by standardised incentive schemes (Pehnt and Brischke 2013).

The growth of renewable energies and the liberalisation of the grid-based energy markets brought substantial changes to the electricity and gas supply sector. The emergence of new actors and competition led to a first transformation of energy companies with the unbundling of power generation, gas production, and grid operation. High price fluctuations as well as capacity shortages, on the one hand, and missing grid infrastructure, on the other, triggered a strong push for demand-side measures. The power crisis in California in 2000 acted as a strong push for research (Faruqui et al. 2001). Major energy research questions concerned themselves with how to reduce costs through a better utilisation of assets and infrastructure, for which the demand side plays with additional flexible demand a more important role. Increasing the number of flexible demand-side participants and reducing their final energy demand induced research activities that focused on the participation of numerous and new stakeholders (Braithwait and Eakin 2002; Department of Energy 2006; FERC 2006).

5.2 *Rebound, Income, and Price Effects and Technical Forecast*

Despite additional energy efficiency policies in the noughties and the high oil prices between 2008 and 2014, consumption of oil products did not fall as expected. For example:

- Car engines became more efficient per unit of power, but cars became bigger and heavier with more powerful engines.
- Airplanes became more energy efficient, but the number of flights and passengers increased as did the distances flown.
- Living space per capita also increased and often offset thermal insulation measures.
- Factories using energy more efficiently became more profitable and competitive encouraging further investment and higher levels of output.

This empirically observable effect, known as the direct rebound effect (which also includes income effects and changes in preferences of private users), was intensively discussed in the noughties (Sorrell 2007; Gillingham et al. 2016). Even if demand of energy services remains unchanged, energy savings across the economy may be less than simple calculations suggest. The question remains what the saved energy budget will be spent on—whether marginal consumption or additionally possible investments induce more or less energy demand. In addition, reductions in energy demand will translate into lower energy prices which encourage increased energy consumption (energy price effects or indirect rebound effects; Sorrell 2007). Not only is this effect important for the design of policy instruments, it is also often overlooked in energy economic models and forecasts (see below).

In addition to these behavioural changes, new technological developments are also difficult to estimate. In order to provide a better scientific basis to project future developments or events, various methods of technological foresight are now combined with one another: For example, the speed at which the technological maturity of individual technologies is reached from the idea to market maturity can be mapped with the help of bibliometric and patent analyses (Jochem et al. 2009). Joint international publications on topics also reveal in which regions developments are taking place and how the players—researchers, manufacturers, and first applicants—network with each other in the early stages of the technology cycle. The number of patents in the individual world regions may indicate the time of market entry and speed of diffusion (Bradke et al. 2009).

It is possible to estimate future cost developments by analysing cumulated production and the associated production costs or market prices, and thus the budgets required and the market opportunities compared to competing technologies (Jakob and Madlener 2004). These empirical findings can be transferred to other technologies by analogy and thus enable well-founded recommendations for the need for initial financial support of new efficient technologies. In cases of mass production, they are projected to "run down their cost curve" for estimating their market diffusion in energy system models.

5.3 Demand-Side Management and Second-Generation Models

Fluctuating electricity production by renewables has started to turn around the role of supply and demand in several countries since the noughties: in the traditional electricity system, production followed the pattern of demand; in the future, however, electricity demand has to realign with fluctuating electricity production by wind, photovoltaic, and hydropower. With the rise of renewable energies and the intensified discussion about climate change, demand-side management and the related research focus on questions of system integration and the security of supply.

Especially in countries with higher shares of renewables and more liberalised markets like in Scandinavia and especially in Denmark, the demand side's contribution to integrating renewables developed into a major topic. An intense discussion between grid operators, policy makers, and research institutions on the potential maximum shares of renewable energies in the power system was the result (see also the discussion on renewable energies in Panny et al. (2024) of this volume), as well as how to balance the power system efficiently (Moller Andersen et al. 2006; Nordel 2004; Nordel Demand Response Group 2006).

Load shifting in very electricity-intensive production processes such as manufacturing electrical steel, aluminium, and chlorine has been around for a long time, but increasing digitisation makes it economically feasible to include smaller industrial and commercial electricity consumers in load management. In several studies and empirical research activities, researchers identified and supported efficient energy use and demand-side management, not only to improve competitiveness of the power system, but also to support the integration into the electricity system and the security of supply (Klobasa 2009).

In line with these considerations, the electricity demand models became much more dynamic in terms of time to match the increasingly fluctuating electricity supply and load shifting options, particularly in countries with substantial seasonal changes of sunshine, wind, or hydropower. In addition, technically based models became more refined and considered future machines and plants, both as optimisation and in simulation models (Quiggin et al. 2012; Sensfuß 2007).

Simulation models went further and endogenised technology diffusion by simulating the investment decisions of actors (e.g., industrial companies or building owners). Although these models are a rather heterogeneous group, most of them represent the age of the technology stock and track individual age classes of, e.g. buildings, cars, or industrial facilities (Fleiter et al. 2011), which yields a more realistic modelling of system inertia (obstacles and market imperfections) and the speed of technological change. Simulation models often use a discrete choice framework that simulates technology choice as a competition among alternative investment options (Elsland et al. 2013; Fleiter et al. 2018; Palzer and Henning 2014).

For example, building sector models typically include various choices of heating systems and consumers decide which one to install based on the total costs of ownership (Stadler et al. 2007). Simulation models are, however, also more experimental and can include elements of non-rational investment behaviour. Daniels and Van Dril (2007), for instance, consider psychological energy price effects and bounded rationality. The CIMS model considers a time preference, heterogeneity of the market and a factor that integrates all other elements of non-rational investment choices (Horne et al. 2005; Murphy et al. 2007; Rivers and Jaccard 2006). In the NEMS model, investments in energy efficiency technologies are determined by payback time thresholds reflecting empirically observed simplified decision rules (Worrell and Price 2001; Energy Information Administration 2009).

Despite their many advantages and a high level of technology detail, bottom-up models also have shortcomings. These include their dependence on detailed technology data and the lacking empirical foundation for data and behaviour assumptions as well as technological optimism. Efforts have been made to estimate decision-parameters empirically (Rivers and Jaccard 2006; Rehfeldt et al. 2019; Beugin and Jaccard 2012). In addition, bottom-up models tend to look at the technological system without considering the interactions with and feedback from the economic system, which motivates researchers to develop hybrid models that draw on both engineering and economics. An important example of such a hybrid model is the Canadian CIMS model (Rivers and Jaccard 2005; Murphy et al. 2007). Other modelling teams couple different types of models in applied studies. Overall, modelling teams increasingly traverse the boundaries between individual disciplines and models incorporate the advantages of different research streams (Herbst et al. 2012; Pfenninger et al. 2018). Applications of modelling tools for policy consultancy often result in combining individual complementary models by, e.g., using bottom-up models to assess technological change in detail and top-down models to provide the overall economic frame (see Fig. 1).

6 The 2010s: Tensions—Slow Progress in Efficient Energy Use, Still Increasing Greenhouse Gas Emissions, Electrification of the Transport Sector, Sector Coupling, and Further Progress in Energy Modelling

The December 2015 Paris Climate Agreement refers to the alarming scientific evidence on global warming. The Paris outcome legitimises more climate action around the world. The question is whether this will happen quickly enough and on a sufficient scale. Certainly, it will not occur without far-reaching government intervention in energy markets and resource efficiency in the next few years, particularly in the largest polluting countries (Clémençon 2016). Energy research took up many of the challenges of the Paris Agreement prior to the 2010's (Jochem 2004); however, the ability to speed up the transformation by politicians and citizens could scarcely be observed by the authors until the middle of 2023, when they completed this publication.

Fracking, the new method to extract additional crude oil and natural gas, was widely accepted in the USA in the 2010s, turning the USA from a net importer into a net exporter of crude oil. The surplus of available oil resulted in the price of oil dropping to around 50 US$ per barrel between 2014 and mid 2020 (see Fig. 1). On the other hand, the findings of climate research still determine energy research and energy policy to a greater extent, but with a moderate impact on the transition needed.

More recent research since around 2010 aims at embedding the barriers of efficient energy use into broader frameworks including:

(a) The conceptualisation of the decision making on energy efficiency investments as a process,
(b) the consideration of psychological factors as well as social dynamics,
(c) broader analyses of the impact of energy-efficient solutions in terms of their co-benefits, but also downsides like rebound, income, and price effects.
(d) And finally, a wider systemic look at the energy-efficient performance of the energy demand and supply side (sector coupling).

The process perspective on decision making is influenced by psychology and aims to show the different stages through which a decision in favour of an energy efficiency investment needs to pass before it is implemented. This includes identifying the need for the investment, compiling information, the actual planning and finally the decision and its implementation. However, this process can end at each of these different phases and usually other actors besides the actual investor also play a role. For example, Globisch et al. (2018) show that the expected reactions from co-workers are important for fleet managers before they invest in electric vehicles. Arning et al. (2020) also point to the crucial role of installers and crafts men in renovation decisions.

The literature on co-benefits highlights the additional effects of energy-efficient solutions like increased thermal comfort in buildings, less noise, improved illumination of production areas, constant product quality, less wastes, or increased real estate value (Reuter et al. 2020).

6.1 Sector Coupling and Integrated Energy

The increasingly ambitious climate protection targets set since the Paris Agreement at national levels require reduced greenhouse gas emissions by reducing final energy demand and by substituting fossil fuels with energy from renewable sources. On a large scale, this can mainly be achieved through electricity from wind, sun, and hydropower, which are subject to intermittency. Questions emerge on how to manage these non-controllable energy sources, how to handle excess electricity generation, and use it in an efficient way in terms of economic, ecologic, and social welfare aspects. To efficiently integrate these variable primary energies, the traditional coupling of the power sector to the residential, transport, industry, and commercial sector has to be adopted to several changing boundary conditions (Schaber et al. (2013), Schaber (2013), Richts et al. (2015)):

- Increasing electricity demand due to substitution of fossil fuels in the transport sector, in industry and commerce (including large heat pumps for district heat systems);
- Short-term electricity storage and longer term storage by thermal heat (including the short-term function of buildings) or hydrogen, ammonia, or methanol;
- Integrating millions of very small photovoltaic and wind generators operated by private households (balcony collectors), small companies, and communes.

This adopted electricity system is called sector coupling (SC). Due to the strong expansion of fluctuating renewables in Central and Northern Europe as well as California since 2010, the discussion about SC started in these two regions. As the shift continues towards the energy transition, in 2017, several German ministries and international energy agencies developed detailed guidelines and information on SC (see BMWi (2016), BMUB (2016), BDEW (2017), IRENA et al. (2018)). In 2020, the European Commission presented a comprehensive EU Strategy for energy system integration (European Commission (2020). A year later, the IEA published a study highlighting the role of SC in ensuring energy security (IEA (2021).

Although the terms "sector coupling" (SC) and "integrated energy" are frequently used in the current energy policy debate, they are often not used clearly or uniformly (Scorza et al. 2018). Several different definitions can be found in Ramsebner et al. (2021). Following one of the first definitions by Wietschel et al. 2018, SC is seen as the "substitution of fossil fuels in conventional technologies with alternative primary energies (e.g., renewables including wind, solar, hydro, biomass, geothermal) in new applications or technologies". This can be done either by directly using electricity, such as.

- in Power-to-Heat PtH, e.g. heat pumps, electro-thermal industrial processes,
- in Power-to-Move PtM, e.g. vehicles driven by electrical motors or by converting electricity into synthetic fuels,
- Power-to-Gas PtG (e.g., hydrogen) as substitution of conventional fossil gases, and
- Power-to-Liquid PtL (e.g., green ammonia, methanol, or e-fuels) as substitution of fossil fuels.

These electricity-based final energies are subsumed as Power-to-X (PtX) energies. In addition, the focus here is on the use of new or alternative technologies and less on classical power applications such as electrical motors, night storage heating, or electric trains and trams. This view of SC focuses on techno-economic issues. The broadly defined aspects of SC also encompass new standards, new business segments, IT issues (including cyber security), and legal as well as regulatory aspects.

The EU Strategy (EU 2020) concludes "that the transition to a more integrated energy system is of crucial importance for Europe. First, for recovery. It proposes a path forward that is cost-effective, promotes well-targeted investments in infrastructure, avoids stranded assets and leads to lower bills for businesses and customers. In short, it is key to accelerating the EU's emergence from the actual economic crisis and for mobilising necessary EU funding as well as private investments. Second, for climate neutrality. Energy system integration is essential to reach increased 2030 climate targets and climate neutrality by 2050. It exploits energy efficiency potentials, enables a larger integration of renewables, the deployment of new, decarbonised fuels, and a more circular approach to energy production and transmission". Whether energy efficiency potentials will be sufficiently exploited in this supply-oriented concept of SC will be questioned in the outlook of this publication.

6.2 Electrification of Road Transport

As already mentioned, a general long-term trend in energy systems with a large impact on society, economy, and policymaking is the gradual change towards direct use of electricity in many applications where fossil fuels are (or were) used. As electricity has been widely available for decades in OECD countries and electric motors are clearly more efficient than combustion engines, electric vehicles in road transport have been researched since the second oil crisis in 1979. Yet, the first fleet trials and vehicle demonstrations did not lead to mass-market introduction. The situation changed around 2010 due to several factors: improvements in battery technologies (lithium-ion batteries offer higher energy density and thus longer ranges), seriousness of climate change with actual policies dramatically reducing tail-pipe emissions of newly sold vehicles beyond the reach of combustion engine vehicles (in particular the 95 g CO_2/km target for new vehicles in Europe).

Against this background of a changing CO_2 landscape, plug-in electric vehicles (PEV) have seen strong support and research. A substantial impact on the existing electricity system is expected as the previously uncoupled sectors of electricity and road transport interact with vehicles frequently connected to the power grid (see the preceding chapter on sector coupling). Early on, the scientific debate has focused on the potential integration of intermittent renewables supplying electric cars and trucks (e.g., Dallinger et al. 2011; Dallinger and Wietschel 2012; Mwasilu et al. 2014; Wang 2021). Many simulations of PEV charging behaviour and their interaction with the grid showed that PEV represent an important additional load but offer only limited power storage capacity. Accordingly, demand-side management and smart charging are the most important aspects of PEV (Peters et al. 2012).

The impact of PEV charging on electricity grids and power generation receives considerable attention in the literature. The additional load implies that some distribution grid extensions or more controlled charging will be needed. Interestingly, although the uptake of PEV requires additional investments on the distribution grid level, the specific grid charges are reduced as the additional electricity demand increases overall grid utilisation and thus lowers its specific costs (Kühnbach et al. 2020). This effect is even higher than the increase in generation costs due to integrating flexible generation with high variable costs (Kühnbach et al. 2020).

Analyses of road transportation systems conducted in the last decade conclude that policies and decision making must be based on a thorough understanding of PEV users and the future market uptake of PEV. Many researchers make important contributions not only to the aforementioned aspects of renewable integration and grid impacts, but also to market diffusion scenarios and the characterisation of PEV early adopters. In a series of national and international publications, they improve existing methods to analyse the future market diffusion of new technologies in an empirically grounded manner (Plötz et al. 2014; Gnann et al. 2015) and help to analyse the national transition towards PEV (Plötz et al. 2013). Early adopters are described in terms of socio-demographics but more importantly for SC, they are shown to also be frequent owners of home PV systems and already use fully renewable electricity contracts (in both aspects showing much higher shares than the general German population) (Scherrer et al. 2019; Preuß et al. 2021; Lee et al. 2019). Likewise, research reveals that today's plug-in hybrid electric vehicle (PHEV) users do not charge their vehicles as frequently as expected (Plötz et al. 2021). Interestingly, there is little research on the efficient electricity use regarding battery charging, wheel driving, storage, and recuperation (Synák et al 2021).

As the transition towards electric passenger cars is already underway, the next open research field for SC in road transport concerns heavy-duty vehicles. Initial results show that battery electric trucks can reduce well-to-wheel emissions from heavy-duty vehicles but represent an inflexible load for electric road systems (Plötz et al. 2019) and megawatt charging (Speth et al. 2022). Researchers will continue to advise policymakers and industry as well as civil society with up-to-date research on the future of electric road transport, e.g. in leading the construction of the first megawatt chargers in Europe (cf. https://www.hochleistungsladen-lkw.de/).

6.3 Energy System Modelling

Over decades of improvements in energy demand modelling, the research questions have also changed dramatically. While early bottom-up models looked at energy efficiency potentials, i.e. how much improved energy efficiency can reduce overall energy demand (e.g., Worrell et al. 2000), the focus then shifted to CO_2 abatement and carbon neutrality, driven in particular by the Paris Agreement and a stronger public push for climate protection by many governments since 2016.

As a result, the models are challenged by the need to include deeper structural and technological changes and regional information to achieve greenhouse gas neutrality (Pfenninger et al. 2014). The focus shifted to topics like sector coupling, electrification, new energy carriers like hydrogen and the potential market diffusion of immature novel technologies. In addition, the simulation of policy instruments like CO_2 markets, standards or subsidy schemes gained importance as policymakers demanded more guidelines on how instruments will impact on future demand and CO_2 emissions (see Fig. 1).

Among other things, these developments require higher temporal and spatial resolution. Energy system models moved from considering individual generic type-weeks towards hourly resolution of the entire year to capture the effects of high wind and solar generation on the system (e.g., Sensfuß et al. 2008). As a consequence, topics such as demand response, e.g. from electric vehicles or heat pumps, have become more important (Boßmann and Staffell 2015; Boßmann et al. 2015; Boßmann and Eser 2016).

With research focusing on SC, entire teams of researchers couple specialised models of individual demand sectors with supply side or systems models with the aim of improving the resolution in the representation of decarbonisation pathways (Sensfuß et al. 2021; Del Crespo et al. 2020). Others integrate the overall system into one optimisation modelling approach that targets minimised overall system costs with perfect foresight (Pfenninger et al. 2014; Plazas-Niño 2022). This is done at the expense of losing technology information to make the optimisation problem solvable (e.g., Henning and Palzer 2014).

At the same time, spatial resolution has also increased drastically from country aggregates to, e.g., NUTS3 regions, or even individual points. A main driver for this is linking the modelling of infrastructures like electricity, gas, heat, or hydrogen transport networks. Only with high spatial resolution can such models consider structural changes in energy demand across regions. While early models calculated energy demand as annual aggregates of one region (e.g., a country), contemporary models aim at hourly resolutions of energy demand and can break down demand spatially from country aggregates to individual NUTS3 regions or even local centers of high energy demand. Some detailed sector models even went further and represented individual agents and their interactions within the simulation (see, e.g., Nägeli et al. 2020 and Steinbach 2016).

Independent of the respective research questions, the movement towards open source models has gained huge momentum over the last decade, driven by public authorities and designers of research programmes like Horizon 2020, with their greater priority for open source and transparency. This is a result of dissatisfaction with "black box" models and the associated difficulties in explaining the results and making the causal chain comprehensible. Modelling teams have reacted and many new open source models have emerged (Pfenninger et al. 2014; Hörsch et al. 2018; Brown et al. 2018), although the opening of proprietary models is still an ongoing process.

It is very likely that modelling approaches in the future will diversify even more and specialise in answering specific research and policy questions. Time will show whether open source models are a way to collaboratively build even bigger models with even more detail. Computing power will certainly continue to increase and will drive corresponding developments in energy system modelling towards greater levels of detail with higher temporal and spatial resolution. New methods like machine learning are likely to play a bigger role.

6.4 Digitalisation: Supporting the Energy System Efficiency

Already at the beginning of 2000, several research papers discussed the links between information technologies and the energy system and identified the potential benefits of closer interactions. Concepts and ideas for sustainable consumption emphasise and indicate individual solutions for different stakeholders to improve the sustainability of the energy system (Schleich et al. 2013). Knowledge about demand and generation is key to stabilise grid infrastructures, but communication links for small-scale assets are still developing. Research topics cover these aspects and several new concepts are elaborated and tested. Researchers support the development and add new dimensions focusing on participation and the acceptance of technologies and their use (Tureczek and Nielsen 2017; see also Heyen et al. (2024) of this anthology). In particular, the planned smart meter roll-out raises several questions about data security and privacy that also affect its implementation into the energy system.

With the increased number of small-scale and decentralised generation assets, the research community's interest turns to their controllability and current status information and large-scale research programmes launched. Within the e-energy programme running from 2008 until 2013, the main goal is to optimise especially the power system using ICT technologies (BMWK 2014). The programme led to technical solutions that are still lacking suitable market solutions and regulatory frameworks. Furthermore, the inclusion of key stakeholders on the demand side and the implementation into real-life settings are identified as major gaps for fast diffusion of the solutions. This leads to new research approaches developed under the concept of smart energy showcases SINTEG[1] (Klempp et al. 2020) and real-life laboratories (REALLABORE[2]), where possible solutions are developed and shown to a wider number of stakeholders in more industrial scale conditions. Five key requirements are identified that are needed in future power systems where digitalisation is seen as a key enabler: Increasing the flexibility of energy supply and demand, integrating flexibility into energy markets and grid operation, optimising and securing the control of flexibilities, testing and validating new solutions in an efficient and fast way, and finally increasing participation and acceptance of energy system users. On an international level user participation and engagement as well as market and grid integration are also identified as key areas where digitalisation can play a crucial role (CODES 2022).

With clear challenges ahead like climate change and the advancing digitalisation of the economy, research concepts are moving more in the direction of a mission-oriented approach, which defines societal goals and clear steps for how and when these should be reached. Related to smart meters and digitalisation of the energy sector the mission-oriented approach sets a clear focus on largest benefits of these technologies to reach climate goals while avoiding or minimising negative environmental impacts. Researchers support this approach with state-of-the-art concepts, including how to best use digital technologies to support energy transmission, improve system operation, and include demand-side options (Klobasa et al. 2019; Singh et al. 2021). Relevant research questions are concerned with how to adapt and change current regulatory conditions (Bekk et al. 2021) and improve demand-side participation (Kühnbach et al. 2022) on a low voltage level as well.

7 Summary and Outlook

First doubts on ever increasing energy demand were expressed in the early 1970s (Chapman et al. 1972). Increasing disputes among energy economists and energy systems analysts could be observed in the energy-related journals in the late 1970s and 1980s about the importance of energy efficiency potentials and their profitability (Hatsopoulos et al. 1978). Existing obstacles and market imperfections called for technology- and target group-specific energy efficiency policies. The importance of

---

[1] SINTEG: Schaufenster intelligente Energie, https://www.bmwk.de/Redaktion/EN/Publikationen/Sinteg/executive-summary-in-english-overview-of-the-key-results-of-five-years-of-sinteg.html.

[2] https://www.energieforschung.de/spotlights/reallabore.

structural changes within the economy to a more service-oriented economy, within industry in favour of non-basic product branches, and even within individual branches to more content of services (e.g., maintenance, ready-mixed concrete), was heavily discussed. The dissent among energy economists and engineers in the 1970s and 1980s can also be understood given the expectation of very inexpensive electricity generated from nuclear power.

First recommendations for energy efficiency research programmes were made to governments, including the development of more detailed energy system modelling, based on models of operations research or simulation. However, since the early 1980s—after the second oil price increase in 1979—the topic of energy efficiency has been increasingly accepted as an "energy source" and detailed energy demand projections have received increasing acceptance. Aspects of obstacles, market imperfections, innovation, and related policies have been taken up by the scenario techniques since the late 1980s, assuming different intensities of energy efficiency policy or efforts and successes of research in efficient energy technologies (e.g., passive houses, waste heat use, electronic control and sensors (Craig et al. 2002)).

Increasing analytical details of efficient energy use, effects of structural change, and saturation on final energy demand are challenging impulses for the development of new methods such as multi-disciplinary hybrid models, complex statistical analytics, patent and bibliometric analyses, multi-criteria assessment methods (see Fig. 1). The social cost of energy use (Hohmeyer 1988), the social benefits of efficient energy investments (such as additional employment and additional exports) led to additional energy-related data and new versions of input–output models (Legler and Jochem 1977; Geller et al. 1992). Final energy demand of several EU countries dropped between the 1990s and 2020 by around 10% despite economic growth.

At the turn of the century, liberalisation of grid-based energies reduced one of the market imperfections and offered the opportunity even to energy supply companies to sell energy-efficient solutions as an energy service. This development realised the statement of some authors of the 1970s that energy efficiency should be considered as "energy source" (Lovins 1976).

Increasing market shares of renewable energy in electricity generation started to convert the role of electricity demand as a driver of electricity generation to the opposite: daily and seasonally fluctuating electricity generation from photovoltaic, wind, and waterpower increasingly determines the patterns of electricity demand. This change induces new technical and organisational innovations in areas of higher variability of electricity demand, hourly fluctuating electricity prices, electricity storage, and more intensive coupling in the energy transformation sector, particularly with heat use and storage, and hydrogen applications in the coming decades. These changes become particularly difficult in northern industrialised countries, if they have little potentials of hydropower, biomass, or geothermal energy. Thus, more efficient energy use may get more attention in the near future—supported by reduced demand of basic goods due to the upcoming circular economy (Jochem et al. 2004). Even the post-growth economy may become a game changer in the long run reducing energy demand in the future in industrialised countries.

The research activities on efficient energy use and energy demand during the last 50 years reflect the framing drivers of energy supply, energy prices, economic development, and policy changes. During the last two decades, climate change policies have given more attention to energy efficiency and conservation. This is likely to be even more important in the next few decades including more resource efficiency reducing the demand of energy-intensive basic products. The progress of knowledge, methods, and empirical data during the last five decades is substantial in understanding remaining profitable energy efficiencies, upcoming new efficiency potentials and in projecting future energy demand. However, the authors have to admit that projecting the transformation of energy systems in the next three decades is extremely difficult:

- The speed of transformation necessary in the light of the Paris target of 1.5 °C maximum surface temperature increase may not be accepted by larger parts of the civil societies in many countries and by oil and natural gas producing countries.
- Market entrance and acceptance of new energy-efficient technologies require a substitution of energy use by capital, a severe obstacle for many private households (moderate incomes), small companies, and organisations (e.g., sport clubs, non-profit organisations).
- There also are high uncertainties considering the long-term performance of new energy-efficient technologies such as high temperature heat pumps, nanomembranes as low temperature separation options, nanocatalysts, electricity-based production of basic products and related interaction with the upcoming circular economy.
- Even if technical options and target-focused policies are clarified as being feasible und accepted, the mere lack of engineers and crafts men in most countries will put a question mark on present target-oriented energy demand projections.

The progress in energy efficiency and the transition of the sectors using fossil fuels for heat generation or road transport deserve closer attention regarding two aspects.

- The abundant options of reducing final energy use in thousands of industrial production processes and even in buildings cause the existence of numerous innovation systems (Wesche et al. 2019). This extreme variety of "energy efficiency" innovation systems leads to little lobbying power in public administration and governments, and also in the group of intermediates (i.e., the banking sector, venture capital, or standardisation; Gallagher et al. 2012). This heterogeneity of innovation systems at the energy demand side reflects the opposite of a few innovation systems of energy supply with high lobbying power (i.e., renewables, green hydrogen, green fuels, possibly also Carbon Capture and Use).

This difference in numbers of innovation systems and lobbying power reflects a high risk that the benefits of energy (and resource) efficiency are substantially underestimated. This inobservance leads to unnecessarily large and costly investments in energy supply (generation, transport, and distribution). Countries which realise and politically counterbalance this uneven situation of innovation systems will have lower energy costs in the coming decades. This will contribute to better competitiveness compared to those countries paying energy (and resource) efficiency not more than lip service (Jefferson 2016).

- Recent social science research on clusters and related narratives that deny climate change or delay the transformation concludes that at least 20% of US inhabitants have strict reservations about the meaningfulness of climate protection (Meyer et al. 2023). This percentage is certainly not much different in the EU, Japan, or emerging countries (Dahlstrom and Rosenthal 2018).

As millions of home owners or small businesses have to make their decisions on energy-efficient use of heat, power, or other energy applications, these societal groups will not only hesitate to take timely decisions on efficient energy use, but also influence political decisions at the various levels of public institutions. Democratic countries are not well prepared to convince those societal groups that deny climate change and delay the necessary transformation. Centrally governed countries with little democratic political structures may have a substantial advantage of realising energy-efficient solutions as an important part of the transition deeper and faster compared to democratic countries. So, communication research is needed to develop methods to convince those societal groups to contribute to efficient energy use and even to lifestyles of sufficiency in high income households (Bertoldi 2022).

Regarding the tough climate protection targets of a maximum increase of between 1.5 and 2.0 °C average surface temperature in the middle of this century compared to 1880, more efficient energy use will have to be substantially supported by:

- Reducing the demand of energy-intensive basic products by means of much more efficient use of final products, buildings, machinery, and plants,
- improved recycling, and more services of renting and sharing (of products and vehicles),
- restructuring global value chains along new energy sources,
- negative CO_2 emissions, and
- the post-growth economy (including sufficiency behaviour) may become a game changer in the long run reducing energy demand in the future in industrialised countries (Vita et al. 2019).

Whether these options of efficiency and changed lifestyles in the industrial and emerging countries will lead to a primary and secondary energy supply based on renewables nuclear power that can be built in the next 25 years remains an open question. While the IEA speeds up its warnings for more efficient energy and resource use (IEA 2022b, 2023), China, India, Indonesia, and Turkey are still planning to build additional coal power stations with a total capacity of more than 100 GW in this decade (Monitor 2023). Whether this growth is justified remains an open question given a steady stream of innovations of efficient energy solutions, given saturation (supported by resource efficient policy) as well as structural effects of domestic consumption and international trade. However, experience may point to the 1970s, when high economic growth rates and nuclear energy were the "dream" of the OECD countries.

References

Abilock H, Bergstrom C, Brady J (1979) MARKAL: a multiperiod, linear-programming model for energy systems analysis. Upton, NY (USA)

Adede AO (1995) The treaty system from Stockholm (1972) to Rio de Janeiro (1992). Pace Environ Law Rev 13(1):33–47

Angerer G (1995) Auf dem Weg zu einer ökologischen Stoffwirtschaft, Teil I: Die Rolle des Recyclings. Gaia 4(2):77–84

Arning K, Dütschke E, Globisch J, Zaunbrecher B (2020) The challenge of improving energy efficiency in the building sector: taking an in-depth look at decision-making on investments in energy-efficient refurbishments. In: Energy and behaviour. Elsevier, pp 129–151

Batelle Institut e.V (1976) Szenarien Chemische Technik. With assistance of Jochem E, Legler H, Lange S. ISI Karlsruhe. Edited by Batelle Institut e.V. Frankfurt am Main

BDEW Bundesverband der Energie-und Wasserwirtschaft e.V (2017) Positionspapier: 10 Thesen zur Sektorkopplung. http://www.bdew.de/service/stellungnahmen/0-thesen-sektorkopplung

Bekk A, Held A, George JF (2021) CO_2-basierte Refinanzierung der Erneuerbaren-Förderung – Welche Konsequenzen ergeben sich für Endverbraucher und Sektorkopplungstechnologien? ifo-schnelldienst – Wie fair ist die Energiewende? Verteilungswirkungen in der deutschen Energie- und Klimapolitik. ifo Schnelldienst 74(06):03–33

Berg CA (1976) Potential for energy conservation in industry. Annu Rev Energy 1:519–534

Berndt ER (1978) Aggregate energy, efficiency, and productivity measurement. Annu Rev Energy 3(1978):225–273

Bertoldi P (2022) Policies for energy conservation and sufficiency: review of existing policies and recommendations for new and effective policies in OECD countries. Energy Build 264:112075

Beugin D, Jaccard M (2012) Statistical simulation to estimate uncertain behavioral parameters of hybrid energy-economy models. Environ Model Assess 17(1–2):77–90. http://www.scopus.com/inward/record.url?eid=2-s2.0-84855351082&partnerID=40&md5=bf603a8d15f9de768ea153f06890dbec

BMUB Bundesministeriums für Umwelt, Naturschutz, Bau und Reaktorsicherheit (2016) Climate Action Plan 2050—Principles and goals of the German govern-ment's climate policy

BMWi (Bundesministerium für Wirtschaft) (1973) First amendment of the energy Programme of the fed. Government. (in German) Bonn

BMWi (2016) Grünbuch Energieeffizienz – Diskussionspapier des Bundesministeriums für Wirtschaft und Energie. BMWi, Berlin

BMWK (2014) Abschlussbericht E-Energy—IKT-basiertes Energiesystem der Zukunft. Bundesministerium für Wirtschaft und Klimaschutz (BMWK). https://www.digitale-technologien.de/DT/Navigation/DE/ProgrammeProjekte/AbgeschlosseneProgrammeProjekte/E-Energy/e-energy.html

Bohn Th, Eich P, Jak W, Meliß M, Oesterwind D, Rath-Nagel S, Turowski R, Voss A, Wagner HJ et al (1977) Options of the development of the energy sector of the Fed. Rep. of Germany – an analysis using a dynamic simulation model. (In German) special reports of the nuclear energy res. Center Jülich no 1, vol 1

Bohnen U (1982) Regression analysis of the development of primary energy consumption as a function of economic growth, energy prices, and temperature. (In German), Oldenburg, Munich

Bossel H, Denton R (1977) Energy futures for the Federal Republic of Germany: three scenarios. Energy Policy 5(3):35–50

Bossel G, Brand P, Denton R, Herz H, Jochem E, Legler H (1976a) Development of a methodology for technology assessment applying dynamic simulation – the example: a changed supply of mineral oil. ISI-Report, Karlsruhe

Bossel H, Klaczko S, Müller N (eds) (1976b) Systems theory in the social sciences. Stochastic and control systems – pattern recognition – simulation – fuzzy analysis – behavioural models. Interdisciplinary Systems Research. Springer Basel, Switzerland

Boßmann T, Eser EJ (2016) Model-based assessment of demand-response measures – a comprehensive literature review. Renew Sust Energ Rev 57:1637–1656. https://doi.org/10.1016/j.rser.2015.12.031

Boßmann T, Staffell I (2015) The shape of future electricity demand: exploring load curves in 2050s Germany and Britain. Energy 90:1317–1333

Boßmann T, Elsland R, Klingler A-L, Catenazzi G, Jakob M (2015) Assessing the optimal use of electric heating systems for integrating renewable energy sources. Energy Procedia 83:130–139. https://doi.org/10.1016/j.egypro.2015.12.203

Bradke H, Marscheider-Weidemann F, Som O, Mannsbar W, Cremer C, Dreher C, Edler J, Ebersberger B, Radgen P, Ruhland S, Krebs A, Jochem E (eds) (2009) Improving the efficiency of R&D and the market diffusion of energy technologies. Springer, Dordrecht Heidelberg London New York. ISBN 978-3-7908-2153-6

Braithwait S, Eakin K (2002) The role of demand response in electric power market design. Edison Electric Institute, Washington, DC

Brand M, Gruber E, Jochem E, Mannsbart W, Meyer T, Witt B (1988) Einfluß der Preisgestaltung leitungsgebundener Energieträger auf die rationelle Energieverwendung in Hessen. Gutachten des Fraunhofer-Institut für Systemtechnik und Innovationsforschung (ISI), ISI Projekt-Nr 10(461):2

Brown T, Hörsch J, Schlachtberger D (2018) PyPSA: python for power system analysis. JORS 6(1):4

Bürki T (1999) The energy model switzerland as experience of swiss companies. (In German) Fed. Office of Energy: Energy 2000. Unit Industry. Benglen, Switzerland

Capros P (1995) Integrated economy/energy/environment models. Proceed. Symp on Electricity, health and the environment. Vienna. 16–19 October 1995. IAEA, Vienna. 343–366

Chapman D, Mount T, Tyrrell T (1972) Electricity demand growth and the energy crisis: an analysis of electricity demand growth projections suggests overestimates in the long run. Science 178(4062):703–708

Chateau B, Lapillonne B (1978) Long-term energy demand forecasting: a new approach. Energy Policy 6(2):140–157. Online available http://www.sciencedirect.com/science/article/B6V2W-48XK6JW-44/2/8496c46a5d19b34aee084384130ba1e2

Chateau B, Lapillonne B (1990) Accounting and end-use models. Energy 15(3/4):261–278

Clémençon R (2016) The two sides of the Paris climate agreement: dismal failure or historic breakthrough? J Environ Dev 25(1):3–24

CODES BE (2022) International review of energy efficiency in data Centres for IEA EBC building energy Codes working group. Northwest National Laboratory, Richland

Craig PP, Gadgil A, Koomey JG (2002) What can history tach us? A retrospective examination of long-term energy forecasts for the United States. Annu Rev Energy Environ 27:83–118

Cuhls K, Dönitz E, Erdmann L, Gransche B, Kimpeler S, Schirrmeister E, Warnke P (2024) Foresight: fifty years to think your futures. In: Edler J, Walz R (eds) Systems and innovation research in transition: research questions and trends in historical perspective. Springer, Heidelberg

Dahlstrom MF, Rosenthal S (2018) Third-person perception of science narratives: the case of climate change denial. Sci Commun 40(3):340–365

Dallinger D, Wietschel M (2012) Grid integration of intermittent renewable energy sources using price-responsive plug-in electric vehicles. Renew Sust Energ Rev 16(5):3370–3382

Dallinger D, Krampe D, Wietschel M (2011) Vehicle-to-grid regulation reserves based on a dynamic simulation of mobility behavior. IEEE Trans Smart Grid 2(2):302–313

Daniels BW, Van Dril AWN (2007) Save production: a bottom-up energy model for Dutch industry and agriculture. Energy Econ 29(4):847–867. https://doi.org/10.1016/j.eneco.2007.02.001
Darmstadter J, Dunkerley J, Alterman J (1977) How industrial societies use energy: a comparative analysis. osti.gov.
DeCanio SJ (1998) The efficiency paradox: bureaucratic and organizational barriers to profitable energy-saving investments. Energy policy 26(5):441–454.
Del Crespo Granado P, Resch G, Holz F, Welisch M, Geipel J, Hartner M et al (2020) Energy transition pathways to a low-carbon Europe in 2050: the degree of cooperation and the level of decentralization. EEEP 9(1):121–136. https://doi.org/10.5547/2160-5890.9.1.PCRE
Department of Energy (2006) Benefits of demand response in electricity markets and recommendations for achieving them. U.S. Department of Energy, Washington, DC
Diekmann J, Eichhammer W, Neubert A, Rieke H, Schlomann B, Ziesing HJ (1999) Energy efficiency indicators – statistical basics, theoretical foundation, and basis of orientation for policy making (in German). Physica, Heidelberg
DIW, EWI, RWI) (ed.) (1978) The future energy demand in the Fed. Rep. of Germany and its supply. (In German). Berlin, Köln, Essen
Durand A, Damian P (2019) Energy efficiency networks: overview of the implementation over the world and lessons learned. https://www.researchgate.net/publication/350637912_Energy_Efficiency_Networks_Overview_of_the_implementation_over_the_world_and_lessons_learned. Accessed 04 Oct 2022.
Dütschke E, Idrissova F, Hirzel S, Nabitz L, Mielicke U, Mai M (2016) Regional energy efficiency networks—what factors make the successfull? ECEEE proceedings of the September 2016 conference. Berlin
Eberle A, Gruber A, Jochem E, von Roon S (2022) Not only technical innovations, but also new formats of groups accelerate the way to sustainability in industry and public organisations. (In German) In: Leal Filho W. (ed) Nachhaltigkeit in den Natur- und Sozialwissenschaften. pp.153–179
Elsland R, Divrak C, Fleiter T, Wietschel M (2013) The Turkish energy efficiency strategy – ex-ante assessment for the residential sector: 7th international conference on energy efficiency in domestic appliances and lighting (EEDAL), Coimbra
Energy Information Administration (2009) Model documentation report: industrial sector demand module of the national energy modeling system, Washington, DC
European Commission (2020) Powering a climate-neutral economy: an EU strategy for energy system integration, COM(2020) 299 final
Faruqui A, Chao HP, Niemeyer V, Platt J, Stahlkopf K (2001) Analyzing California's power crisis. Energy J 22(4):25–52
FERC (2006) Assessment of demand response & advanced metering—staff report. Federal Energy Regulatory Commission (FERC), Washington, DC
Finon D, Midttun A (eds) (2004) Reshaping European gas and electricity industries. Elsevier
Fisher AC, Rothkopf MH (1989) Market failure and energy policy: a rationale for selective conservation. Working Paper 474. University of California and LBL, Berkeley
Fleiter T, Worrell E, Eichhammer W (2011) Barriers to energy efficiency in industrial bottom-up energy demand models – a review. Renew Sust Energ Rev 15(6):3099–3111. Online verfügbar unter. http://www.sciencedirect.com/science/article/pii/S1364032111001286
Fleiter T, Rehfeldt M, Herbst A, Elsland R, Klingler A-L, Manz P, Eidelloth S (2018) A methodology for bottom-up modelling of energy transitions in the industry sector. The FORECAST model. Energ Strat Rev 22:237–254. https://doi.org/10.1016/j.esr.2018.09.005
Gallagher KS, Grübler A, Kuhl L, Nemet G, Wilson C (2012) The energy technology innovation system. Annu Rev Environ Resour 37:137–162
Garnreiter F, Jochem E, Schön M (1986) Product structural change and technological progress as parameters of the energy consumption in four energy-intensive industries. FhG-Rep 10(4):271–283. (in German)

Geller H, Attali S (2005) The exprience with energy efficiency policies and programmes in IEA countries – learning from the critics. IEA Information Paper, Paris
Geller HS, DeCicco JM, Laitner S (1992) Energy efficiency and job creation: the employment and income benefits from investing in energy conserving technologies. American Council for an Energy-Efficient Economy, Washington, DC
German Bundestag (1991) Protecting the earth. A status report with recommendations for a new energy policy. In: Third report of the Enquete commission of the 11th German Bundestag "preventive measures to protect the Earth's atmosphere", vol 2. Economica Verlag, Bonn
Gillingham K, Rapson D, Wagner G (2016) The rebound effect and energy efficiency policy. Rev Environ Econ Policy 10(1):68–88
Globisch J, Dütschke E, Schleich J (2018) Acceptance of electric passenger cars in commercial fleets. Transp Res Part A Policy Pract 116:122–129. https://doi.org/10.1016/j.tra.2018.06.004
Gnann T, Plötz P, Kühn A, Wietschel M (2015) Modelling market diffusion of electric vehicles with real world driving data–German market and policy options. Transp Res A Policy Pract 77:95–112
Gruber E, Brand M (1991) Promoting energy conservation in small and medium-sized companies. Energy Policy 19(4):279–287
Gruber E, Jochem E (2007) Local learning networks on energy efficiency in industry—successful initiative in Germany. Appl Energy 84:806–816
Gruber E, Garnreiter F, Jochem E, Kuntze U, Mannsbart W, Meyer-Krahmer F, Overkott S (1982) Evaluation of energy conservation programmes in the EC countries. Report for the Commission of the C, DG XVII. Karlsruhe
Hake JF, Jagodzinski P, Kuckshinrichs W, Markewitz P, Martinsen D, Walbeck M (1994) IKARUS: an energy-economy model to reduce energy-related greenhouse gas emissions in Germany. CONF-940426. Air and waste management association, Pittsburgh, PA
Hatsopoulos GN, Gyftopoulos EP, Sant RW, Widmer TF (1978) Capital investment to save energy. Harvard Bus Rev 56:2
Hennicke P, Ramesohl S, Barthel C, Starzer DIO, Schmid W, Ostertag DVK, ... Klok J (1998) Interdisciplinary analysis of successful implementation of energy efficiency in the industrial, commercial and service sector. Final Rep. Copenhagen, Karlsruhe, Kiel, Vienna, Wuppertal
Henning HM, Palzer A (2014) A comprehensive model for the German electricity and heat sector in a future energy system with a dominant contribution from renewable energy technologies—part I: methodology. Renew Sust Energ Rev 30:1003–1018
Herbst A, Toro F, Reitze F, Jochem E (2012) Introduction to energy systems modelling. Swiss Soc Econ Stat 148(2):111–135
Heyen NB, Friedewald M, Jochem E, Reiß T, Thielmann A (2024) Assessing technological innovations: from early warning to the governance of socio-technical transformations. In: Edler J, Walz R (eds) Systems and innovation research in transition: research questions and trends in historical perspective. Springer, Heidelberg
Hirst E (1991) Improving energy efficiency in the USA: the federal role. Energy Policy 19(6):567–577
Hohmeyer O (1988) Social costs of energy consumption. Dissertation. Springer, Berlin, Heidelberg
Hohmeyer O (1991) Impacts of external costs on the competitive position of wind energy in the Federal Republic of Germany. In: External environmental costs of electric power. Springer, Berlin, Heidelberg, pp 213–225
Horne M, Jaccard M, Tiedemann K (2005) Improving behavioral realism in hybrid energy-economy models using discrete choice studies of personal transportation decisions. Energy Econ 27(1):59–77. https://doi.org/10.1016/j.eneco.2004.11.003
Hörsch J, Hofmann F, Schlachtberger D, Brown T (2018) PyPSA-eur: an open optimisation model of the European transmission system. Energ Strat Rev 22:207–215
IEA (Intern. Energy Agency) (2023) World energy investment 2023, Paris
IEA (International Energy Agency) (2021) Secure energy transitions in the power sector. International Energy Agency, Paris

IEA (International Energy Agency) (2022a) World energy outlook 2022. OECD, Paris
IEA (International Energy Agency) (2022b) 7th annual global conference on energy efficiency – the value of urgent action on energy efficiency. Paris
IEA (International Energy Agency) (n.d.) Users Technological Collaboration Program (TCP) – Social License to Automate investigates the social dimensions of user engagement with automated technologies in energy systems. https://userstcp.org/social-license-to-automate-task
IPCC (2001) Climate change 2001 – mitigation: contribution of working group III to TAR of the IPCC. Cambridge University Press, Cambridge
IPCC (2001a) Climate change 2001: synthesis report: a contribution of working groups I, II, and III to the third assessment report of the IPCC. Cambridge University Press, Cambridge, pp 214–224
IRENA, IEA, & REN21 (2018) Renewable energy policies in a time of transition
Jakob M, Madlener R (2004) Riding down the experience curve for energy-efficient building envelopes: the Swiss case for 1970–2020. Int J Energy Technol Policy 2(1–2):153–178
Jefferson M (2016) A global energy assessment. Wiley Interdiscipl Rev: Energy Environ 5(1):7–15
Jhirad D, Mintzer IM (1992) Electricity: technological opportunities and management. Confronting climate change: risks, implications and responses, p 171
Jochem E (1991) Long-term potentials of rational energy use – the unknown possibilities of reducing greenhouse gas emissions. Energy Environ 2:31–44
Jochem E (ed) (2004) Steps towards a sustainable development—a white book for R&D of energy-efficient technologies. ETH Zurich and Novatlantis, Zurich
Jochem E et al (2009) Improving the efficiency of R&D and the market diffusion of energy technologies. Physica. Heidelberg
Jochem E, Gruber E (1990) Obstacles to rational electricity use and measures to alleviate them. Energy Policy 18(4):340–350
Jochem E, Hauser U, Holzer G, Mentzel T, Nickel J, Oanea A, Wiesner J (1978) Energy consumption and potentials of efficient energy use in three industrial branches in Baden-Württemberg. (in German). ISI-Report. Karlsruhe
Jochem E, Sathaye J, Bouille D (eds) (2000a) Society, behaviour, and climate change: advances in global change research, Dordrecht, Boston, London. ISBN 0-7923-6802-9
Jochem E, Adegbulugbe A, Aebischer B, Bhattacharjee B et al (2000b) End-use energy efficiency. Chapter 6 in: UNDP/WEC/DESA (edit.), World Energy Assessment. UNDP, New York
Jochem E, Schön M, Angerer G, Ball M, Bradke H, Celik B, Eichhammer W, Mannsbart W, Marscheider-Weidemann F, Nathani C, Walz R, Wietschel M (2004) Resource efficiency. In: Efficiency potentials in production and use of energy-intensive basic products (in German). Fraunhofer IRB, Stuttgart, p 378. ISBN 3-8167-6622-6
Klempp N, Heilmann E, Pelka S, Köppl S, Bekk A (2020) Strategisches Gebotsverhalten auf der FlexPlattform – ein Engpass für die Weiterentwicklung des Netzengpassmanagements. Energiewirtschaftliche Tagesfragen
Klobasa M (2009) Dynamische Simulation eines Lastmanagements und Integration von Windenergie in ein Elektrizitätsnetz. Fraunhofer-IRB-Verl, Stuttgart. (ISI-Schriftenreihe "Innovationspotenziale")
Klobasa M, Pelka S, Plötz P, Vogel L (2019) Artificial intelligence for the integrated energy transition. Fraunhofer ISI, Karlsruhe
Kraus M (1988) Root cause failure analysis of projected-actual value differences of energy demand projections in the Fed. Rep. of Germany between 1950 and 1980 (in German). Dissertation. KIT, Karlsruhe
Kühnbach M, Stute J, Gnann T, Wietschel M, Marwitz S, Klobasa M (2020) Impact of electric vehicles: will German households pay less for electricity? Energ Strat Rev 32:100568
Kühnbach M, Bekk A, Weidlich A (2022) Towards improved prosumer participation: electricity trading in local markets. Energy 239:122445. https://doi.org/10.1016/j.energy.2021.122445
Lantzke U (1980) The role of nuclear energy in the year 2000. The OECD Observer 107:42

Lebert B (1977) Analysis of short- and long-term energy price elasticities of energy demand in the Fed. Rep. of Germany (in German). IfW (Inst. für Weltwirtschaft). Kieler working papers no 59. Kiel

Lee JH, Hardman SJ, Tal G (2019) Who is buying electric vehicles in California? Characterising early adopter heterogeneity and forecasting market diffusion. Energy Res Soc Sci 55:218–226

Legler H, Jochem E (1977) The relationship between energy consumption, economic growth and employment (in German). Blätter für deutsche und intern Politik 22(3):270

Levine MD, Adamson B (1988) Energy use in buildings: The U.S. experience and lessons for China. Proceedings Chinese-American symposium on energy markets and the future of energy demand. Nanjing, China, June 22–24, 1988. LBL. University of California

Levine M, Koomey J, McMahin J, Sanstad A, Hirst E (1995) Energy efficiency policies and market failures. Annu Rev Energy Environ 20:535–555

Lovins AB (1976) Energy strategy: the road not taken? Foreign Aff 55(1):65–96

Meadows DH, Meadows DL, Randers J, Behrens WW (1972) The limits to growth: a report for the Club of Rome'project on the predicament of mankind. Universe Books, New York

Meyer H, Peach AK, Guenther L, Kedar HE, Brüggemann M (2023) Between calls for action and narratives of denial: climate change attention structures on twitter. Media Commun 11(1):278–292

Møller Andersen F, Grenaa Jensen S, Larsen H V, Meibom P, Ravn H, Skytte K, & Togeby M (2006) Analyses of demand response in Denmark (No. RISO-R—1565 (EN)). Risoe National Lab

Möller K-P, Ströbele W (1978) Surplus capacities in 1985? Arguments of the energy policy – electricity demand. (In German) Wirtschaftsdienst. ISSN 0043-6275, Weltarchiv, Hamburg, 58, 10, pp 515–521

Monitor GE (2023) China permits two new coal power plants per week in 2022. Centre for Research on Energy and Clean Air (CREA) and the Global Energy Monitor (GEM)

Morovic T, Schön M (1987) Energy conservation indicators. Springer, Berlin, Heidelberg, New York

Morovic T, Schön M (1989) Energy conservation indicators II. Springer, Berlin, Heidelberg, New York

Morrison WE, Readling CL (1968) An energy model for the United States: featuring energy balances for the years 1947 to 1965 and projections and forecasts to the years 1980 and 2000, vol 8384. US Bureau of Mines, Washington, DC

Mueller HF (1957) The role of mineral oil in the Sovjet energy sector (in German). Berliner Wissenschafts-Verlag, Osteuropa, pp 495–504

Murphy R, Rivers N, Jaccard M (2007) Hybrid modeling of industrial energy consumption and greenhouse gas emissions with an application to Canada. Energy Econ 29(4):826–846. https://doi.org/10.1016/j.eneco.2007.01.006

Mwasilu et al (2014) Electric vehicles and smart grid interaction: a review on vehicle to grid and renewable energy sources integration. Renew Sust Energ Rev 34:501–516

Nägeli C, Jakob M, Catenazzi G, Ostermeyer Y (2020) Towards agent-based building stock modeling: bottom-up modeling of long-term stock dynamics affecting the energy and climate impact of building stocks. Energ Build 211:109763. https://doi.org/10.1016/j.enbuild.2020.109763

Neu AD (1978) Decoupling of economic growth an energy consumption: a strategy of energy policy? (in German). Kieler Diskussionsbeiträge, no 52, IfW (Inst. f. Weltwirtschaft) Kiel

Nitsch J (2000) Renewable energies in the 21st century – additive or alternative? Manuscript (in German). Int. solar forum. Freiburg. July 5 to 9. DLR Stuttgart

Nitsch J, Schott D, Kleiß H (1981) Development of systems of secondary energies in the Fed. Rep. of Germany until 2000 (in German). Arbeitsgemeinschaft für Großforschungseinrichtungen (ASA). Köln

Nordel D (2004) Rules for congestion management Evaluation of availability of capacity and possibilities for increased counter trade (No. NEI-DK--4332)

Nordel Demand Response Group (2006) Enhancement of demand response – final status report, Denmark

Palm J, Backman F (2020) Energy efficiency in SMEs: overcoming the communication barrier. Energ Effic 13:809–821

Palzer A, Henning HM (2014) A comprehensive model for the German electricity and heat sector in a future energy system with a dominant contribution from renewable energy technologies–Part II: Results. Renew Sust Energ Rev 30:1019–1034

Panny J, Held A, Winkler J, Breitschopf B, Jochem E, Walz R (2024) From niche to mainstream: exploring innovation and progress of renewable energy development. In: Edler J, Walz R (eds) Systems and innovation research in transition: research questions and trends in historical perspective. Springer, Heidelberg

Pehnt M, Brischke L-A (2013) Energy savings fonds and guaranty of efficiency. Report (in German). Friends of the Earth Germany. Pp 18–19

Peters A, Doll, C, Kley F, Möckel M, Plötz P, Sauer A, et al (2012) Konzepte der Elektromobilität und deren Bedeutung für Wirtschaft, Gesellschaft und Umwelt. Innovationsreport für das Büro für Technikfolgenabschätzung beim Deutschen Bundestag, Arbeitsbericht, p 153

Pfenninger S, Hawkes A, Keirstead J (2014) Energy systems modeling for twenty-first century energy challenges. Renew Sust Energ Rev 33:74–86. https://doi.org/10.1016/j.rser.2014.02.003

Pfenninger S, Hirth L, Schlecht I, Schmid E, Wiese F, Brown T et al (2018) Opening the black box of energy modelling. Strategies and lessons learned. Energ Strat Rev 19:63–71. https://doi.org/10.1016/j.esr.2017.12.002

Plazas-Niño FA, Ortiz-Pimiento NR, Montes-Páez EG (2022) National energy system optimization modelling for decarbonization pathways analysis: a systematic literature review. Renew Sust Energ Rev 162:112406. https://doi.org/10.1016/j.rser.2022.112406

Plötz P, Gnann T, Kühn A, Wietschel M (2013) Markthochlaufszenarien für elektrofahrzeuge Study commissioned by the National Academy of Science and Engineering and Working Group, 7

Plötz P, Gnann T, Wietschel M (2014) Modelling market diffusion of electric vehicles with real world driving data—part I: model structure and validation. Ecol Econ 107:411–421

Plötz P, Gnann T, Jochem P, Yilmaz HÜ, Kaschub T (2019) Impact of electric trucks powered by overhead lines on the European electricity system and CO_2 emissions. Energy Policy 130:32–40

Plötz P, Moll C, Mock P, Bieker G (2021) From lab-to-road: real-world emissions of plug-in-hybrid electric vehicles. Environ Res Lett 16:054078

Preuß S, Kunze R, Zwirnmann J, Meier J, Plötz P, Wietschel M (2021) The share of renewable electricity in electric vehicle charging in Europe is higher than grid mix. Fraunhofer ISI Working Papers Sustainability and Innovation no S 11/2021

Prömmel A (1978) Increased energy efficiency in existing buildings. Habitat Int 3:569–575

Quiggin D, Cornell S, Tierney M, Buswell R (2012) A simulation and optimisation study: towards a decentralised microgrid, using real world fluctuation data. Energy 41(1):549–559

Radgen P, Tönsing E (1996) Energieeinsparmöglichkeiten im Industriebereich in Europa bis zum Jahr 2050. Endbericht. Untersuchung für das Zentrum für Europäische Wirtschaftsforschung (ZEW). FhG-ISI Karlsruhe, Mannheim

Ramesohl S (1999). Opening the black box-what can be learned from socio-economic research for energy policy analyses? In *The IEA international workshop on technologies to reduce greenhouse gaseEmissions: Engineering-economic analyses of cnserved energy*

Ramsebner J, Hass R, Ajanovic A, Wietschel M (2021) The sector coupling concept: a critical review jasmine. WIREs Energy Environ 10(4):e396

Rehfeldt M, Globisch J, Fleiter T (2019) Fuel choice in industrial steam generation: empirical evidence reveals technology preferences. Energ Strat Rev 26:100407. https://doi.org/10.1016/j.esr.2019.100407

Rennings K (2000) Redefining innovation — Eco-innovation research and the contribution from ecological economics. p 319–332

Reuter M, Patel M, Eichhammer W, Lapillonne B, Pollier K (2020) A comprehensive indicator set for measuring multiple benefits of energy efficiency. Energy Policy 139:4

Richts C, Jansen M, Siefert M (2015) Determining the economic value of off-shore wind power plants in the changing energy system. Energy Procedia 80:422–432. https://doi.org/10.1016/j.egypro.2015.11.446

Rivers N, Jaccard M (2005) Combining top-down and bottom-up approaches to energy-economy modeling using discrete choice methods. Energy J 26(1):83–106. http://www.scopus.com/inward/record.url?eid=2-s2.0-14844340859&partnerID=40&md5=d178ef095d4887cc3a603a5fc505ddea
Rivers N, Jaccard M (2006) Useful models for simulating policies to induce technological change. Energy Policy 34(15):2038–2047. https://doi.org/10.1016/j.enpol.2005.02.003
Sanstad AH, Howarth RB (1994) 'Normal' markets, market imperfections and energy efficiency. Energy Policy 22(10):811–818
Schaber K (2013) Integration of variable renewable energies in the European power system: a model-based analysis of transmission grid extensions and energy sec-tor coupling. Diss. Technical University Munich
Schaber K, Steinke F, Hamacher T (2013) Managing temporary oversupply from renewables efficiently: electricity storage versus energy sector coupling in Germany. 22. International energy workshop, Paris
Scherrer et al (2019) Early Adopter von E-Fahrzeugen: Ladeleistungen, Eigenerzeugung und Einstellungen zum Lademanagement. Energiewirtschaftliche Tagesfragen 69(11):23–26
Schipper LJ, Hollander J, Levine MD, Craig PP (1979) The National Energy Conservation Policy Act: an 'Evaluation'. Nat Resour J 19:765–785
Schipper L, Scholl L, Price L (1997) Energy use and carbon emissions from freight in 10 industrialized countries: an analysis of trends from 1973 to 1992. Transp Res Part D-Transp Environ 2:57–76. https://doi.org/10.1016/S1361-9209(96)00014-4
Schleich J, Klobasa M, Gölz S, Brunner M (2013) Effects of feedback on residential electricity demand—findings from a field trial in Austria. Energy Policy 61:1097–1106. https://doi.org/10.1016/j.enpol.2013.05.012
Schröter M, Weißfloch U, Buschak D (2009) Energieeffizienz in der Produktion—Wunsch oder Wirklichkeit? Energieeinsparpotenziale und Verbreitungsgrad energieeffizienter Techniken. Mitteilungen aus der ISI-Erhebung. No. 51, Karlsruhe
Scorza SA, Pfeiffer J, Schmitt A, Weissbart C (2018) Kurz zum Klima: "Sektorkopplung" – Ansätze und Implikationen der Dekarbonisierung des Energiesystems. ifo Schnelldienst 71(10):49–53
Sensfuß F (2007) Assessment of the impact of renewable electricity generation on the German electricity sector: an agent-based simulation approach (Dissertation, KIT Karlsruhe)
Sensfuß F, Ragwitz M, Genoese M (2008) The merit-order effect: a detailed analysis of the price effect of renewable electricity generation on spot market prices in Germany. Energy Policy 36(8):3086–3094. https://doi.org/10.1016/j.enpol.2008.03.035
Sensfuß F, Lux B, Bernath C, Kiefer C, Pfluger B, Kleinschmitt C, et al (2021) Long term scenarios projecting the transformation of the energy Systems of Germany. Off Shore Potentials of Wind Energy (in German). ISI Karlsruhe
Singh M, Jiao J, Klobasa M et al (2021) Making energy-transition headway: a data driven assessment of German energy startups. Sustain Energy Technol Assess 47, art. 101322:11
Sioshansi FP (1991) The myths and facts of energy efficiency: survey of implementation issues. Energy Policy 19(3):231–243
Smeers Y (1997) Computable equilibrium models and the restructuring of the European electricity and gas markets. Energy J 18(4):1–31
Sorrell S (2007) The rebound effect: an assessment of the evidence for economy-wide energy savings from improved energy efficiency. Report Sussex Energy Group. UK Energy Research Centre ISBN 1-903144-0-35
Sorrell S, Schleich J, O'Malley E, Scott S (2004) The economics of energy efficiency: barriers to cost-effective investment. Edward Elgar, Cheltenham
Speth D, Plötz P, Funke S, Vallarella E (2022) Public fast charging infrastructure for battery electric trucks—designing a network for Germany. Under review
Stadler M, Kranzl L, Huber C, Haas R, Tsioliaridou E (2007) Policy strategies and paths to promote sustainable energy systems –the dynamic invert simulation tool. Energy Policy 35:597–608
Steinbach J (2016) Model based analysis of policy instruments promoting renewables and energy efficiency in buildings (in German), dissertation

Sweeney JL (1981) World oil. Preliminary results from the energy modelling forum-study. Kvartalsskrift, Bergen Bank, Norway
Synák F, Kucera M, Skrucany T (2021) Assessing the energy efficiency of an electric car. Komunikacie, Univ. of Zilana 23(1):A1–A13
Tureczek AM, Nielsen PS (2017) Structured literature review of electricity consumption classification using smart meter data. Energies 10(5):584
Velthuijsen JW (1995) Determinants of investment in energy conservation. Diss. Rijksuniversiteit Groningen (Netherlands)
Vita G, Lundström JR, Hertwich EG, Quist J, Ivanova D, Stadler K, Wood R (2019) The environmental impact of green consumption and sufficiency lifestyles scenarios in Europe: connecting local sustainability visions to global consequences. Ecol Econ 164:106322
Walz R (1994) The electricity industry in the United States and the Fed. Republic of Germany. Comparison including co-generation and efficient use of electricity. Diss. Physica. Heidelberg
Wang Y (2021) RES to drive EVs. Renew Energy 171:85–86
Weber L (1997) Some reflections on barriers to the efficient use of energy. Energy Policy 25(10):833–835
Wesche JP, Negro SO, Dütschke E, Raven RPJM, Hekkert MP (2019) Configurational innovation systems—explaining the slow German heat transition. Energy Res Soc Sci 52:99–113. https://doi.org/10.1016/j.erss.2018.12.015
Wietschel M, Plötz P, Pfluger B, Klobasa M, Eßer A, Haendel M, Müller-Kirchenbauer J, Kochems J, Hermann L, Grosse B, Nacken L, Küster, M, Pacem J, Naumann D, Kost C, Kohrs R, Fahl U, Schäfer-Stradowsky S, Timmermann D, Albert D (2018) Sektorkopplung—Definition, Chancen und Herausforderungen. ISI Working Paper Sustainability and Innovation.Band: S 01/2018. Fraunhofer-Institut für System- und Innovationsforschung ISI, Karlsruhe
Worrell E, Price L (2001) Policy scenarios for energy efficiency improvement in industry. Energy Policy 29:1223–1241
Worrell E, Martin N, Price L (2000) Potentials for energy efficiency improvement in the US cement industry. Energy 25(12):1189–1214
Zweifel P, Filippini M, Bonomo S (1997) Electricity tariffs and electricity consumption in private households. New knowledge from Switzerland (in German), vol 138. Physica, Berlin, Heidelberg

Eberhard Jochem is a senior scientist at the Fraunhofer Institute of Systems and Innovation Research ISI. Internationally acknowledged expert in technical and socio-economic research, mainly in the fields of energy, material efficiency, and climate change. Major contributions to the development and implementation of a sustainable energy and resources policy. Member of various national and international scientific organisations and advisory committees during the last five decades. Author of several books and numerous publications in the fields of energy, resource efficiency, and environment with regard to R&D, innovation, obstacles and policy issues, and energy economics. Promoter of energy efficiency and climate protection networks in industry.

Harald Bradke studied Mechanical Engineering. PhD in Energy Technology and worked as a research assistant at the University of Siegen, Chair of Energy Engineering. Since 1989 senior scientist at the Fraunhofer Institute for Systems and Innovation Research ISI, Head of the Competence Centre Energy Policy and Energy Systems from 1996 to 2011, Head of the Competence Centre Energy Technology and Energy Systems since 2012.

Elisabeth Dütschke has a background in psychology and completed her doctoral thesis in social sciences. She joined Fraunhofer ISI in 2009 as a senior scientist and later became the coordinator of the Business Unit Actors and Social Acceptance in the Transformation of the Energy System in 2019. Since 2023, she has served as the Head of the Competence Centre Energy Policy and Energy Markets. Dr. Dütschke is also an Associate Editor of the Journal Energy Research & Social

Science. Her primary focus is on the societal dimension in the context of the transition of the energy system.

Marian Klobasa studied engineering at the University of Karlsruhe. From September 2002 to December 2011 researcher in the Competence Centre Energy Policy and Energy Systems at Fraunhofer Institute for Systems and Innovation Research ISI. Since January 2012 research associate in the Competence Centre Energy Policy and Energy Markets. Since June 2014 Coordinator of the Business Unit Demand Response and Smart Grids in the Competence Centre Energy Technology and Energy Systems.

Martin Wietschel studied Industrial Engineering and Management at the University of Karlsruhe. From 2002 to 2011, member of the Competence Centre Energy Policy and Energy Systems at the Fraunhofer Institute for Systems and Innovation Research ISI, since January 2012 Deputy Head of the Competence Centre Energy Technology and Energy Systems. Since 2007 Coordinator of the Business Unit Energy Economy. From 2005 to 2011, lecturer at the ETH Zurich and since 2008 adjunct professor at the University of Karlsruhe (now Karlsruhe Institute of Technology KIT). Since February 2020, Head of Competence Centre Energy Technology and Energy Systems.

Patrick Plötz is Coordinator of Business Unit Energy Economy at the Fraunhofer Institute for Systems and Innovation Research ISI. Patrick Plötz has studied Physics in Greifswald, St. Petersburg, and Göttingen. He holds a doctorate degree in Theoretical Physics from the University of Heidelberg and has completed his habilitation in Economics at the Karlsruhe Institute of Technology (KIT) and has been a private lecturer at KIT since 2020. Dr. Plötz has over 10 years of experience in energy research, having served as a researcher at Fraunhofer ISI since 2011.

Tobias Fleiter began work at the Fraunhofer ISI in the Competence Centre Energy Technology and Energy Systems in 2007. He has a master's degree in industrial engineering from the University of Flensburg and obtained his PhD in 2012 at Utrecht University on the topic "The adoption of energy-efficient technologies by firms". Tobias Fleiter has a longstanding experience in coordinating research and consulting projects for companies, governmental institutions, and research funds. He has published peer-reviewed articles on industrial energy efficiency, decarbonisation, and modelling.

Understanding the Co-Evolution of Research and Water Protection Policies: From Single Technologies to Systemic Integrated Approaches for the Sustainable Use of Water

Thomas Hillenbrand, Harald Hiessl, Frank Marscheider-Weidemann, Jutta Niederste-Hollenberg, Christian Sartorius, and Rainer Walz

Abstract Over the last 50 years, water research has been strongly driven by the poor condition of water bodies underpinned by growing environmental awareness in society. This was characterised by interactions between the state of the environment, policy making and water research. Three phases can be distinguished: In the first phase (1970s and 1980s), the focus of activities in the establishment of environmental policy was on improving the precarious state of the environment in the water sector (first phase "establishment"). From 1990s to around 2010, further environmental pressures and new findings on the causes and changed framework conditions led to a need for further action. The interactions with other issues and a systemic perspective gained in importance (second phase "expansion"). In the years since 2010, the pressure to act has increased once again, particularly due to climate change. The need for systemic change is becoming increasingly clear (third phase "transformation"). The development is also characterised by an internal dynamic that is constantly increasing the scope and complexity of water research. The focus has shifted from individual technologies to systemic solutions. The need for transformation in the water sector is reflected, for example, in the national water strategy.

Our thanks go to Eberhard Böhm and Felix Tettenborn.

T. Hillenbrand (✉) · F. Marscheider-Weidemann · J. Niederste-Hollenberg · R. Walz
Fraunhofer Institute for Systems and Innovation Research ISI, Karlsruhe, Germany
e-mail: Thomas.Hillenbrand@isi.fraunhofer.de; Frank.Marscheider-Weidemann@isi.fraunhofer.de; jutta.niederste-hollenberg@isi.fraunhofer.de; rainer.walz@isi.fraunhofer.de

H. Hiessl · C. Sartorius
Formerly Fraunhofer Institute for Systems and Innovation Research ISI, Karlsruhe, Germany

J. Edler, R. Walz (eds.), *Systems and Innovation Research in Transition*, Sustainability and Innovation, https://doi.org/10.1007/978-3-031-66100-6_10

1 Background/Objective

The field of water research is very broad. It encompasses many areas, which show no obvious relation to innovation research, such as basic natural-science related research on the fundamentals of the natural water system. However, our chapter focuses on water research which is clearly linked to innovation issues, in particular research which deals with the specific segments of water systems as part of the built environment. This segment is key for determining the interaction between the socio-economic system and the natural water system. From the perspective of the socio-economic system, water resources provide input for production and consumption in form of water supply. But water bodies also have the function of absorbing output from the socio-economic system in the form of wastewater and inflows from agricultural land and soil surfaces. A sustainable use of water requires that there is no overuse of both functions, which would lead to a deterioration of the natural system. The last 50 years, however, have been characterised by the fact that these functions were being increasingly challenged. For the water systems there are three dimensions to the challenges in particular: (a) water quality problems due to emissions from the socio-economic system into the water bodies; (b) overuse of water resources; (c) problems in management of storm water caused by the increased sealing of soil surfaces and changing precipitation patterns.

The water sector is peculiar in that it is subject to a triple regulatory challenge for innovation: The first challenge relates to the external benefits of R&D and requires regulation to deal with the protection of intellectual property rights and the incentives for research. This challenge is the same in all other innovation domains. The second challenge arises from the external nature of environmental pollution, which is key to the challenges arising for water systems. It follows from this second challenge that the demand for water innovations crucially depends on regulation and environmental policy (Rennings 2000). The third regulatory challenge arises from the economic nature of (public) water supply and wastewater treatment systems as monopolistic bottlenecks, which require economic regulation. These economic regulations also have specific effects on the demand for innovation in regulated areas (Walz 2007). Demand, however, is an important factor for innovation activities, which is interacting with the supply side of innovations (Edler 2016). Thus, our chapter is also of particular interest as a case study on the interaction of pressure to act, research and regulation in an innovation system which is subject to this triple regulatory challenge.

Our chapter deals with two research questions. First, we want to analyse how the pressure to act interacts with innovation-related water research and regulation or policies. Second, we want to find out more about the logic of the long-term development of research and innovation in the water sector. The underlying analysis focuses on the national development in Germany. Cross-connections, especially with developments at the EU level, are taken into account. For both research questions, we start with the following hypotheses.

Our first hypothesis states that the interplay between the pressure to act, which is triggered by the state of the environment, the knowledge gained through research work and the identification and implementation of regulation and policies, contributed in Germany to the emergence of three different phases. These three phases can be summarised as follows:

- Particularly in the 1970s, but also to some extent in the 1980s, the focus of activities in the context of establishing environmental policy was on improving the precarious state of the environment in the water sector. Major causes of a large proportion of environmental pollution were obvious, so that measures could be directly derived and implemented (First Phase "establishment").
- In the subsequent phase (1990s to approx. 2010) further environmental pressures, new findings on the causes and the strongly changing framework conditions led to an additional need for action. The measures resulting from this pressure highlighted the need for a systemic perspective that took into account the intensive interactions with other thematic fields and areas of action (Second Phase "expansion").
- In the years since 2010, the pressure to act has intensified once again, also, and especially against the backdrop of the climate change impacts that are now becoming increasingly clear. The need for comprehensive, systemic changes in the water sector is now becoming widely accepted (Third Phase "transformation").

The hypothesis for the second research question states that starting from the approaches which addressed the need to act in the 1970s with the initial strong problem-oriented focus on technologies and regulations, water policy has evolved into more integrated approaches. This is consistently increasing the complexity of the underlying innovation system by integrating more actors and more perspectives. Initially, the approach was very much motivated by the precarious state of the aquatic environment and the search for technical solutions. However, with increasing insights into the complexity of water management, the role of water research and a systemic approach based on it increased as a basis for the formulation and implementation of environmental policy measures.

In order to tackle the research questions, the connections between the development of the environmental states in the area of water, water policy and water research are shown and explained in more detail for each of these three phases. Due to the complexity of the water sector, it is not possible to provide a fully comprehensive view. Instead, specific lines of development are selected as examples that are of particular importance and representative of the water sector in terms of the changes that have taken place.

The chapter is structured chronologically through the decades. The three phases are analysed in three sections. The individual sections are subdivided according to the following aspects: (a) pressure to act; (b) concrete important changes in research activities, which are related to the change in the phase; (c) important implemented measures and regulations. The chapter concludes with a discussion and an outlook, with regard to our two research questions: (a) logic of interaction between pressure to act, innovation-related water research, and policies; (b) logic of change in research over time. In order to support our line of argument with empirical data on development of research, we analysed databanks which show information of the projects funded in Germany. To back up our arguments, we also looked at the development of innovation indicators such as publications and patents in order to figure out, how our specific research topics linked to sustainable use of water developed within the overall broad topic of water research.

2 Phase 1: Establishing Water as Key Issue in Environmental Policy

2.1 Pressure to Act

In 1971, the first environmental programme of the federal government was adopted. It included fundamental aspects such as the "polluter pays" and the precautionary principle as well as the topic of water as seen from various facets. The catastrophic condition of the water bodies was reflected in very high levels of organic matter and nutrient concentrations in the water bodies. Improving water quality by reducing these loads was seen as a key imperative (Sachverständigenrat für Umweltfragen (SRU) 1974), and the pressure to act increased quite substantially in terms of the need for wastewater treatment. The first environmental programme therefore called for a cleaning up of the German water bodies by 1985 (BMI 1971). Various activities started in subsequent years—both in the field of research and with regard to the establishment of the legal and institutional framework, such as the founding of the German Environment Agency in 1974.

2.2 Important Research Activities

In the 1970s, research activities in the field of water were greatly intensified. This was reflected in the increase in research projects and funding in the early 1970s. Various funding institutions were active (Eberle 1978) with different focuses (cf. Merkel and Reiff 1983; Bauer 1977; Eberle 1978):

- the federal government, primarily through the then BMFT as well as the BMI, which was responsible for environmental affairs prior to establishment of the BMU;
- the federal states, which are responsible for supporting the construction of sewage treatment plants, with funding of very specific application-oriented research,
- more academic-oriented research funded by the DFG and institutional university funds,
- projects and institutes supported by the water supply companies themselves.

The diversity of institutions active in the field of water research, which developed in Germany in the first phase, has remained a special characteristic in this country until the present day (see Fig. 1). The most important funding agency over time has been the BMBF, which funded larger research projects in particular compared to the numerous research projects of the Bundesländer, which typically have been rather small.

The research projects in this first phase focused primarily on technical approaches to water treatment and wastewater purification. This reflected the pressure to act quickly: The primary focus was on the development and optimisation of single technologies. The analysis of water research projects included in the BMBF funding catalogue clearly shows that research in this phase was organised primarily as individual projects. Furthermore, most research was performed in a disciplinary manner, with a move towards more interdisciplinary research occurring only slowly (Sachverständigenrat für Umweltfragen (SRU) 1974). As the importance was recognised at the beginning of the 1980s of embedding research into the practical implementation context, the involvement of practical partners and the technical implementation of the findings from the research projects were a requirement for some of the funding measures. This also resulted in the number of collaborative projects increasing steadily, a trend which can be seen also in the BMBF funding catalogue.

Important developments during the first phase were, for example, the improvement of biological purification or treatment processes and the use of new processes (activated carbon, ion exchangers, membrane technology, ozone, etc.) for drinking water and wastewater purification, the elimination of nitrogen/nitrate and phosphorus, or the improved treatment of industrial wastewater. At the end of the 1970s and in the 1980s, the increased procedural and process engineering effort due to the increased requirements also led to research efforts in the areas of process stability with the necessary accompanying measurement, control and regulation technology. In addition, improved wastewater treatment led to an increase in the amount of sewage sludge and the resulting often anaerobic sludge treatment with sewage gas production. In-plant generation of electricity, combined heat and power generation and improving the energy efficiency of the facilities became important R&D topics. Furthermore, first research and demonstration projects were performed that fulfilled the requirements of the integration of practical partners, so that practice-oriented guidelines for action were developed as a result (e.g., Kunz and Müller 1986; Hillenbrand and Kunz 1989).

Another important research focus during this time was the improvement of the analysis and monitoring of environmental pollutants in bodies of water, in drinking water or wastewater, as well as the ecological assessment of the effects of pollutants. It was here that the foundations were laid for more far-reaching requirements in terms of protecting health and water. Thus, from the perspective of a policy cycle of environmental policy (Böcher and Töller 2012), water research did not only support the design and implementation of policy strategy, but it also contributed towards identification and agenda setting.

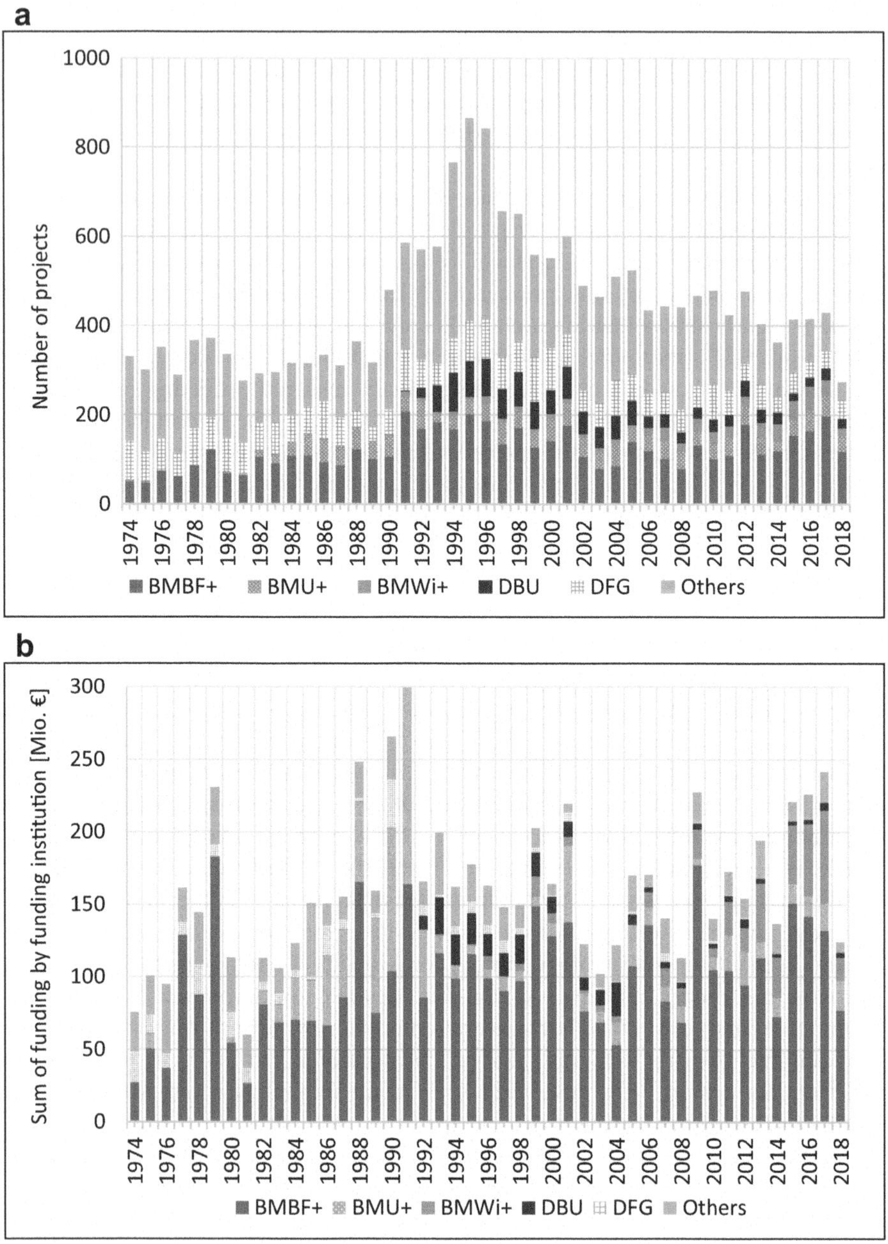

Fig. 1 Representation of (**a**) number of research projects and (**b**) funding volumes in the water sector, subdivided by funding institutions. Explanations: Evaluations of UFORDAT, without EU projects; BMBF+, BMU+, BMWi+: each with predecessor institutions in the period under consideration; others: federal states, foundations, companies, other ministries, etc.; funding totals adjusted for inflation with reference year 2020. Data for 2018 incomplete. Source: Fraunhofer ISI

Due to the favourable climatic conditions with its ample supply of water resources in Germany, no urgent need for research to improve water efficiency or reduce water demand was seen in the 1970s, and studies on the future development of water consumption were carried out only to a limited extent. The forecasts of the population's water consumption and the associated wastewater generation were extrapolated from the history of water consumption in a highly simplified manner and assumed a significant increase in per capita water consumption. For example, in the early 1970s, household water consumption was predicted to reach 200 litres per capita per day (lpcd) by the year 2000 (Battelle-Institut e.V. 1972), while it actually dropped to less than 130 lpcd. These forecasts served as a basis for the design of water infrastructure systems (especially water supply and sewerage networks) in settlement areas. Only a few studies (e.g., Merkel and Reiff 1983), motivated primarily by possible cost savings, started at the beginning of the 1980s to study the multiple use of water and the reduction of water losses.

2.3 *Measures and Regulations*

The high political pressure to act caused by the poor water quality led to regulatory measures at the national as well as the international level. In Germany, parallel to the further expansion of drinking water and sewer networks, a total of more than 8000 municipal wastewater treatment plants with mechanical-biological stages were built or expanded with the primary objective of reducing the input of oxygen-consuming organic and easily degradable wastewater constituents into bodies of water. Minimum requirements for wastewater treatment were formulated through the amendment of the Federal Water Act (Wasserhaushaltsgesetz) in 1976 with the introduction of the state of the art through §7a as a requirement for wastewater treatment and the concretisation through the first Wastewater Management Regulation (Schmutzwasser-Verwaltungsvorschrift) of 1979. Based on research results, the necessity of limiting nutrient inputs became increasingly clear for reasons of water and ocean protection and additional restrictions on nitrogen and phosphorus were imposed in 1989 as part of the so-called "10-point programme" from 1986 (Töpfer et al. 1988).

On the basis of the emission principle, industrial wastewater discharges in particular were more strongly regulated with the help of sector-specific wastewater management regulations. In addition, restrictions began to be imposed on individual areas of application through a chemicals policy (Detergents and Cleaning Agents Act (Wasch- and Reinigungsmittelgesetz) in 1975, a Fertilisers Act (Düngemittelgesetz) in 1977, a Chemicals Act (Chemikaliengesetz) in 1980, and aEU Directive 76/464/EEC on the discharge of certain dangerous substances into the aquatic environment (Sachverständigenrat für Umweltfragen (SRU) 1987). The stricter requirements were supported by economic incentives within the framework of the Wastewater Discharge Act (Abwasserabgabengesetz) passed in 1976 with levies for the discharge of various pollutants.

The success of the measures could be verified, for example, by the water quality map published for the first time in 1975: The quality classes I (unpolluted to very slightly polluted) to IV (excessively polluted) show the results of biological-ecological inventories of watercourse sections based on the saprobic system, with the help of which the biological water quality can be assessed using the organisms found in the water as bioindicators. Water quality maps illustrated the improvements achieved in the late 1970s and in the 1980s and also were quickly used in the new federal states after reunification, illustrating short-term improvements as a result of the investments in water infrastructure made there (cf. Kautt 1996).

3 Phase 2: Expanding Existing Approaches

3.1 Pressure to Act

Phase 1 had led to initial regulations and measures being taken for major sources of nutrient and pollutant inputs (wastewater management regulations, ChemG, WRMG, nitrate guidelines, etc.). Despite the resulting improvement achieved in the quality of surface waters, there was still a clear need for further action in order to achieve a good hygienic, ecological and chemical state in which the aquatic ecosystems deviate only slightly from their natural condition. This would allow for water to be used without restriction, for example, for bathing water, fishing water or for the production of drinking water (Sachverständigenrat für Umweltfragen (SRU) 1987; UBA 2006). Furthermore, various chemical accidents (Seveso in 1976; Bhopal in 1984; Sandoz in 1986) demonstrated the need to look beyond organic matter and nutrients as the only sources of pollution.

In the second phase, it also became increasingly clear that trade-offs between different goals existed. Traditionally, the sewage sludge from wastewater treatment had been used as agricultural fertiliser, and phosphorus was kept within the nutrient cycle. However, the conflict of goals between the nutrient character of the phosphorus contained in sewage sludge and the pollutants also contained in sewage sludge led to a decrease in direct agricultural utilisation. Thus, the question arose whether the phosphorus could be technically recovered and used as fertiliser (Sachverständigenrat für Umweltfragen (SRU) 2002).

In addition to these remaining challenges, the focus changed in phase 2, both in water research and in the political objectives and implementation measures. Changes in important framework conditions played a very important role (cf. Hillenbrand and Hiessl 2007):

- The original forecasts of drinking water consumption (cf. Battelle-Institut e.V. 1972) forecast an increase in the specific water demand of households up to 200 lpcd, on which the planning of the water infrastructure had been largely based until then. But instead of a steady increase, there was a stagnation or even a decline in water demand. In the industrial sector, the reasons for this decline were cost reduction measures, changed production processes, and stricter environmental requirements. In the private sector (decline to below 130 lpcd), the use of more efficient household appliances also played an important part. These efficiency improvements were often motivated by the associated energy and cost savings. However, reduced water use could also lead to problems in the operation of the water and wastewater networks, e.g. the deposition of solids from the wastewater in the sewer network.

- These problems were exacerbated by demographic change, which had very different regional effects in Germany, especially after reunification. The decline in population figures which was in part sharp, partially coupled with a decline in industrial activities, led to an under-utilisation of the water infrastructure and thus to technical operating problems. It also resulted in a deterioration of the economic framework conditions which was in part drastic, due to the decline in revenues to cover a fixed cost structure of infrastructure assets with long lifespans and depreciation periods (Hillenbrand et al. 2011). This led to debates about increasing water and sewage fees.
- Due to the global intensification of the water issue, the importance of the export relevance of water technologies also increased against the background of the globally recognised excellent state of the art of German technology. This resulted in additional incentives for research activities in cooperation with German technology providers, in order to further strengthen the advantage of German exporters (cf. Kluge and Schramm 2016).
- In addition, the in part poor condition of the existing water infrastructure in the settlement areas, including the associated hazard potential and the need for rehabilitation to be derived from it, became clear from the mid-1980s onwards. For example, on the one hand, the risk to groundwater from exfiltration of wastewater increased. Leaky sewage systems, on the other hand, cause extraneous water quantities that put a strain on the sewage treatment plants both in terms of the purification processes and the amount of wastewater produced.
- The threat to water infrastructure from possible attacks and the link to security issues came into particular focus against the backdrop of perceived threats from terrorist organisations in the early 2000s. Thus, even though not debated in public a lot, innovations for protecting water infrastructure systems against terrorist attack were seen as necessary.

To sum up, the pressure to act in Phase 2 is shown, on the one hand, as a continuation of existing challenges and on the other hand, as challenges related to the water infrastructure itself. Thus, expanding existing approaches and increasing the scope of solutions became central for the pressure to act in Phase 2.

3.2 Important Research Activities

Water Protection

With reduced emissions from key point sources, induced by the initial regulations and measures such as wastewater management regulations, ChemG, WRMG, nitrate guidelines, more precise statements on the sources of inputs, input pathways and effects became necessary. Thus the development of corresponding methods and findings became important research tasks. The various chemical accidents (Seveso 1976; Bhopal 1984; Sandoz 1986) also resulted in calls for more information on substances used in industry, their quantities and their emissions into the environment. In addition, the effects of accidents and shock or continuous loads on bodies of water, and thus also on water uses (e.g., after bank filtration and treatment as drinking water), were increasingly investigated scientifically. In corresponding research projects on forecasting and assessing water quality or quantifying the effect of measures, a more comprehensive view of water bodies and their environment had become necessary. This was stimulated in particular by international activities, such as the Toxic Release Inventory (TRI) in the USA (1986) as the first national emission inventory, and by Chap. 19 of Agenda 21 on Establishment of pollutant emission registers ("Pollutant Release and Transfer Register—PRTR"). Thus, research began in Germany on the inventorying of pollutant inputs. The results showed the clearly increased importance of diffuse sources of pollution and input pathways, e.g. heavy metal inputs via urban areas or nitrogen inputs from agriculture (Böhm and Hillenbrand 2000; Fuchs et al. 2002; Hillenbrand et al. 2005).

The various inventories also supported the comprehensive view of the body of water, taking into account its entire catchment area: at such a river basin management level, the influence of the important global change processes can be identified and cost-efficient measures can be developed. Such inventories also supported the new institutional approach of the European Water Framework Directive to manage emissions with a look at the entire catchment area instead of looking at single processes only (see below). This resulted in various projects which were devoted to how to institutionalise such integrated water management (Richter et al. 2013; Helmholtz-Zentrum für Umweltforschung—UFZ 2015). The resulting projects also led to even higher degrees of bringing together different research institutions in large research projects. The BMBF's GLOWA research focus,[1] for example, consisted of research consortia which linked different systems modellers with technical expertise. In the GLOWA project for the river Elbe catchment area, this led, on the one hand, to an integration of modelling the diffusion of technical innovations in the municipal and industrial wastewater sector with modelling of the socio-economic framework conditions. On the other hand, the modelling of the effects of the innovations was embedded within natural-science based models, which show the effects of

[1] BMBF-Forschungsschwerpunkt GLOWA https://www.fona.de/en/measures/funding-measures/archive/global-change.php.

resulting emission reductions on water quality (Wechsung et al. 2013; Sartorius et al. 2011; Hillenbrand and Böhm 2008). Furthermore, research started to look at systemic issues of water basin management, such as setting up water management plans and cost-effectiveness analysis (Böhm et al. 1999; Böhm et al. 2002).

In conjunction with improved analysis and assessment methods, research results simultaneously clarified the multitude of relevant pollutants and the effects they trigger. While pollution with heavy metals had already been focussed on at the end of the 1970s, organic microcontaminants (sometimes also referred to as micropollutants or later also trace substances) only moved into the spotlight of research efforts in phase 2. This can also be seen by looking at the development of topics of publications or focus of funded research projects (Figs. 2 and 3). Studies were conducted on the whereabouts of pharmaceuticals in the environment, among other things, which demonstrated their occurrence in groundwater and drinking water for some representatives of this group of substances (Bergmann et al. 2011).

After the simple landfilling of sewage sludge as well as its agricultural utilisation had been increasingly restricted, the importance of sewage sludge incineration grew significantly. In order to counteract the associated loss of phosphorus contained in sewage sludge, processes were developed from the beginning of the 2000s on, with

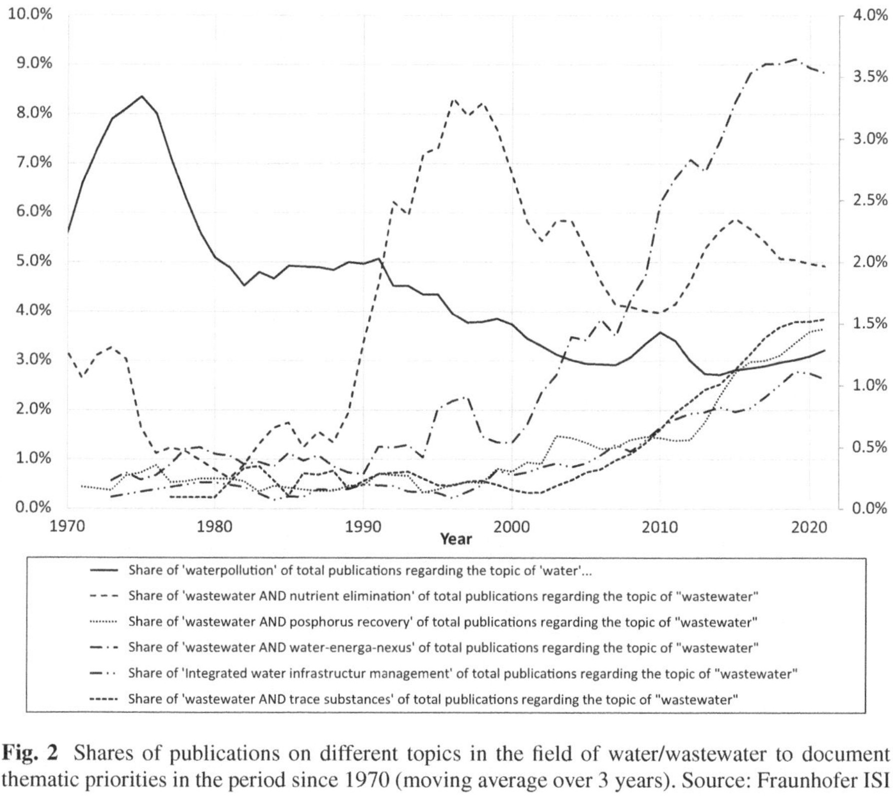

Fig. 2 Shares of publications on different topics in the field of water/wastewater to document thematic priorities in the period since 1970 (moving average over 3 years). Source: Fraunhofer ISI

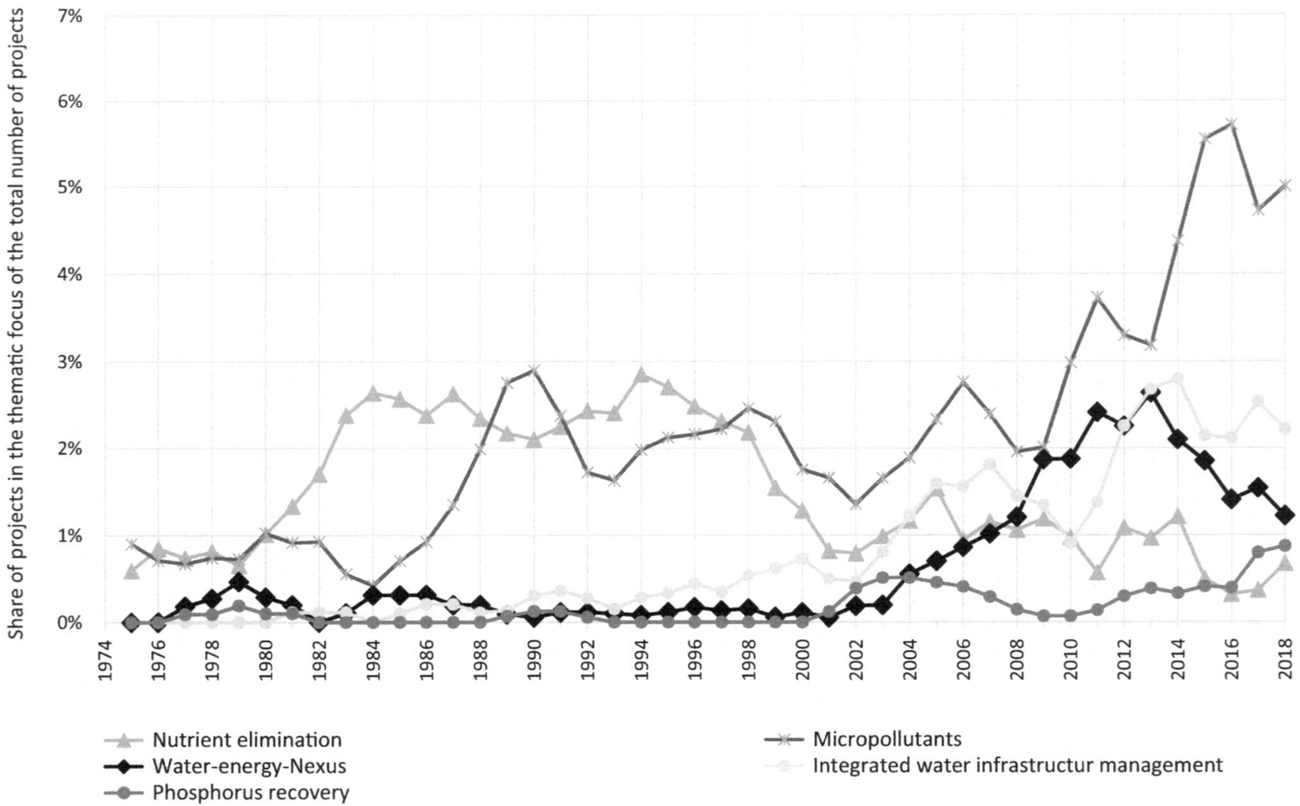

Fig. 3 Shares of research projects on selected topics of the total amount of projects in the field of water/wastewater since 1974 (data basis: UFORDAT; moving average over 3 years). Source: Fraunhofer ISI

which phosphorus can be recovered from wastewater, sewage sludge or sewage sludge ash (cf. BMBF funding call in 2004). Investigations showed that recovery from ash is more effective than recovery from sewage sludge or wastewater. At the same time, a conflict of objectives between cost efficiency and environmental effectiveness became apparent: the higher quality of the p-products was usually accompanied by relatively high costs, whereas more cost-effective processes produced p-products with a limited fertilising effect only and higher pollutant contents. This conflict of objectives was ultimately responsible for the fact that under the conditions prevailing in the early 2010s, research results concluded that phosphorus recovery could not be implemented in an economically viable matter (Sartorius et al. 2012).

The remaining pressure to reduce emissions also led to further development of existing approaches. In the case of drinking water treatment, for example, an important task was to develop treatment processes, even for smaller facilities, in cases where nitrate and pesticide concentrations were too high. In the industrial and commercial sector, comprehensive, integrated approaches were developed, that aimed at avoiding the transfer of harmful substances across all media, at closing the cycle or reusing treated wastewater, and also at recovering valuable substances from wastewater. As early as 2003, studies were carried out analysing industries which used persistent, toxic chemicals. The aim of these studies was to enable companies to identify environmentally hazardous substances in products and processes and replace them with lower-risk, technically efficient solutions (Ahrens et al. 2003).

The large number of problematic substances and different sources of input also showed the limits of end-of-pipe technologies. Investigations on the basis of successive substance flow and substance input modelling highlighted the need for a combined approach of emission and immission principles and the need for more source-related efforts to prevent environmental pollution (Hillenbrand et al. 2014). This resulted in calls for implementing improved agricultural practices with regard to the input of nutrients and pesticides, reducing the atmospheric deposition of pollutants or developing substitution options for emission-relevant uses of substances.

Water Infrastructure for Water Supply and Sanitation

The significantly increased pressure in the area of (urban) water infrastructure systems for water supply and wastewater disposal led to an intensive search for improved procedures for the assessment and restoration of pipeline and sewer networks. The involvement of practical partners in collaborative research projects was a prerequisite for the rapid implementation of research results. With regard to wastewater discharge, the subject of sustainable rainwater management (decoupling, percolation, treatment, charging structures, etc.) was also increasingly addressed. In particular, in the second half of phase 2, this also resulted in a substantial increase in collaborative projects dealing with integrated water infrastructure management (see Fig. 3).

At the same time, research into improving energy efficiency within the water infrastructure was intensified in different directions. Work on anaerobic sludge stabilisation was advanced quite significantly in the 1990s. The approaches were expanded in the early 2000s to include the goal of hydrogen production at sewage treatment plants and the energetic use of sewage gases in fuel cells (cf. significant increase in the importance of the topic of energy in Fig. 3).

The changing framework conditions also led to more fundamental reflections and scenario considerations for the future development of water infrastructure. The framework concept "Urban Ecology" (Stadtökologie) developed by the BMBF in 1990 focused on recommendations for action in the design and use of water and bodies of water in cities and urban regions, which included the objective of near-natural rainwater management in settlement areas. At the same time, long-term sustainable decentralised or semi-centralised system concepts as well as novel or resource-oriented sanitation concepts (NASS) were developed and tested (Hiessl et al. 2005; Deutsche Vereinigung für Wasserwirtschaft, Abwasser and Abfall e. V. (DWA) 2008; Hiessl et al. 2012; Londong et al. 2011). Possibilities for sector coupling with the energy or waste sector were also developed. In addition to the energy technologies described above that can be used at the wastewater treatment plant, the use of the waste heat contained in the wastewater from the sewer network also plays an important role. After initial research had been carried out as early as the 1980s, broader implementation-oriented projects started in the 2000s. Wastewater heat from grey water, the lightly contaminated domestic wastewater partial flow, e.g. from the shower, can be recovered closer to where the heat is generated and therefore with a higher degree of efficiency. More intensive research activities on this began at the end of the second phase. The prerequisite for this is the separate discharge of black and grey water, as favoured in the concept of resource-efficient infrastructures (Vetter et al. 2011; Menger-Krug et al. 2012).

The increasing prices for water supply and discharge of wastewater also induced research activities. One string of research looked into ways of improving the efficiency of the water utilities by investigating the role of technical standards or various forms of benchmarking tools (Böhm et al. 1998; Zschille et al. 2010; Hoffjan et al. 2011; Holländer 2011). Another string debated the role of privatisation and competition in the water sector (Mankel and Schwarze 2000; Michaelis 2001; Sachverständigenrat für Umweltfragen (SRU) 2002; Koch 2003; Brunner and Riechmann 2004; Wackerbauer 2009). This research concluded, by and large, that a modernisation of the industry is necessary, though not a general restructuring (Gawel and Bedtke 2015). The threat to water infrastructure from possible attacks and the link to security issues also influenced research. Against the background of the more comprehensive requirements, water management systems and measures were assessed in terms of their vulnerability as critical infrastructure and their sustainability (van Leuven 2011; Petermann et al. 2011; Schneidmadl et al. 2000; Deutsche Vereinigung für Wasserwirtschaft, Abwasser and Abfall e. V. (DWA) 2008; Hillenbrand et al. 2009).

The expansion of existing approaches and a stronger focus on the infrastructure systems led not only to interdisciplinary collaborative projects but also required the involvement of important stakeholder groups, including those outside the water sector. Furthermore, socio-economic issues, e.g. regarding the acceptance of innovative approaches and their effectiveness, were increasingly included in the projects (Hiessl et al. 2005; Hiessl et al. 2012; Kluge and Libbe 2010).

3.3 Measures and Regulations

In conjunction with the research efforts aimed at improved wastewater treatment processes which can (also) be applied in the industrial and commercial sector, additional sector-specific minimum requirements were imposed via annexes to the Wastewater Management Ordinance (Abwasserverwaltungsvorschrift) (or, from 1997, as annexes to the Wastewater Ordinance (Abwasserverordnung)) and its amendment (Sachverständigenrat für Umweltfragen (SRU) 2004). The technological advancement in treatment technologies achieved in research projects contributed to the introduction of more stringent regulations by showing that technical solutions exist which enable the standards to be fulfilled. Increased emphasis was also placed on the purification of wastewater partial flows and, if necessary, their pre-treatment. The European IPPC Directive (96/62/EC—Directive on Integrated Pollution Prevention and Control) introduced an integrated, cross-media approach for the first time for selected, particularly environmentally relevant industrial sectors. The aim was to achieve the highest possible level of protection for the environment on the basis of the best available techniques (BAT). According to Article 15 (3) of this Directive, the main pollutant emissions should be recorded and published every 3 years. The first inventory was published in 2003 for the reference year 2001; 90% of the emissions from IPPC installations were to be recorded within.[2] From 2007 on, the EPER was replaced by the E-PRTR (Regulation (EC) 166/2006), which significantly extended the obligation to provide information.

To implement the research results with regard to the relevance and entry pathways of chemicals into the environment, the requirements were also extended via chemicals policy—both at the international and European as well as at the national level (EU Directive on the placing of plant protection products on the market ((91/414/EEC); EU Biocide Directive (98/8/EC) on the placing of biocidal products on the market; Chemicals Prohibition Ordinance (Chemikalienverbotsverordnung) for implementing the Chemicals Act (Chemikaliengesetz), etc.). In addition, information tools were developed to inform consumers about environmentally hazardous substances (e.g., in 1993 the first Blue Angel award for detergents with the criteria complete biodegradability and limited toxicity to aquatic organisms was established).

In contrast, the establishment of immission-related limit values (quality targets) for hazardous substances in bodies of water, which was already envisaged in EU Directive 76/464/EEC, was delayed considerably. Quality objectives became legally

[2] cf. https://cwm.unitar.org/publications/publications/cbl/prtr/pdf/cat3/eper_de.pdf.

binding for the first time in Germany with the Quality Objectives Ordinance (Qualitätszielverordnung) being used to implement the EC Water Protection Directive 76/464/EEC in 2001.

Shortly before that, in 2000, the EC Water Framework Directive came into force (Directive 2000/60/EC). This changed the implementation of German water management fundamentally. The Water Framework Directive requires water management in river basins, an integrated management of ground and surface water, and the definition of ecological, chemical and quantitative environmental targets with the involvement of the public in planning processes (cf. Leinweber 2008). The goal was and is to align all uses that have an impact on ground and surface waters with the management objectives. This conceptually reoriented, integrated approach took place in parallel to changes in research activities in the area of water protection and water management described. Nevertheless, before its implementation the Water Framework Directive had been already debated intensively. Thus, even though the Water framework Directive was implemented in 2000, it can be assumed that its influence on the direction of research already started much earlier towards the beginning of Phase 2.

Over time, it became increasingly clear what an important role agriculture played in the condition of bodies of water. After this could be scientifically proven by corresponding inventories of the pollution situation, the pressure to act in this area increased. The influence of the 1977 Fertiliser Act (Düngemittelgesetz) was minor. However with the Nitrate Directive at the European level (91/676/EEC: Combating water pollution caused by nitrates from agriculture (Nitrate Directive) and the resulting Fertiliser Ordinance of 1996, improvements were achieved, even though the overall balance for Germany still shows high N and P surpluses (see UBA 2006, p. 88).

With regard to nutrient management, the majority of scientific experts agreed as early as at the end of the 2000s that the recycling of sewage sludge in agriculture had no future and that phosphorus recovery would instead represent the state of the art by 2030 (Sartorius et al. 2012). At this stage, however, the political will for widespread implementation was lacking, due to the fact that this would most likely have resulted in an increase in sewage fees.

The discussions, especially in eastern Germany, on the sharply increased water and wastewater charges led not only to research, but also to some changes. Based on the results of Böhm et al. (1998), changes were initiated in the standard-setting associations, which were intended to strengthen the innovation-promoting influence of technical regulations. Furthermore, water utilities were introducing some measures to increase efficiency whilst still remaining within the existing regulatory framework (Gawel and Bedtke 2015).

4 Phase 3: The Current Status—The Need for Transformation in the Water Sector

4.1 *Pressure to Act*

The pressure to act on the water sector increased significantly in the 2010s, especially with regard to the impact of climate change, which also affects the water system (Sachverständigenrat für Umweltfragen (SRU) 2020). Increasing heavy precipitation leads to more frequent overloading of the sewer networks, which were not designed for such precipitation events. As both heavy precipitation and the duration of dry periods increased and, at the same time, the actual water consumption of households decreased significantly in comparison with assumptions on which their design was based, the sewer networks' ability to function was no longer ensured to the extent required.

The greatly increased pressure to act, also at the political level, is likewise closely linked to the extreme weather events that are widely perceived and discussed in the public domain as an indicator of the anticipated future impacts of climate change on the water cycle. The extreme periods of drought and heat in the summers from 2018 to 2020, as well as the increase in heavy precipitation events, and extreme flash floods, flooding or high water events such as in the summer of 2021, have made the need for action clear to the wider public.

At the same time, the effects of change processes from other areas (e.g., energy transition, circular economy) were and are to be integrated. As well as the continuing poor state of the water infrastructure in some areas, demographic changes, which depend on the region show either significant population growth or decline combined with the perspective of an increasingly ageing population (associated with a significant increase in the consumption of medicines and increased inputs of pharmaceutical residues into the bodies of water), all continue to play an important role. The increased environmental awareness among the population also makes greater participation of the public or the relevant groups of actors necessary. The PFT scandal, the heavy contamination of groundwater with nitrates and the widespread, to some extent increasing, contamination of water bodies with micropollutants or trace substances from the most diverse sources (pharmaceuticals, pesticides, biocides, etc.) are important additional drivers towards comprehensive change in water management—changes which now increasingly involve the relevant stakeholders.

4.2 Important Research Activities

Water and Resource Protection

The ability to better analyse the effect of pollutants in the environment has been increasing over time. It is now possible not only to analyse those pollutants that show effects in the aquatic environment even in very low concentrations, but also those that are so mobile that they can be detected in boundaries ground and drinking water. This is due to the improved analytical and assessment methods that have been developed in research projects. With the use of more advanced, very specific analytical methods (non-target analytics), detection of additional, potentially environmentally relevant substances was possible. However, there is still a need for research concerning the exact classification and evaluation of the characteristics. In order to identify input sources and to make estimates of elimination measures and their costs, new or improved methods were also used to better quantify the input pathways in conjunction with input modelling.

As a result of the findings regarding the need for action to reduce water pollution through trace substances and pathogens, the shares of research and demonstration projects and publications in this field increased (Figs. 2 and 3). On the one hand, the project's focus was on the development of reduction technologies, e.g. the so-called fourth purification stage for improved wastewater treatment in the municipal sector. On the other hand, research increasingly targeted also comprehensive strategies which included communication measures and raising awareness of the target groups (cf. Löwe et al. 2012, BMBF funding measures RiSKWa: http://riskwa.de resp. DECHEMA 2016).

Extensive work regarding selected micropollutants showed that sufficient emission reductions following the polluter pays and precautionary principle are only possible through a combination of source- and application-oriented as well as downstream measures (Hillenbrand et al. 2016; cf. Figure 4). These results formed the basis for the subsequently launched stakeholder dialogue on the federal government's trace substance strategy. It aimed at developing comprehensive approaches to solutions involving all relevant stakeholders. This development also signifies a transformational shift in paradigm in the reduction of emissions. The challenge is no longer solely assigned to the water sector. New actor constellations together with new forms of producer and user responsibilities are emerging.

Expanded requirements also affect research efforts in the area of resource and energy efficiency. The data on innovation indicators clearly show a substantial increase in the field of energy-water nexus in Phase 3 (see Figs. 2 and 3). In terms of energy and resource efficiency, efforts are being made in line with the energy transition and changes on the energy market. Corresponding research projects aim at energy self-sufficient sewage treatment plants, or sewage treatment plants as resource centres (Deutsche Vereinigung für Wasserwirtschaft, Abwasser and Abfall e. V. (DWA) 2018; Ministerium für Umwelt, Landwirtschaft, Natur- and Verbraucherschutz des Landes Nordrhein-Westfalen (MULNV) 2018).

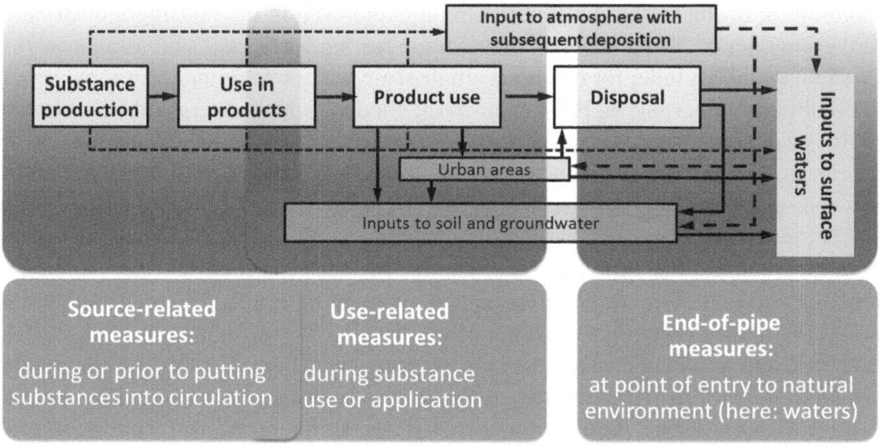

Fig. 4 Starting points for emission reduction measures. Source: Hillenbrand et al. 2017

Niederste-Hollenberg et al. (2021) show the potential of the wastewater industry to achieve climate protection goals and also discuss integrating the wastewater industry into the energy system in order to contribute to regional balancing energy at the same time as using innovative technologies, such as hydrogen production. These scenarios are part of a systemic transition that includes a consistent sector coupling for the synergetic utilisation of all potentials. These developments signify a transformational change on two levels: First, on a technical level, it enlarges the system boundaries and links water and energy technologies together. Second, it shows the overriding impact of climate change as a cross-cutting influence for different innovation domains. In addition to climate adaptation, which greatly influences the challenges of water systems to handle changing precipitation patterns, climate mitigation is now also increasingly shaping solutions in wastewater treatment.

Water Infrastructure for Water Supply and Sewage Disposal

With regard to the recycling of phosphorus contained in wastewater, comprehensive concepts for its practical application are being developed on the basis of other separate technologies that have been developed in the meantime and are now available (see BMBF funding measure Regional Phosphorus Recycling "RePhoR"[3]).

Changes in water and the water infrastructure are an essential element of the climate adaptation measures in the urban environment. Improving the urban climate has become a key issue as a result of the overall changes in climate conditions. This gives rise to the paradigm of blue-green infrastructures or the concept of "water-sensitive urban design", in which the integration of water issues into urban

[3] https://www.bmbf-rephor.de.

planning, in traffic and green space planning, and even in the housing industry are core components. New concepts also include resource-oriented sanitation systems (NASS), which include, for example, small-scale water cycles and heat recovery at a neighbourhood level (von Horn et al. 2013; Niederste-Hollenberg and Hillenbrand 2020), or decentralised system concepts, in which the traditional disadvantages of decentralised approaches can be overcome through the use of innovative technologies and operator concepts (Eggimann et al. 2015; Niederste-Hollenberg et al. 2017).

The new system concepts essentially enable new options for the further development of existing water infrastructure systems. In scenario processes, corresponding development options were concretised and assessed in terms of their sustainability and vulnerability (Hiessl et al. 2005; Hiessl et al. 2012; Hillenbrand et al. 2016). However, the associated changes were so extensive, both in terms of the actors involved, as well as the necessary technologies and the organisational and legal framework conditions, that comprehensive transition processes were required, which various research projects have started to investigate (Hillenbrand et al. 2019; Hillenbrand et al. 2018; Schramm and Winker 2023; Winker et al. 2017; Winker 2017).

4.3 *Measures and Regulations*

New forms of governance to reduce the emissions of micropollutants have been triggered by the need to act in this field and by research results indicating that reduction potentials also involve other groups of actors outside the water sector. Based on the research results which showed a necessity for a very broad approach to solving the trace substance problem, a stakeholder dialogue "Federal Trace Substance Strategy" was launched by the Federal Ministry for the Environment in 2016. The results produced show that new instruments involving the relevant stakeholders are needed both in assessing the relevance of substances and in developing comprehensive emission reduction concepts (Hillenbrand et al. 2017, 2019b). The results also led to the establishment of the Federal Centre for Trace Substances at the German Environment Agency in 2021.

The implementation of the cross-sectoral, resource-oriented and integrated approach to water infrastructure, which was developed through research projects, among other things, requires extensive changes to the framework conditions. In the WHG (Wasserhaushaltsgesetz—water resources act), for example, the 2010 amendment prioritised the local management of precipitation. In 2014, an amendment to Annex 1 of the Wastewater Ordinance (Abwasserverordnung) stipulated that wastewater facilities were to be operated in an energy-efficient manner and that the energy potential arising from wastewater disposal was to be used, as far as technically possible and economically justifiable. At the same time, the DWA (German Association for Water, Wastewater and Waste) developed the first set of rules for new sanitation systems (NASS, DWA-A 272: Principles for the planning and implementation of new sanitation systems (neuartige Sanitätssysteme; NASS)) with the aim of implementing resource-oriented concepts.

With regard to phosphorus recovery, in the 2010s more and more federal states completely withdrew from the use of agricultural sewage sludge due to the problem of pollutants. Against the background of the classification of phosphorus as a resource in need of protection in the German Resource Efficiency Programme (ProgRess) in 2011 by the BMU, the question increasingly arose as to the whereabouts of the phosphorus contained in the ash. A clear answer to this question was provided by the amendment to the Sewage Sludge Ordinance (Klärschlammverordnung) in late 2017, which stipulates phosphorus recovery for large sewage treatment plants (>100,000 inhabitant equivalents.) by 2029 and by 2032 for somewhat smaller ones (>50,000 inhabitants equivalents). Implementation requires inter- and transdisciplinary work involving plant operators, potential users of the products emerging in the future, public administration staff, politicians and other members of society. Ultimately, there must be a restructuring and substantive changes in essential parts of all areas involved, which also justifies the term transformation process in this case.

In 2023, the BMU's water strategy included the goal of adapting water infrastructure to climate change (BMU 2023). To this end, further development of water infrastructures is envisaged, with the implementation of a water-sensitive city (e.g., sponge city, multifunctional land use during heavy rainfall) and the use of resources-oriented sanitation systems as well as the connection of water, energy and material cycles. Corresponding objectives are now also being embedded at the level of the federal states. For example, the future plan water ("Zukunftsplan Wasser") of the state of Hesse was developed with a variety of different measures aiming to ensure the long-term protection of water resources and supply, including the support of groundwater formation through retention and seepage; or the protection of bodies of water from pollutant inputs (Hessisches Ministerium für Umwelt, Klimaschutz, Landwirtschaft and Verbraucherschutz (HMUKLV) 2022). The DWA position paper "Water-conscious development of our cities (Wasserbewusste Entwicklung unserer Städte)" published in 2021 also contains the demand for a water-conscious design of cities in which blue-green infrastructures are strengthened, resources are used efficiently and land is used multifunctionally (Deutsche Vereinigung für Wasserwirtschaft, Abwasser and Abfall e. V. (DWA) 2021). These developments demonstrate the need for comprehensive change and transformation in water management in Germany, which is meanwhile widely recognised.

5 Discussion and Conclusion

The development in water management since the 1970s has shown three phases in which pressure to act, research and regulation and policies all interacted with each other. Figure 5 shows the main developments over time. In the first phase, especially in the 1970s, specific individual measures were required as a direct response to the

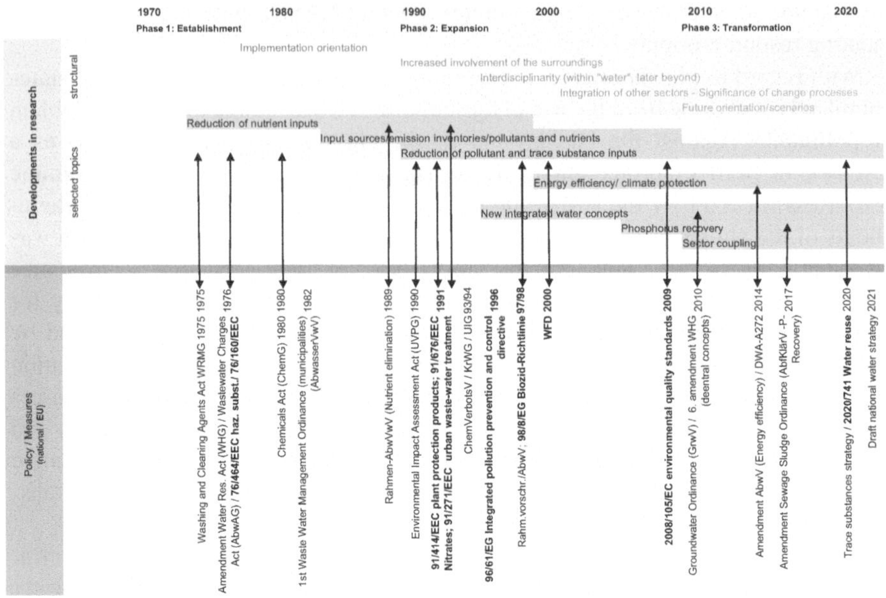

Fig. 5 Priorities and developments in research in comparison with policy measures in the water sector since 1970. Source: Fraunhofer ISI

critical environmental situation of water bodies. The essential cause–effect relationships were largely known, and the necessary measures could therefore be derived directly from them. Accordingly, the focus of research activities lay on the development of the necessary technologies, e.g. for improved wastewater treatment. Due to the implementation requirements, it was of great importance to ensure the practical relevance of the work already while the research activities were taking place. For this reason, an increasing number of projects were funded that could guarantee this practical relevance by involving practical partners. Regulation played a central role in the broad implementation of the measures.

In the subsequent second phase, which spanned the 1990s and 2000s until around 2010, the focus of research expanded significantly: the investigations to remedy the persisting deficits in water protection revealed significantly more complex interrelationships and a comprehensive need for action, which required interdisciplinary, cross-media and integrated solution approaches. The contents and structures in research changed accordingly: the importance of collaborative projects increased significantly, not only to involve practical partners to implement the results as directly as possible, but also to ensure interdisciplinary and integrated processing of the (new) questions. Research results showed the great importance of environmental factors and change processes. Work with scenario processes was started in order to identify development possibilities owing to the uncertainties associated with the development of the above factors and against the backdrop of the long duration of use and substantial sunk costs in the area of grid-based infrastructure systems and the associated high technological (and organisational) path dependencies. The comprehensive need for action was also evident at the level of measures. Particularly outstanding examples are the European IPPC Directive with its parallel consideration of the various environmental media and the EU Water Framework Directive with its overarching objective of "good status" for all bodies of water and the demand for comprehensive river basin management that also takes the water body's environment into account.

From 2010 onwards, one can speak of a third phase in which the pressure to act, which had already been identified in the work performed and in the scenario considerations in the second phase, intensified significantly. The increase in severe precipitation events or the very hot and dry summers also showed the effects of climate change on water management in Germany. The requirements with regard to resource and energy efficiency are increasing as is the need for action with regard to water pollution with nutrients and trace substances. The questions derived from this were taken up in research and as a result showed a comprehensive need for change in water management and the necessity for integrated concepts at different levels: the interaction of bodies of water and their surrounding environment have to be coupled with source-oriented and downstream measures to reduce trace substance pollution of bodies of water. Furthermore, the water infrastructure systems have to be integrated into the urban environment much stronger. Against this background, additional instruments were used with regard to measures being implemented (participatory approaches, stakeholder processes and education and awareness campaigns). The pressure on the entire water sector to adapt, as evidenced by research findings and reinforced by changes in key framework conditions, has led to a gradual recognition of the need for transformation in the water sector. This necessity is now reflected in various regulations and political objectives such as the national water strategy (BMU 2023).

The diverse challenges suggest that a holistic approach in combination with cross-sectoral coupling with areas that also face challenges (energy transition, resource efficiency, urban development, etc.) will determine the next phase. Sustainable solutions will only be possible through cross-system concepts.

The analysis for the three phases also revealed the following pattern of interrelation: The pressure to act influences not only regulation and policy, but also research. Thus, water-related research is a field where contributing to societal challenges is not new, but has been practised for a long time. The impact of policies and regulations influences the pressure to act in accordance with their impact on the state of the environment. Research, on the other hand, also influences the pressure to act through its influence on identifying new problems and agenda setting (Fig. 6).

The interactions between research, on the one hand, and regulation and policy responses, on the other, take place in both directions and through different channels. Regulatory agencies influence research in the first instance through the funding of research projects. In particular, the research projects of the Federal Ministry for the Environment and the projects funded by the federal states are driven directly by supporting future regulation and implementing policies. Sufficient leeway is important in the design of the regulations, for example, with regard to the use of different

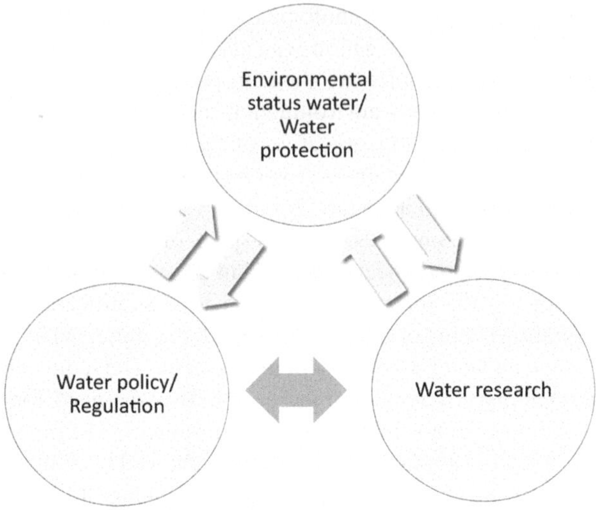

Fig. 6 Interplay between environmental status, policy and research in the water sector. Source: Fraunhofer ISI

technical solution concepts. Thus, research projects are designed in such a way that planned new regulations can be met technically. Other research topics, which have increased from phase one to three are the evaluation of policies and issues such as planning procedures and acceptance, which are also of increasing importance for policymakers. Furthermore, regulation also influences research by establishing new institutional rules. In particular, the EU Water Framework Directive has driven an approach of looking into the integrated management of water bodies and strongly influenced the topics and organisation of research projects. However, as it takes time for regulations and policies to be adopted and implemented, the influence of regulation and policies on research usually begins with the debates on the regulation.

If the influence of regulation on research leads to successful results, then research establishes an influence on regulation. There are several links to achieve this impact path: First, the traditional paradigm of the implementation of regulation follows a best available technology approach: New technical solutions, which allow for reduced emissions limits to be reached, can be implemented quickly during implementation even without the formal modification of existing laws and ordinances. Second, research can also influence regulation and policies in the medium and long-term if it identifies shortcomings and new forms of policy solutions. The need for solutions to tackle micropollutants during phase 3 outside the water sector is one of the key success stories, where research also triggered a new policy approach. However, there is no guarantee that this sort of interplay always takes place. The institutional move towards integrated water basin management in phase 2 also improves the institutional setting for making use of new policy instruments such as emissions trading. During Phase 2, this policy instrument was implemented in parallel in the field of climate mitigation. However, despite these developments, emissions trading has hardly been recognised in German water research and policy at all.

If we look at the development of water research over the course of all three phases, we can see certain developments which also indicate a development of the underlying research and innovation system. Several changes have contributed to research getting more systemic and more interdisciplinary:

- technical systems are becoming more integrated, e.g., across different infrastructure areas,
- policy strategies are increasingly looking beyond technical solution concepts and their feasibility,
- climate change and approaches such as the water and energy nexus are linking problems and sectors and require integration with other local domains such as urban development,
- new approaches require sectorial expertise within a water basin, but also require to move beyond the water sector by integrating producer and user perspectives,
- technical perspectives have to be integrated with social science perspectives, not only because acceptance and communication issues gain in importance, but also because technical solutions are supplemented with social innovations.

Even though the challenge of transformation calls also for questioning the organisational and institutional setting of the water sector, there is currently neither a debate about changing the innovation culture within the water sector nor one about its restructuring. Although the need to move beyond incremental innovations is called for, e.g., by the national water strategy, the discussion about the general need to increase innovation dynamics in the water industry and on ways to achieve this still seems to be rather hesitant. There is also only limited debate about the need to create successful transition strategies for socio-technical regime change within the water sector. Indeed, there are some developments in innovation strategies and transformation needs which are similar in the innovation domain of renewable energy and in water systems. In the case of renewable energy, such a need has led to an intensive debate about policy instruments and analysis, and the heuristics of the innovation system and multi-level-perspectives are important components of research (see Panny et al. 2024 in this anthology). In the water domain, however, there are only a few exceptions in this direction, (e.g., Hillenbrand et al. 2013; Heiberg and Truffer 2022; Kiparsky et al. 2013; Hohmann and Truffer 2022), and approaches such as technological innovation system analysis or multi-level perspectives have not been shaping research in the water sector. It remains to be seen to what extent the need for action to transform the water system will also lead to these key heuristics of innovation research being included in the water research portfolio.

References

Ahrens A, Böhm E, Heitmann K, Hillenbrand T (2003) Guidance for the use of environmentally sound substances. For producers and professional users of chemical products relevant to the aquatic environment. Part one: five steps for the evaluation of environmental risks, Berlin

Battelle-Institut e.V. (1972) Wasserbedarfsentwicklung in Industrie, Haushalten, Gewerbe, öffentlichen Einrichtungen und Landwirtschaft: Prognose des Wasserbedarfs in der Bundesrepublik Deutschland bis zum Jahr 2000. Prognose des Wasserbedarfs in der Bundesrepublik Deutschland bis zum Jahr 2000 / Battelle-Institut e.V., Frankfurt am Main. Bundesministerium des Innern, Bonn

Bauer K (1977) Wasserforschung muss stark zielgerichtet sein. Beratende Ingenieure:20–24

Bergmann A, Fohrmann R, Weber F-A (2011) Zusammenstellung von Monitoringdaten zu Umweltkonzentrationen von Arzneimitteln. UBA-FB, vol 1525. Umweltbundesamt, Dessau-Roßlau

BMI (1971) Materialienband zum Umweltprogramm der Bundesregierung. In: Drucksache, vol 6. Deutscher Bundestag, Bonn, p 2710

BMU (2023) National Water Strategy, Berlin. Online: https://www.bmuv.de/fileadmin/Daten_BMU/Download_PDF/Binnengewaesser/nationale_wasserstrategie_2023_en_bf.pdf

Böcher M, Töller AE (2012) Eigendynamik und Zufall als Triebkräfte der Umweltpolitik? Ein Ansatz zum Erklären umweltpolitischer Ergebnisse. Zeitschrift für Umweltpolitik und Umweltrecht 35(4):450–479

Böhm E, Hillenbrand T (2000) Emissionsinventar Wasser für die Bundesrepublik Deutschland: Forschungsbericht 29621524/01. Umweltforschungsplan des Bundesministeriums für Umwelt, Naturschutz und Reaktorsicherheit: Wasserwirtschaft, vol 2000,53. Umweltbundesamt, Berlin

Böhm E, Hillenbrand T, Hiessl H, Walz R (1998) Effektivität und Effizienz technischer Normen und Standards für kommunale Umweltschutzaufgaben am Beispiel der kommunalen Abwasserentsorgung. Abschlussbericht, Abschlussdatum 26.03.1998. Fraunhofer-Institut für Systemtechnik und Innovationsforschung. Karlsruhe (Umweltforschungsplan des Bundesministers für Umwelt, Naturschutz und Reaktorsicherheit Wasserwirtschaft, UBA-FB-98–180)

Böhm E, Hillenbrand T, Walz R, Borchardt Dietrich (1999) Maßnahmenplan Nachhaltige Wasserwirtschaft – Handlungsschwerpunkte für einen zukunftsorientierten Umgang mit Wasser in Deutschland. Ed Umweltbundesamt (Texte, 55/1999)

Böhm E, Hillenbrand T, Liebert J, Schleich J, Walz R (2002) Kostenwirksamkeit von nachhaltigen Maßnahmen im Gewässerschutz. Abschlussbericht des FhG-ISI im Auftrag des Umweltbundesamtes. Umweltbundesamt, Berlin

Brunner U, Riechmann C (2004) Wettbewerbsgerechte Preisbildung in der Wasserwirtschaft: Vergleichsmarktkonzepte, -methoden und Erfahrungen aus England & Wales. Zeitschrift für öffentliche und gemeinwirtschaftliche Unternehmen 27:115–130

DECHEMA (2016) RiSKWa – risk management of emerging compounds and pathogens in the water cycle: handbook of good practice. BMBF funding measure, Frankfurt am Main

Deutsche Vereinigung für Wasserwirtschaft, Abwasser und Abfall e. V (2008) Neuartige Sanitärsysteme, 2008th ed. DWA-Themen. DWA, Hennef

Deutsche Vereinigung für Wasserwirtschaft, Abwasser und Abfall e. V. (DWA) (ed) (2018) Energieeffiziente und Ressourcenschonende Wasserwirtschaft (ERWAS) Perspektivpapier. Online verfügbar unter. https://www.fona.de/medien/pdf/erwas_perspektivpapier_2018_ansicht.pdf

Deutsche Vereinigung für Wasserwirtschaft, Abwasser und Abfall e. V. (DWA) (ed) (2021) DWA-Positionen – Wasserbewusste Entwicklung unserer Städte. Online verfügbar unter. https://de.dwa.de/files/_media/content/01_DIE_DWA/Politikinformationen/Positionspapiere/Positionspapier_Wasserbewusste_Entwicklung_unserer_St%C3%A4dte_2021_Netz.pdf

Eberle SH (1978) Öffentliche Förderung der Wasserforschung. In: Wasser '77: Die technisch-wissenschaftlichen Vorträge auf dem Kongreß Wasser Berlin 1977. Band 1, vol 1. Colloquium Verlag, Berlin, pp 455–462

Edler J (2016) The impact of policy measures to stimulate private demand for innovation. In: Handbook of innovation policy impact. Edward Elgar Publishing, Cheltenham, UK, pp 318–354

Eggimann S, Truffer B, Maurer M (2015) To connect or not to connect? Modelling the optimal degree of centralisation for wastewater infrastructures. Water Res 84:218–231. https://doi.org/10.1016/j.watres.2015.07.004

Fuchs S, Scherer U, Hillenbrand T, Marscheider-Weidemann F, Behrendt H, Opitz D (eds) (2002) Emissions of heavy metals and lindane into river basins of Germany: research report 200 22 233. Environmental research of the Federal Ministry of the environment, nature conservation and nuclear safety, vol 2002,55. Umweltbundesamt, Berlin

Gawel E, Bedtke N (2015) Effizienz und Wettbewerb in der deutschen Wasserwirtschaft zwischen "Modernisierung" und "Regulierung". Zeitschrift für öffentliche und gemeinwirtschaftliche Unternehmen 38:97–132

Heiberg J, Truffer B (2022) The emergence of a global innovation system—a case study from the urban water sector. Environ Innov Soc Trans 43:270–288

Helmholtz-Zentrum für Umweltforschung – UFZ (2015) Integriertes Wasserressourcen-Management: Von der Forschung zur Umsetzung. 5. Aufl. Wurmberg. Online. https://www.wasserressourcen-management.de/_media/150408_IWRM_Broschuere_deutsch_barrierefrei.pdf

Hessisches Ministerium für Umwelt, Klimaschutz, Landwirtschaft und Verbraucherschutz (HMUKLV) (2022): Zukunftsplan Wasser – Wasserwirtschaftlicher Fachplan Hessen. Hessisches Ministerium für Umwelt, Klimaschutz, Landwirtschaft und Verbraucherschutz. Online. https://hessen.de/sites/hessen.hessen.de/files/2022-10/zukunftsplan_wasser.pdf

Hiessl H, Toussaint D, Becker M (2005) AKWA 2100: Alternativen kommunaler Wasserver- und Abwasserentsorgung. Innovationspotenziale. Fraunhofer-IRB-Verl, Stuttgart

Hiessl H, Hillenbrand T, Klug S, Lange M, Vöcklinghaus S, Flores C, Weilandt M (2012) Nachhaltige Weiterentwicklung kommunaler Wasserinfrastrukturen—Strategischer Planungsprozess unter Einbindung aller wesentlichen Akteure. Energie-, Wasser-Praxis 63:13–16

Hillenbrand T, Böhm E (2008) Entwicklungstrends des industriellen Wassereinsatzes in Deutschland. KA: Abwasser, Abfall 55 (2008), 8, p. 872–882

Hillenbrand T, Hiessl H (2007) Sich ändernde Planungsgrundlagen für Wasserinfrastruktursysteme. Teil 2: Technologischer Fortschritt und sonstige Veränderungen. KA: Korrespondenz Abwasser, Abfall 54:47–53

Hillenbrand T, Kunz P (1989) Optimierung des Energiebezugs von Kläranlagen. KA: Korrespondenz Abwasser, Abfall 36:439–445

Hillenbrand T, Toussaint D, Böhm E, Fuchs S, Scherer U, Rudolphi A, Hoffmann M (2005) Einträge von Kupfer, Zink und Blei in Gewässer und Böden. Analyse der Emissionspfade und möglicher Emissionsminderungsmaßnahmen. Dessau. Online. http://www.umweltbundesamt.de/sites/default/files/medien/publikation/long/2936.pdf

Hillenbrand T, Sartorius C, Hiessl H (2009) Anwendungspotenziale für neue Systemkonzepte in Deutschland. In: Abwasserrecycling – Chancen und Risiken. 2. Internationales symposium Abwasserrecycling 2009, Braunschweig, pp 143–159

Hillenbrand T, Niederste-Hollenberg J, Holländer R, Lautenschläger S, Galander C (2011) Demografischer Wandel – Auswirkungen und Lösungsansätze für die Abwasserinfrastruktur. KA: Korrespondenz Abwasser Abfall 58:1132–1138. https://doi.org/10.3242/kae2011.12.002

Hillenbrand T, Hiessl H, Klug S, Freiherr von Lüninck B, Niederste-Hollenberg J, Sartorius C, Walz R (2013) Herausforderungen einer nachhaltigen Wasserwirtschaft – Innovationsreport. Verhandlungen des Deutschen Bundestages Drucksachen, 18/2085. Bundestag, Berlin

Hillenbrand T, Tettenborn F, Menger-Krug E, Fuchs S, Toshovski S, Metzger S, Tjoeng I, Wermter P, Kersting M, Abegglen C (2014) Kombination quellenorientierter und nachgeschalteter Maßnahmen zur Emissionsminderung von Mikroschadstoffen aus dem urbanen Bereich. KA: Korrespondenz Abwasser, Abfall 61:1019–1028

Hillenbrand T, Tettenborn F, Toshovski S, Fuchs S, Metzger S, Tjoeng I et al. (2016) Maßnahmen zur Verminderung des Eintrages von Mikroschadstoffen in die Gewässer – Phase 2. Unter Mitarbeit von Yannis Geiger. Fraunhofer-Institut für System- und Innovationsforschung. Karlsruhe (Umweltforschungsplan des Bundesministeriums für Umwelt, Naturschutz, Bau und Reaktorsicherheit, UBA-FB-002362)

Hillenbrand T, Tettenborn F, Bloser M (2017) Policy-Paper Empfehlungen des Stake-holder-Dialogs »Spurenstoffstrategie des Bundes« an die Politik zur Reduktion von Spurenstoffeinträgen in die Gewässer. Hg. v. BMU/UBA. Bonn. Online. https://www.dialog-spurenstoffstrategie.de/spurenstoffe-wAssets/docs/PolicyPapier_FINAL.pdf

Hillenbrand T, Eckartz K, Hiessl H, Hohmann C, Niederste-Hollenberg J (2018) Transition urbaner Wasserinfrastruktursysteme – notwendig und machbar? KA: Korrespondenz Abwasser, Abfall 65:121–129. https://doi.org/10.3242/kae2018.02.003

Hillenbrand T, Hohmann C, Bieker S, Niederste-Hollenberg J, Tettenborn F (2019) Kopplung technischer Infrastruktursysteme am Beispiel Wasser: Potenziale, Hemmnisse und deren Überwindung. Transforming Cities 2019(4):75–79

Hillenbrand T, Tettenborn F, Bloser M (2019b) Ergebnispapier – Ergebnisse der Phase 2 des StakeholderDialogs »Spurenstoffstrategie des Bundes« zur Umsetzung von Maßnahmen für die Reduktion von Spurenstoffeinträgen in die Gewässer. Ed. BMU/UBA. Online. https://www.dialog-spurenstoffstrategie.de/spurenstoffe-wAssets/docs/ergebnispapier_stakeholder_dialog_phase2_bf.pdf

Hoffjan A, Lechtenberg C, Schuchardt LD (2011) Methoden zur Effizienzmessung in der deutschen Wasserversorgung-Eine Auswertung internationaler Studien. Zeitschrift für öffentliche und gemeinwirtschaftliche Unternehmen 34:325–345

Hohmann C, Truffer B (2022) The infrastructure transition canvas: a tool for strategic urban infrastructure planning. Nature-Based Solutions 2:100039

Holländer R (2011) Kosten und finanzielle Nachhaltigkeit in der kommunalen Wasserwirtschaft. Zeitschrift für öffentliche und gemeinwirtschaftliche Unternehmen 34:396–414

Kautt M (ed) (1996) Wasserforschung: Ergebnisse der Förderung des BMBF in den Jahren 1990 bis 1996. Forschungszentrum Karlsruhe GmbH, Technik und Umwelt, Projektträgerschaft für Wassertechnologie und Schlammbehandlung (PtWT), Karlsruhe

Kiparsky M, Sedlak DL, Thompson BH, Truffer B (2013) The innovation deficit in urban water: the need for an integrated perspective on institutions, organizations, and technology. Environ Eng Sci 30:395–408. https://doi.org/10.1089/ees.2012.042

Kluge T, Libbe J (eds) (2010) Transformationsmanagement für eine nachhaltige Wasserwirtschaft: Handreichung zur Realisierung neuartiger Infrastrukturlösungen im Bereich Wasser und Abwasser. Difu-Sonderveröffentlichung, Difu, Berlin

Kluge T, Schramm E (2016) Wasser 2050: Mehr Nachhaltigkeit durch Systemlösungen. Oekom, München

Koch T (2003) Vorgaben des europäischen und bundesdeutschen Rechts für eine Neuorganisation der kommunalen Wasserversorgung. Zeitschrift für öffentliche und gemeinwirtschaftliche Unternehmen 26:241–258

Kunz P, Müller A (1986) Ergebnisse von Stromverbrauchsmessungen in kleineren und mittelgrossen Kläranlagen: Ein Beitrag zur differenzierteren Betrachtung des Energiehaushalts von Kläranlagen; Arbeitspapier. KA: Korrespondenz Abwasser, Abfall 33:406–414

Leinweber T (2008) Umsetzung der EU-Wasserrahmenrichtlinie in Deutschland: Bestandsaufnahme, Monitoring, Öffentlichkeitsbeteiligung und wichtige Bewirtschaftungsfragen, 2008th ed. In: Materialien zur Umweltforschung, Nr. 39. Sachverständigenrat für Umweltfragen, Berlin

Londong J, Hillenbrand T, Niederste-Hollenberg J (2011) Demografischer Wandel: Anlass und Chance für Innovationen in der Wasserwirtschaft. KA: Korrespondenz Abwasser, Abfall 58(2):152–158. https://doi.org/10.3242/kae2011.02.005

Löwe H, Schade J, Höckele V (2012) Förderschwerpunkt "Nachhaltiges Wassermanagement" (NaWaM) im Rahmenprogramm "Forschung für nachhaltige Entwicklungen" (FONA). Bonn, Berlin

Mankel B, Schwarze R (2000) Wettbewerb in der Wasserversorgung: Konzepte, Modelle, Effekte. Zeitschrift für öffentliche und gemeinwirtschaftliche Unternehmen 23(4):418–427

Menger-Krug E, Niederste-Hollenberg J, Hillenbrand T, Hiessl H (2012) Overview of material and energy flows in German wastewater infrastructures. GWF – Wasser/Abwasser 153:96–104

Merkel W, Reiff G (1983) Forschung und Entwicklung (F und E) in der Wasserversorgung. GWF – Wasser/Abwasser 128:243–253

Michaelis P (2001) Wasserwirtschaft zwischen Markt und Staat: Zur Diskussion um die Liberalisierung der deutschen Wasserversorgung. Zeitschrift für öffentliche und gemeinwirtschaftliche Unternehmen 24(4):432–450
MULNV (ed) (2018) Energie- und Abwasser-Handbuch NRW
Niederste-Hollenberg J, Hillenbrand T (2020) Von der Siedlungsentwässerung zum Wassermanagement. fbr-Wasserspiegel:3–8
Niederste-Hollenberg J, Hillenbrand T, Ungermann A (2017) Zentraler Betrieb dezentraler Anlagen. z*dez – Erfahrungen aus drei Jahren Betrieb. KA. Wasserwirtschaft, Abwasser, Abfall 64:977–984
Niederste-Hollenberg J, Schirmer G, Borger J, Winkler J, Zheng L, Fritz M et al. (2021) Klimaschutz- und Energieeffizienzpotenziale in der Abwasserwirtschaft – aktueller Stand und Perspektiven. Online verfügbar unter. https://www.umweltbundesamt.de/sites/default/files/medien/5750/publikationen/2021-04-30_texte_50-2021_klimaschutz_abwasserwirtschaft.pdf
Panny J, Held A, Winkler J, Breitschopf B, Jochem E, Walz R (2024) From niche to mainstream: exploring innovation and progress of renewable energy development. In: Edler J, Walz R (eds) Systems and innovation research in transition: research questions and trends in historical perspective. Springer, Heidelberg
Petermann T, Bradke H, Lüllmann A, Poetzsch M, Riehm U (2011) Was bei einem Blackout geschieht: Folgen eines langandauernden und und großräumigen Stromausfalls. Nomos Verlagsgesellschaft, S.l.
Rennings K (2000) Redefining innovation: eco-innovation research and the contribution from ecological economics. Ecol Econ 32:319–332
Richter S, Völker J, Borchardt D, Mohaupt V (2013) The water framework directive as an approach for integrated water resources management: results from the experiences in Germany on implementation, and future perspectives. Environ Earth Sci 69:719–728. https://doi.org/10.1007/s12665-013-2399-7
Sartorius C, Hillenbrand T, Walz R (2011) Impact and cost of measures to reduce nutrient emissions from wastewater and storm water treatment in the German Elbe river basin. Reg Environ Chang 11:377–391. https://doi.org/10.1007/s10113-010-0140-6
Sartorius C, von Horn J, Tettenborn F (2012) Phosphorus recovery from wastewater--expert survey on present use and future potential. Water Environ Res 84:313–322. https://doi.org/10.2175/106143012x13347678384440
Schneidmadl J, Fuchs S, Hillenbrand T (2000) Vergleichende Ökobilanz unterschiedlicher Abwasserkonzepte. KA- Wasserwirtschaft, Abwasser, Abfall 47:681–696
Schramm E, Winker M (2023) Transformation der Wasserinfrastruktur im Siedlungsbestand als Herausforderung: Pfade, Instrumente, Entwicklungsbedarf. In: ISOE Materialien Soziale Ökologie, vol 68, 68th edn. Institut für sozial-ökologische Forschung ISOE GmbH, Frankfurt am Main
SRU (1974) Umweltgutachten 1974
SRU (1987) Umweltgutachten 1987
SRU (2002) Umweltgutachten 2002: Für eine neue Vorreiterrolle
SRU (2004) Umweltgutachten 2004: Umweltpolitische Handlungsfähigkeit sichern
SRU (2020) Umweltgutachten 2020: Für eine entschlossene Umweltpolitik in Deutschland und Europa
Töpfer K, Schäfer HB, Brösse U (1988) Das Umweltprogramm zum Schutz von Nord- und Ostsee. Wirtschaftsdienst <Heidelberg> 68(9):443–450
UBA (ed) (2006) Wasserwirtschaft in Deutschland- Teil 2: Gewässergüte. Berlin/Dessau
van Leuven LJ (2011) Water/wastewater infrastructure security: threats and vulnerabilities. In: Clark RM, Hakim S, Ostfeld A (eds) Handbook of water and wastewater systems protection. Springer, New York, New York, NY, pp 27–46
Vetter C, Gropp M, Erhardt M, Menger-Krug E, Hillenbrand T (2011) Wärmerückgewinnung aus Grauwasser mit dem Pontos HeatCycle. In: Fachvereinigung Betriebs- und Regenwassernutzung: Wasserautarkes Grundstück, Darmstadt, pp 63–77

von Horn J, Maurer M, Londong J, Lautenschläger S, Steinmetz H, Hillenbrand T (2013) Welche neuartigen Sanitärsysteme (NASS) sind für Deutschland besonders Erfolg versprechend? KA: Korrespondenz Abwasser, Abfall 60:673–683
Wackerbauer J (2009) Struktur und Entwicklung der Wasserversorgung in Deutschland im Vergleich zu anderen europäischen Ländern. Zeitschrift für öffentliche und gemeinwirtschaftliche Unternehmen
Walz R (2007) The role of regulation for sustainable infrastructure innovations: the case of wind energy. IJPP 2:57. https://doi.org/10.1504/IJPP.2007.012276
Wechsung F, Hartje V, Kaden S, Venohr M, Hansjürgens B, Gräfe P (eds) (2013) Die Elbe im globalen Wandel: Eine integrative Betrachtung, Ökologie, vol 9. Weißensee-Verl, Berlin
Winker M (ed) (2017) Wasserinfrastruktur: Den Wandel gestalten: Technische Varianten, räumliche Potenziale, institutionelle Spielräume. Edition Difu – Stadt, Forschung, Praxis, vol 16. Deutsches Institut für Urbanistik, Berlin
Winker M, Trapp JH, Libbe J, Schramm E (eds) (2017) Wasserinfrastruktur: den Wandel gestalten: Technische Varianten, räumliche Potenziale, institutionelle Spielräume, 1st edn. Edition Difu – Stadt, Forschung, Praxis, Bd. 16. Deutsches Institut für Urbanistik, Berlin
Zschille M, Walter M, von Hirschhausen CR (2010) Ineffizienz und Strukturunterschiede in der deutschen Wasserversorgung. Zeitschrift für öffentliche und gemeinwirtschaftliche Unternehmen

Thomas Hillenbrand studied chemical engineering. PhD at the University of Karlsruhe, Department of Civil Engineering, Geo and Environmental Sciences, on the "Analysis and Evaluation of new Urban Water Infrastructure Systems". Working at the Fraunhofer Institute for Systems and Innovation Research in the Competence Centre Sustainability and Infrastructure Systems, since 2008 as Coordinator of Business Unit "Water Resources Management". Member of different specialist groups of the German Association for Water, Wastewater and Waste (DWA) (KA-1.4 "Systemintegration" (speaker), KA1: "Neue Sanitärsysteme", KA 1.8 "Forschungs- und Entwicklungsbedarf", ES-1.2 "Stoffeintraege in Entwaesserungssysteme").

Harald Hiessl holds a Diploma Degree in Hydrology from Freiburg University (1980). Doctoral Degree (Dr.-Ing.) in Water Resources Engineering from Technical University Karlsruhe (1985). Feodor-Lynen Post-Doc Fellowship of the Alexander-von-Humboldt-Foundation at the Dept. of Hydrology and Water Resources and Visiting Assistant Professor at the Optical Sciences Centre, both at the University of Arizona, Tucson, AZ, USA (1986–1988). Various positions at the Fraunhofer Institute for Systems and Innovation Research ISI, Karlsruhe: Project Manager and Group Manager (1988–1996), Head of Dept. Environmental Technology and Environmental Economics (1996–2004), Head of Competence Centre of Sustainability and Infrastructure Systems (2005–2008), and Deputy Director of Fraunhofer ISI (2007–2017). He retired at the beginning of 2018.

Frank Marscheider-Weidemann, born in 1960, studied chemistry at the University of Hamburg where he completed his doctorate degree in technical chemistry on the improvement of the products of the Hamburg pyrolysis process. From 1992 to 1993 he worked as a postdoctoral research assistant at the Institute for Technical and Macromolecular Chemistry. Since 1994 to present, Frank Marscheider-Weidemann has been working for the department "Sustainability and Infrastructures" at the Fraunhofer ISI. He focusses on evaluating innovative processes and products.

Jutta Niederste-Hollenberg is a civil engineer specialising in urban water management and environmental technology. In her research at Fraunhofer ISI, she deals with issues of sustainable water management, in particular sustainable urban water infrastructure systems and micropollutants in water bodies. She also works on transformation and policy processes and sustainability assessment. She is a member of various DWA committees, Vice President of the fbr and a member of the BDZ Scientific Advisory Board.

Christian Sartorius studied biology and economics and received PhDs in Biochemistry and Evolutionary Economics from the universities in Saarbrücken and Jena. Since 2001 at the Technical University Berlin and since 2004 with Fraunhofer ISI, he developed assessment criteria and investigated conditions for implementing more sustainable technologies in the fields of energy supply and use, water supply and wastewater systems and the efficient use of material resources.

Rainer Walz is Deputy Director of the Institute and Head of the Competence Centre Sustainability and Infrastructure Systems at the Fraunhofer Institute for Systems and Innovation Research ISI in Karlsruhe. He obtained his PhD and "habilitation" in economics. In 1991, Prof. Dr. Walz joined the Fraunhofer ISI. Prior employment includes the University of Wisconsin and the Enquête Commission "Protecting the Earth's Atmosphere" of the German Bundestag. He is also an adjunct professor of economics at the Karlsruhe Institute of Technology (KIT) and member of the EU expert group "Economic and Social Impact of Research".

Drivers and Patterns of Change in Systems and Innovation Research

Jakob Edler and Rainer Walz

Abstract This chapter draws overall conclusions across all the contributions to this volume. Its first aim is to characterise each of the nine analysed systems and innovation research fields (SIR fields) in terms of their patterns and dynamics of change. How and why have the respective research questions and approaches evolved, and how has the role of each field vis-à-vis policy changed? The second aim is to draw overall conclusions concerning the dynamics of change across the fields. To achieve these two aims, the chapter first develops a change model of system and innovation research that is then applied to all fields. This model defines four drivers of change in the fields, i.e. policy processes, contextual developments, theoretical developments and developments in data and methods. It also takes into consideration the relationship of the SIR fields to traditional academic fields that were established prior to the evolution of our fields. Although we find some dynamics specific to each field, we also identify commonalities and similar patterns across the fields, mirroring the overarching zeitgeist changes of the last 50 years. The chapter closes with a few speculations and normative claims regarding the future development of our nine fields.

J. Edler (✉)
Fraunhofer Institute for Systems and Innovation Research ISI, Karlsruhe, Germany

Manchester Institute of Innovation Research (MIoIR), University of Manchester, Manchester, United Kingdom
e-mail: jakob.edler@isi.fraunhofer.de

R. Walz
Fraunhofer Institute for Systems and Innovation Research ISI, Karlsruhe, Germany
e-mail: Rainer.Walz@isi.fraunhofer.de

J. Edler, R. Walz (eds.), *Systems and Innovation Research in Transition*, Sustainability and Innovation, https://doi.org/10.1007/978-3-031-66100-6_11

1 Introduction

In this chapter, we want to explore whether there are commonalities and differences between the research fields analysed in the previous chapters of this book, and whether we can identify general patterns in their development. As a first step, we develop and apply a simplified systemic model of (research) field dynamics to help the analysis. This model has emerged inductively by analysing how these nine research fields covered in this book have developed over time. In the second step, we use this model to reflect on the development in each research field. In the third step, we draw conclusions. We use the experience from the nine research fields to reflect on the drivers behind the development of SIR research fields in general and to consider how research in the various fields of systems and innovation research (SIR) could and should develop in the future.

2 Systemic Model of (Research) Field Dynamics

Research provides new knowledge and helps to spur innovation. Thus, the heuristics used to analyse knowledge creation and innovation are also useful starting points for our systemic model.

Modern innovation research emphasises that knowledge creation and innovation are social processes driven by the interaction between different actors within an innovation system. They are characterised by co-evolutionary processes between the different elements in such a system, as well as by a cumulative process. In hindsight, we can conceptualise the development of research fields to follow a certain general—simplified—pattern. At the beginning, the traditional, long established research fields were perceived to be insufficient to tackle new needs. Thus, those needs were addressed with new research paradigms, a specific mode of research and a new set of actors, which led towards establishing new areas of research, subsequently developing into new research fields. This process has been cumulative and recursive, knowledge stocks thus created built on previous ones by combining existing knowledge with new insights, and in doing both challenge, influence and re-confirm the nascent research field. Over time, this process led to new established research fields with their own established epistemological structures and actor landscape. Therefore, SIR developed in a co-evolutionary processes with the external demands and an internal logic driven by the cumulative nature of knowledge and development of specific epistemic practices.

One of the specific and defining features of our SIR fields is the fact that they from the beginning have been strongly linked to policy making. Throughout their history, they have provided conceptualisation and empirical evidence to support, underpin or assess policies. Thus, in all nine fields, we see strong patterns of co-evolution between research activities and policy practice. Although this co-evolution differs in each field, and its nature changes over time, it is still possible to detect patterns concerning the mutual influence of policy and research. Analysing the histories of the nine research fields, and in order to take a comparative and systematic view of these dynamics, we propose to conceptualise this co-evolution in a four-dimensional systemic model. We assert that each of the four dimensions can be the source for the demand, orientation and shape of research in a given research field.[1]

As we are concerned with application and policy-oriented research fields, the most obvious dimension is the definition of what a *policy concern* is. The formulation or adaptation of policy problems requests sound, scientific advice that often necessitates adjusted conceptual underpinning and new forms of empirical analysis. A second dimension is formed by the new opportunities and challenges stemming from *contextual developments*, i.e. technological, economic, environmental, societal or (geo)political dynamics. Those developments can be of very different levels of gravity and severity. Some of the recognised, major societal challenges, such as climate change concerns may exert a somewhat dominant influence over time both on policy and on the academic community with re-enforcing dynamics across all dimensions of the model. Thirdly, from time to time, there are *new theoretical or conceptual developments* that have the power to bring about profound change in research fields. Fourthly, new *methodological developments* and the availability of new data sources can generate new questions and new types of empirical evidence. In particular the latter two dimensions are linked to established academic fields, which influence the behaviour of individual scientists, their career paths and options, as well as patterns of scientific exchange.

In addition to being influenced by these external dimensions, research field development is also driven by its internal logic. The above-mentioned cumulative nature of knowledge means that research questions and methodologies are also influenced by previous results within the research field, which interact with new developments in the external dimensions. Those dynamics influencing the development of the field are not uni-directional. Rather, the development of the field itself feeds back to the four dimensions of the field. Finally, and very importantly, the epistemic community of the SIR is not developing out of thin air. Rather, it draws upon theories, concepts, methods and data from established academic fields, which opens various options of mutual influence between new and established fields over time. Figure 1 depicts the conceptual model of SIR field development.

It can be expected that the dynamics and patterns of change vary, that each dimension can be the source, or the bottleneck, of further development of the demand for and supply of research for policy making in our fields. Also, there is seldom just one connection, one arrow in place in this model. Instead, dynamics from one dimension spill over to the other three. However, the analytical model

[1] When writing this chapter, this model was particularly inspired by analysing the first versions of the chapter on innovation indicators by Frietsch et al. and the chapter on water research by Hillenbrand et al. in this anthology, who pointed to the importance of elements from the four dimensions, and the feedback from research to policy making and economic and environmental developments. Subsequently, Frietsch et al. structured their analysis by using and further developing this model in a version tailored to the specificities of their research field.

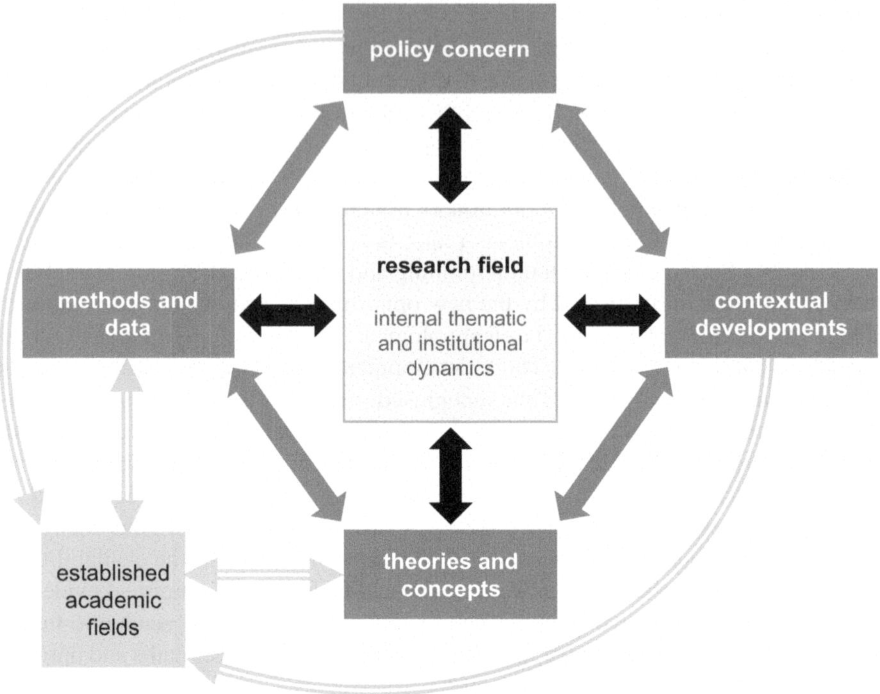

Fig. 1 Systems and innovation research change model. Source: own representation

allows us to look for the origin of developments and the timing and relative weight of the cross-influence between dimensions. It also allows us to detect the absence of influence, i.e. the lack of any policy consequence of, for example, new data availability or new scholarly concepts. In the following, we apply this model to reflect in brief on the nine contributions to this volume and to try and detect any overarching patterns.

3 Patterns of Change in the Research Fields

3.1 Understanding Paradigm Changes in Science, Technology and Innovation Policy: Between Science Push and Policy Pull

It is now widely recognised that the analysis and conceptualisation of STI policy has developed in three phases, with each phase characterised by a major STI policy paradigm. As these paradigms are additive rather than substituting each other, policy ambitions and policy instrumentation have broadened as a result. Academics have tried to understand and conceptualise policy for science and innovation and have drawn on theories and concepts concerning the specific role of the state. Thus, this area of SIR is a very good example of the interplay between theoretical paradigms, on the one hand, and policy concerns and instrumentation, on the other hand, throughout its five decades of history.

The first phase was dominated by the linear model of innovation and market-driven exploitation. The basic, dominant idea was that the state should support basic research and selected technologies in order to foster economic competitiveness and military security. One could argue that, in this phase, economists, who dominated the STI policy thinking at the time, were reacting to developments in policy ambition. Economists like Solow and Arrow explained and justified the role of the state for science and technology as it was unfolding, rather than driving that change. However, academics also suggested and supported the development of framework conditions and instruments in line with the dominant paradigm of supporting science production for subsequent technological and economic exploitation through market forces.

The relationship between scholars and policy making started to change in the 1970s and especially in the 1980s and 1990s. While the endogenous growth theory and economic geography developed as a response to the traditional macro-economic theory and amplified the relevance of knowledge and technology for economic and societal development, the influence on STI policy making was dominated by a new, complementary, albeit rather small class of evolutionary and institutional economists. As in other SIR areas of research, the (national) innovation systems approach and its institutional and policy-related implications began to prevail. These epistemic contributions were well received by politicians and policy makers seeking ways out of the stagflation crisis, a new phenomenon in the 1970s. The awareness of a deep crisis and concerns about the loss of competitiveness vis-à-vis new global players such as Japan catalysed both academic conceptualisation and the eagerness with which policy makers embraced systems thinking and its policy implications. Re-defining the role of the state in the system and broadening the instrumentation used were thus the result of the interplay between an academic paradigm change and the existence of an unprecedented crisis demanding novel solutions.

In the third and most recent phase, yet another relationship between the scholarly community and policy makers drove the development of both policy and of the scholarly community. One driver was again a crisis; this time a failure of policy to deliver on the ambitious economic goals of innovation enhancement announced in Europe and at a European level. Another driver was the normative turn in STI policy. A novel sense of urgency in society and policy making called for new ways to mobilise science and technology to address pressing societal challenges at national, European and UN level (SDGs). This finally opened the door to policy concepts that had already been immanent in the innovation systems thinking of the 1980s and early 1990s and that had already demanded the mobilisation of STI towards problems and its coordination with other policies to that end. In the late 2010s, the time was ripe for this policy paradigm. It was then popularised and further developed by innovation scholars and—importantly—by scholars in transition studies, who realised the leverage STI policy can and should have on transformations. The current turn towards mission-based and transformational policies is therefore the result of a co-generation of policy and broader governance approaches, in parts driven by explicit joint efforts,[2] in parts by efforts from policy-making organisations to take stock of and further develop these policies collectively.

At this point of the development, in the early 2020s, the community of STI scholars and policy practice are facing a dilemma. Evolutionary economists, STI policy scholars and transition scholars have, consciously and unconsciously, combined their efforts in the direction of the new normative paradigm of STI policy making. While this is promising, it has also led to demands for far-reaching changes in governance, the requirements for the state in terms of capability and institution building as well as policy coordination. Again, as with the Lisbon agenda of system enhancement for leading edge global competitiveness, there is a danger that governance and in particular STI policy practice will be overwhelmed by these demands. The role of the STI policy community at this point would have to be to offer support in very pragmatic, formative terms.

An additional concern that was already apparent in the 1980s but largely ignored is now coming to the fore, the need to consider STI development and the application of innovations within the context of sectoral policies. We still witness a dual compartmentalisation here: the division of responsibilities in ministries between STI policy, on the one hand, and sectoral, solution-oriented policies, on the other hand. It is important to note that this compartmentalisation is mirrored to a large extent by persistent separation of the scholarly community, where the convergence of innovation economists, STI policy scholars and transition scholars is still not sufficient.

Reflecting on the development of this research field over the last five decades also makes one think about the limitation of STI policy scholars in terms of focussing too narrowly on STI policy organised in innovation, science or economic ministries. Even as early as the post-World War II years have there been all sorts of purpose-oriented programmes supporting science, technology and innovation in sectoral policies. The value of scientific knowledge and in particular of new technologies and their application has always been part of policy making in sectoral policies such as energy, health or mobility. Against this background a more ambitious link between STI policy and sectoral policies should be possible.

3.2 Analysing the Nature and Dynamics of Innovation: Innovation Monitoring and Innovation Indicators

The research field of indicators occupies a very special position in innovation and systems research. While a stand-alone academic community has emerged here featuring a major journal (Scientometrics), European networks and a range of established conferences, science and innovation monitoring indicator development has

[2] The most prominent example is the transformative innovation policy consortium TIPC (https://tipconsortium.net/).

been at the service of other research fields from the very beginning. These fields include innovation studies, science studies, science and innovation policy studies, technology management studies, business economics and macro-economic analyses, and technological and systems analysis in sectors such as energy, mobility and so on.

In the same way, quantitatively measuring the key dimensions of innovation systems has been of great importance for policy making throughout the last five decades. Here, we see a mutually reinforcing dynamic, as new indicators have led to new policy demands, and new policy aspirations have led to new indicators. For example, if the state assumes responsibility for the performance of the science system, it becomes imperative to measure and map its internal dynamics, input, performance and connectivity to economic actors and to measure the relative contribution of science to economic dynamics. Likewise, if the support for innovations serves first and foremost to drive national economies, as was prevalent in the 1980s and 1990s, it becomes necessary to develop, test and apply a range of indicators and methodologies to analyse the contribution these innovations make to economic dynamics.

The critical importance of quantitatively monitoring the contribution of science and innovation to economic competitiveness and, more generally, to societal welfare has thus put indicator development at the forefront of policy concerns. This importance is backed by the actions of international organisations such as the OECD and the EU, which have supported the development of innovation indicators and ensured international comparability through standardisation and the formation of transnational epistemic and policy-making communities.

Further, this field is also characterised by a particularly pronounced interplay with the other three dimensions of our conceptual model. In the last five decades, the development of the field has been driven by the advent and dominance of new theoretical paradigms, such as the endogenous growth model and the linear model of innovation that stress input and output indicators, or the innovation system paradigm, which requires the definition and operationalisation of indicators to measure capabilities and connectivity. Due to these close links, major changes in the dominant model of technological development have necessitated adaptations to the focus of measurement. For example, the increasing prevalence of science-based technologies, the growing dominance of platform-based business models and the technologies enabling them, or the rising importance of the service economy have all required the adaptation of existing indicators or the development of new ones. In addition, methodological advances based on the availability of new data or the accessibility of new computational power have led to a novel perspective of the innovation system, as they allow, for example, more profound, inductive analyses at the micro level.

While these complex interactions have, all in all, contributed to the development of policy making, not all of the methodological aspirations have been successful. For example, the idea to quantitatively measure and map the development of a technology through technometrics did not succeed. Instead, it has proven necessary to mobilise additional methods from neighbouring fields with expertise in the specific technological area for technological system analysis. Similarly, early applications of big data analysis have not yet led to new theoretical concepts that can capture and assess the dynamics measured. The biggest contribution of this field has been to co-evolve so profoundly with all the dimensions of our model and to enable an operationalisation of the new theoretical paradigms in policy making.

The field's further development will be equally challenging if it is to support analyses of the contribution of science and technology to broader transformations in the wake of transformational policies. STI indicators have been used to analyse technological systems for some time already, but monitoring the transformation of systems and the relative role of S&T in this system transformation will require further methodological advances, as well as a novel joint discourse of yet diverse epistemic communities and new combinations of policy communities. For this development, it is of the utmost importance that all producers and users of indicators constantly remind themselves what the chapter on indicators in this book emphasises, i.e. that indicators are a means to an end and must not take on a life of their own. Indicators will always need contextualisation, which is even more challenging in the context of monitoring system dynamics and the related role of science, technology and innovation.

3.3 Foresight: From Detecting Futures for Strategising to Participatory Approaches for Governance

As the chapter on this research field shows, foresight has had a twofold function in systems and innovation research. It started as a means to reduce uncertainty in order to develop strategies. This was then complemented by collective reflection on possible futures that served to better understand the directions and drivers of change and to structure a discourse based on this about what is normatively desirable and how decisions can influence future developments. This field is a fascinating example of how changing demands in society and policy have interacted with intellectual and methodological developments in the epistemic communities.

Foresight was first developed as a technique for planning and strategising by the military and large corporations. The aim was to identify possible futures including estimates of plausibility using sophisticated foresight techniques such as systems modelling and technology forecasting as well as Delphi studies.

Policy-oriented foresight beyond military planning first began in the 1960s and intensified in the 1970s, in the era of "planification" and the dawn of cybernetic systems thinking. This was motivated by broad societal concerns and enabled by an increasingly active epistemic community and advanced modelling methods. The activities of the Club of Rome highlighted and supported the new ambition of predicting the future development of entire systems rather than specific technologies or limited problem areas. It does not seem to be a coincidence that foresight developed at the same time as other research fields of SIR, which sought to understand the systemic dimension and dynamics of technological development and were also driven by the normative appeal to inform policy and support the governance of system development.

As in other areas of systems and innovation research, the explicit and formalised innovation systems approach of the later 1970s and 1980s advanced this ambition even further. Foresight was increasingly applied to help understand and shape those national systems. The 1980s and 1990s saw mutually reinforcing dynamics between the conceptualisation and empirical analysis of systems dynamics, on the one hand, and concrete foresight activities, on the other hand, that solidified the very idea of national systems as the focus of policy concern.

Reflecting on and arguing about possible futures, their drivers and consequences also affected the policies aimed at influencing systems. Foresight thus became a policy tool in two respects: first, to explore the role of polices for alternative futures and second, to engage in broad stakeholder discourses on the very nature of existing systems and the factors influencing their future development. The foresight practitioners and academics therefore positively reinforced innovation systems thinking.

Subsequent developments in the field of foresight were driven by new data and methods as well as new epistemic and political aspirations. First, foresight in and for STI policy has been influenced by new data sources and methods. In the 1970s, new patent and publication data made it possible to detect technology development at an early stage and assess the contribution of technologies to the competitiveness of the—mainly national—systems. In more recent years, big data analytics and artificial intelligence tools have enabled the processing of much larger sets of unstructured data and thus increased the ability to detect early signals of trends and develop far advanced topic models to inform trend analysis and discourse.

Second, and more importantly for the development of the field, foresight has become a participatory and more reflexive practice. Having realised the limited possibility to predict or determine futures, scholars and practitioners began to appreciate the value of collective sense-making about possible future developments. This was pushed by a move towards recognising societal concerns and matters of the social responsibility of science and technology sparked by approaches like Responsible Research and Innovation and the orientation towards societal challenges. Forecast and modelling was complemented, and even substituted in parts, by participatory practices. Inclusive reflection and co-construction of alternative futures became a policy imperative, especially at European level, and traditional technology and scientific experts were increasingly complemented by broader stakeholder groups, in particular those societal groups affected by technological and scientific developments.

The growing methodological sophistication and diversification of foresight has been accompanied by the emergence of an increasingly diverse foresight community that is both inter- and transdisciplinary in nature. However, in conjunction with the problem-driven and application-oriented nature of foresight, this has also resulted in very slow institutionalisation into an academic field.

With the UN SDGs and transformational STI policy now at the top of policy agendas, foresight is, ironically, at a crossroads. It is called upon to support system change both by identifying the role of specific technologies or social innovation for transformation (predictive) and by a broad discourse on possible system trajectories, their material and normative tensions and how political and societal action can shape future developments. As this role of foresight unfolds in highly normative contexts, it will be important to define the role of different kinds of foresight exercises very explicitly and to address normative apprehension transparently—without falling back into the trap of the 1970s to claim prediction and systems engineering through advanced modelling even of behaviour.

3.4 Evaluating Public Research and Innovation Policies: A Short History of Co-Evolution

The main force driving developments in evaluation research has been the interplay between the theoretical development in innovation studies and changing practices and demands of policy makers. A succession of theoretical concepts such as market failure concepts (neoclassical economics), innovation systems approach (evolutionary economics) and mission orientation (based on a range of neo-interventionist approaches) have interacted with policy ambitions and instrumentation and defined research questions. Evaluation practitioners have developed operational concepts to measure the intended (and unintended) effects of policies that changed as a result of this co-evolution. Methods and data sources did not play a major role here, although there are instances of the co-development of policy intentions (connectivity) and new methods (network analysis).

In the 1960s and 1970s, public policy for science, technology and innovation was about steering the economy and increasing welfare systematically. Consequently, evaluation concerned the contribution of R&D programmes to economic growth and technological development and thus the accountability of policy and funding. Early methods in policy evaluation were based on input–output calculations of the innovation process (at the macro level). In parallel, questions were addressed regarding the implementation of programmes (at least in Germany, implementation research). From the mid to late 1980s, the dominant theoretical background was the innovation systems approach, focusing on the need for connectivity and capacity building in systems. This resulted in a broadening of methods from changes of outputs to encompass a number of indicator approaches (patents, publications, citations). This was accompanied by a further diversification of the research questions, which then also included the innovation activities of firms (behavioural additionality) and capacity building in the economy. Policy makers then wanted information about the networking effects of actors and how their relationships with each other changed as a consequence of complex, multi-measure, multi-actor programmes. In the 1990s, a wave of evaluations analysed system-level changes in structures and performance based on policy mixes. Further developments included questions about broader societal impact that became more important in the 1990s and 2000s and which have recently expanded to cover the contribution of research and innovation policy to specific missions and to system transformation. This has led to a search for and the application of new methods to assess the impact of STI policy on the development of transformations or the achievement of missions.

The demand for evaluation has evolved from giving a very crude account of the impact of interventions to a more sophisticated analysis of different types of impact. With its increasing complexity, however, evaluation has also become a device used to support learning, to accompany, monitor and influence policy implementation. Evaluators have become partners of policy makers, research and innovation funders as well as target group stakeholders. Looking ahead, it is likely that the need to understand the contribution of STI and STI policy will lead to further attempts to understand its impact on system developments and to model and monitor impact and transition pathways. The evaluation of STI policy will be even more closely linked to wider stakeholder groups and other disciplines. How this development will play out is not yet clear, but we may very well see much more formative evaluation efforts as well as a broadening of evaluation towards ex-ante systems modelling and systems monitoring. Finally, as evaluation now analyses policies that target system-level dynamics, in some instances, evaluators have begun to influence the societal and political debate about the very nature of the research and innovation system. This includes value judgements about the relative importance of actors and activities across systems and about policy support for the system.

3.5 Assessing Technological Innovations: From Early Warning to the Governance of Socio-Technical Transformations

This field of research has been driven less by the concerns of policy makers and more broadly by societal concerns and parliamentarians. Institutionalised technology assessment (TA) was initiated by US parliamentarians as a counterweight to the decision-making of the executive and the judicative. Growing societal concerns have led to a rising demand for TA. TA has been called upon to assess what to expect from a technology in the future and, in doing so, to support decision-making, regulation and also management with regard to the future development and deployment of technologies. More recently, another major modification is the shift towards systemic questions, such as what is the role of technologies in system transformation and how can systems be transformed more generally?

TA has also evolved into an approach to democratise the development of technology and to make it "more responsible" by involving multiple stakeholder groups (constructive technology assessment). TA has therefore expanded its methodological approaches and taken on the role of active moderator of a process that seeks to develop a vision of what a technology should deliver and ways to make that vision more likely to come true.

There are many different factors influencing the research questions in this field. In the context of our four-dimensional model, however, the main driving force is the interplay between basic attitudes towards technology (contextual development) that are translated into policy concerns and different kinds of academic communities.

Within TA, we can observe the dynamic interaction between defining societal and political problems, on the one hand, and mobilising different academic communities and their epistemological and methodological approaches, on the other hand. As the societal zeitgeist has passed through fluctuating waves of technology pessimism and optimism, so have the needs of parliamentarians, and increasingly of policy makers and societal groups to assess the future impacts of emerging technologies. With these changing needs, different kinds of academic communities have been called upon to participate in TA, themselves driving further changes in TA approaches and methodologies.

In simplified terms, the relationship between TA and society has passed—highly stylised—through three distinct phases, distinguished by a shift of roles and foci of different scientific communities in each. These changing roles reflect the dynamic interplay between problem perceptions and the different scientific communities, each with their own identity, methodological skill set and normative expectations. A first phase saw a particular strong role of modelling and technology experts (expert TA), while in a second phase process and moderation experts became more prominent (participatory TA). In a third phase we saw a stronger combination of technological experts and those with strong expertise in participatory processes, marking a broader shift towards identifying the possible technological and behavioural contributions to solutions for pressing societal problems (pragmatic TA).

Initially, technology experts were seen as the main sources for TA. However, as the users of TA and researchers became critical of the early technocratic modelling approaches employed, participation broadened to include interdisciplinary and some transdisciplinary approaches. Participatory TA was based on insights from the scholarly community of Science and Technology Studies (STS) and constructivist thinking, while later deliberation theories (Habermas) contributed to a more measured, mixed model. Scientific approaches to "the nature of technology assessment and deliberation" made a real difference to the identity and role of TA, as did advances in participatory methods. As a result, activists, especially those from the Science and Technology Studies community, started to question the predictive and warning role of TA and turned towards co-constructivist approaches (CTA). Finally, with the inclusion of the role of future technologies for system transformation, other academic communities from innovation systems and innovation studies as well as transition studies have been increasingly called upon and shown an interest in TA.

3.6 Understanding Paradigm Change in Industrial Production

This field of research has been evolving over the last 50 years through a succession of five successive technological and organisational production paradigms. Two of these paradigms have been driven by technology: computer and automation (CAM, or Industry 3.0), and cyber-physical systems and Internet of Things (Industry 4.0). The other three have been more organisational and comprise the quality of working life, lean management and perfected human–machine interaction.

The chapter on industrial production provides additional background, in particular, that research in this field has been driven very strongly by paradigmatic shifts, and identifies three such phases. Relating the paradigmatic shifts to the elements of our model allows us to interpret the development of production research in the logic of our systemic model of research dynamics.

Technological and economic developments, but also policy concerns have been key factors in research development:

- Most obviously, the production paradigms Industry 3.0 and Industry 4.0 represent technological developments in themselves.
- Changes in society's perspective of work from simply being a way to earn money to emphasising the well-being effects of working have also influenced research as well as policy concerns and working regulations.

The interaction of technological and economic developments with policy concerns has been the most influential driver of production research. However, the nature of this interaction has changed over time from a rather contradictory to a more complementary one: In the technological paradigm of phase 1, concerns about the quality of work and the need to increase productivity characterised research, driven by the conflict between greater productivity and labour conditions. The interaction of technological and economic development and policy concern was therefore contradictory in nature. In the second phase, globalisation meant that maintaining national competitiveness became an overarching political concern, and research reacted by analysing lean management as an organisational innovation that could enhance competitiveness in addition to industry 3.0. Therefore, the interaction of economic and technological development and policy concerns was complementary in influencing research. Finally, the third phase is characterised by the coupling of the physical and digital world, and with the emerging debate about taking a more human-centric approach to Industry 4.0, once again research is being shaped by the contradiction between technological and economic developments and societal concerns.

The development of the research field has also interacted with the institutional research environment. A specific research community developed, dominated by large collaborative projects between different research partners—many of them from non-university institutions—and companies. Methodologically, research predominantly took place as accompanying research of case studies, in which a paradigmatic innovation was applied as a pilot. However, globalisation and the increasing need to evaluate the competitiveness of domestic industry in comparison with competitors have led to a broadening of the types of methodology and projects and large-scale surveys undertaken by research institutes have been added to the research portfolio. The links between production research and socioeconomic theories are less clearly defined. New economic geography and the evolutionary theory of trade developed in the 1980s and 1990s have had the most obvious influence. These supported the need to continuously upgrade the capabilities for competitiveness, to "climb-up the ladder to stay ahead".

3.7 Exploring Innovation and Progress in Renewable Energy Development: From Niche to Mainstream

The chapter on renewable energy offers plenty of evidence for how the interaction of economic (geo)-political and environmental developments with policy concerns and measures drives the development of research. The first oil price crisis and the debate about finite non-renewable energy sources together with concerns about the security of supply amplified by the oil embargo in the early 1970s kick-started the research on renewable energy. In this first phase, research focused on technological solutions, in particular for heating, responsible for consuming the biggest share of imported oil. The second phase was marked by decreasing oil prices, which lowered the pressure to substitute oil. Political concern focused on the electricity sector in the aftermath of the Chernobyl accident, and subsequently on the need to reduce CO_2 emissions, especially those from burning coal. The targets set and measures taken defined the research in this second phase, which was focused on analysing the support schemes to foster the diffusion of renewable energy. Implementing these support schemes resulted in the increasing diffusion of renewables and rising electricity prices, which turned policy concerns towards the expansion of renewables with lower costs. This resulted in research focusing on market integration and the functioning of innovation systems to bring down technology costs. Finally, with climate-related risks becoming more apparent and the corresponding political targets of achieving net zero emissions, research is increasingly concerned with the transformation of the energy system towards renewable energy.

The research on renewable energy also shows the importance of the interactions with methodological developments. Advances in systems analysis and computing applications enabled the first energy system models, which identified the need for and the possibilities to move towards renewable energy sources. The political challenge of managing the increasing costs of renewables support schemes led to new ways to combine energy and economic models to analyse the economic impacts of renewable energy expansion. The favourable overall economic impacts demonstrated by these models supported the continuation of the support schemes and even more ambitious target setting. Analysing support schemes and acceptance issues of energy system transformation required interdisciplinary research approaches, which combined methodologies from economics and social sciences with more engineering-based concepts.

There are also very strong links between renewable energy research and the institutional and epistemic community. Renewable energy research co-evolved with the establishment of non-university research centres and institutes, which still dominate the technical research on renewable energy technologies. These non-university research institutions were also the first to pursue interdisciplinary research and provided the manpower and institutional homes for those driving the expansion of research. Finally, with renewables now becoming the main source of energy in the electricity sector, this field of research has become an established part of the university landscape as well. Interaction with social science theories first became apparent with the influence of neoclassical environmental economics and transaction economics on the analysis of support schemes. However, research on renewable energy has also led to further advances in innovation heuristics in a co-evolutionary process. Indeed, innovation studies in the field of renewable energy have greatly advanced the concept of technological innovation systems.

3.8 Analysing Energy Demand and Modelling of Energy Systems: From Little Knowledge to Differentiated Know-How

Similar to renewable energy research, the chapter on analysing energy demand and modelling of energy systems provides plenty of evidence for how the interaction of economic, (geo)political and environmental developments with policy concerns and measures has driven this research field over time. Three examples illustrate how substantial this influence has been:

- The first oil price crisis and concerns about the security of energy supply amplified by the oil embargo in the early 1970s instigated research challenging the existing paradigm that energy demand increases in line with economic growth. Technical analyses of energy efficiency potentials led to a new paradigm in analysing energy demand.
- In the 1990s, a new wave of research to identify additional barriers was triggered by limited energy efficiency improvements in practice despite the proven existence of large low-cost potentials to reduce energy demand.
- The growing importance of fluctuating renewable electricity resulted in research turning towards new ways to influence energy demand such as demand-side management and has ultimately led to broadening the scope of research by including sector coupling.

The research on energy demand also shows the importance of the interactions with methodological developments. Advances in systems analysis and computing enabled the development of energy systems models that could combine techno-economic analyses of energy efficiency potentials with structural changes within the economy. This made it possible to construct energy demand scenarios, which became more sophisticated over time. In turn, these scenarios influenced policy making and have formed the basis for policy decisions on CO_2 reduction targets ever since the recommendations of the German Enquête Commission on Climate change for Germany's first target to reduce CO_2 emissions. Increasing interdisciplinarity and integrating methodologies from economics and the social sciences have also helped to improve the methodologies in this field. This has opened the door to analysing topics such as energy efficiency behaviour and new instruments to alleviate obstacles, which have contributed to expanding the portfolio of energy policies.

The research on analysing energy demand and energy efficiency also co-evolved with the institutional community. In particular, energy system models co-evolved with the formation of a scientific modelling community, which allowed for scientific exchange and fostered continuous methodological improvement. There has also been some influence of theoretical developments on energy demand research, most visibly transaction and institutional economics, together with motivation theories, which have provided the theoretical background to analysing obstacles to energy efficiency.

The development of the research on energy demand is also influenced by the internal dynamics in the field. Research results have sparked new challenges leading to a broadening of the disciplinary background. For instance, the technical feasibility of energy efficiency options gave rise to questions of their cost-effectiveness. Extending analyses from the status quo to include future-oriented studies has required the integration of structural change and foresight into the research portfolio. The existence of low-cost energy efficiency options led to research on the barriers to energy efficiency and measures to overcome them. This internal logic has meant that research has become inevitably more complex and more interdisciplinary over time.

3.9 Understanding the Co-Evolution of Research and Water Protection Policies: From Single Technologies to Systemic Integrated Approaches for the Sustainable Use of Water

Water research was initiated and is still strongly driven by the deterioration of our bodies of water, underpinned by growing environmental awareness in society. This has caused increasing concern among policy makers and researchers and has triggered activities to reduce emissions and adapt water infrastructure to new challenges. At the same time, there are very close links between water research and policy making. Regulatory measures prescribe maximum emission thresholds which are technically feasible using current sewage treatment technologies. Any technological advances that make further emission reductions feasible directly influence regulatory standard setting, and goals to strengthen regulatory standards directly guide the research needed to provide the technical basis for that. This is backed by how the research in this field is funded at least in Germany, mostly by the federal and state governments, which are also responsible for policy making.

The co-evolution of research and policy making is not only influenced by the state of the environment, but also has repercussions on the level of pollution. This in turn leads to new priorities in policy making and a new phase of water research. This interplay between the environment, policy making and water research has been the main driver of water research over the last 50 years, and three phases can be distinguished: In the first phase, policy making and research emphasised the quick reduction of emissions by applying or improving single technologies. This reduced the emissions from sewage treatment plants, but was not enough to clean up water bodies, and the focus therefore widened to include diffuse emission sources as well. This contributed to the second phase of water research, which built on more integrated solutions in addition to improving single technologies in wastewater treatment plants. The increasing number of options called for a more systemic approach, which was supported by policy making that itself was more concerned with the management of water bodies rather than only the emissions of individual sewage treatment plants. At the same time, economic challenges in the form of rising costs for sewage treatment and the maintenance of water supply infrastructure further emphasised the need to look for least-cost options from a system perspective. Finally, climate change with the resulting changes in precipitation patterns and higher frequencies of both droughts and flooding triggered the third phase of research, which calls for greater integration of sewage treatment and water supply and a transformation of the entire sector.

There have been also interactions between water research and methodological improvements. Increasing abilities to measure pollutants and collect such data helped to identify new environmental challenges. Water research addressed these challenges, and systematically used this opportunity to develop new information tools such as emission inventories, which greatly facilitated the policy debates about target setting in water policy, e.g. with regard to micropollutants. These new challenges reinforced the need for more research on new emission control technologies and ultimately led to a broadening of the actors involved in water issues, such as the pharmaceutical industry, which is called upon to provide innovations to reduce the environmental burden of their products.

The development of water research is also characterised by internal dynamics, which constantly increases the scope and complexity of water research. The focus has shifted from single technologies to more systemic solutions integrating multiple different technologies. This has also had repercussions on the disciplinary nature of water research. Although it is still dominated by engineering-related scientists, social sciences are also playing a bigger role with issues like the acceptance of systemic solutions and the transformation of the sector gaining momentum. It has also had repercussions in the form of a growing scientific community with its own associations, which has institutionalised water research as an academic field and opened up new career perspectives for the scientists involved.

Other elements in our model show a lower level of interaction. This is especially true for the role of social scientific theories in shaping water research. The impetus of regulatory economics, which so strongly influenced the liberalisation of the energy markets, has only marginally affected water research. One explanation might be that, even though decentralised technologies have become more important, they are still not in a position to phase out monopolistic bottlenecks. Another, perhaps surprising lack of interaction is with innovation system heuristics and approaches. Despite an increasing number of actors and the need to analyse their interaction, and despite the importance attached to analysing the transformation of the sector, approaches such as Technological Innovation Systems and the Multi-Level Perspective have hardly been used in water research so far.

4 Conclusions

4.1 Dynamics of Research Fields and Patterns of Change

Before drawing the final, overarching conclusions, we need to reiterate the impetus for and aim of this volume. It was motivated by wanting to understand how scientific research has been analysing and supporting the development of innovation systems and of socio-technical systems, their interplay and their governance over the last 50 years. In order to do so, we first selected nine research fields in the area of systems and innovation research. While the delineation of the area of systems and innovation research and the selection of our nine fields are somewhat arbitrary and subjective, there are common qualities to the fields that define the area of systems and innovation research. All the fields defined in this volume are interested in phenomena that can only be understood using a systemic approach; any analysis of specific elements has always been put into the context of the functional system it is referring to; and as we can observe, this has happened with increasing intensity over time. All fields deal with the role and impact of innovations, of novelties that are put to practical use, albeit often on very different levels and in very different forms. Finally, all the fields have developed out of a normative impetus or concern to support the governance of systems, and they are all clearly application-oriented. Consequently, all the fields have co-evolved with the systems, in which they are embedded and, more concretely, with the governance and policy in these systems.

The final questions to answer in this concluding chapter are: How have the fields and their roles changed over time? Are there any overarching patterns beyond the idiosyncrasies highlighted in the nine contributions to this volume? If so, what do those patterns tell us about the specific nature and responsibilities of innovation and systems research and its future challenges?

Three Phases of Zeitgeist

It is possible to identify a few, very high level and severely simplified patterns across the fields. First of all, the demands on and roles of the fields have changed with the respective zeitgeist, a term we use slightly differently to the Oxford dictionary definition,[3] as the defining spirit or mood of a particular period of history nurtured by dominating ideas and beliefs of the time. Of course, those phases of changing zeitgeist are in themselves ambiguous, different ideological and ideational streams, some of them even contradictory, some co-exist and some compete. But very crudely, despite marked differences even between developed OECD countries, overall a few dominant phases of zeitgeist can be identified, which also shaped the activities and developments in our fields. A first phase in the 1960s and parts of the 1970s was characterised by strong economic growth and a general sense of optimism

[3] https://www.oxfordreference.com/display/10.1093/oi/authority.20110803133418753.

towards technological and societal developments. This sense of progress was buoyed up by the growing strong sense of the governability of systems and thus of a strong and pro-active role of the state. During that time, the field of innovation indicators and monitoring, for example, provided concepts and evidence to make sense of innovation dynamics and to optimise the allocation of research inputs, while STI policy concepts were developed that sought to improve the understanding of policy makers on how to steer innovation activities and sub-systems.

Decreasing economic growth rates culminating in stagflation in the late 1970s that posed severe challenges to economic and welfare systems, accompanied by growing environmental concerns, led to a zeitgeist shift. At the end of the 1970s and beginning of the 1980s, major Western countries, led by the US, UK and—to some extent—Germany, developed their own variants of a combination of neo-liberal and neo-conservative ideologies. In this eclectic ideological combination, the role of the state was—grosso modo—weakened and the role of the market enhanced. The political landscape became increasingly polarised with intensifying societal conflicts over the responsibilities of the state for economic development, social welfare and environmental concerns.

As a consequence of both the ideological turn and growing environmental and social concerns, the demands on and roles of our nine fields were ambiguous. On the one hand, there was a fading of the general optimism that new technologies are always favourable. The role of technology assessment broadened, with a number of countries following the US model of TA. Existing technological paths were questioned, such as the expansion of electricity systems based on centralised nuclear power stations. On the other hand, science and technology were increasingly seen as the engines driving economic growth, and the innovation systems approach was popularised through comparisons of national systems and their innovation-driven competitiveness. The field of innovation policy emerged and the paradigm shift in manufacturing industries was strongly supported by policy measures, all of which resulted in new demands in areas such as innovation system monitoring and analysis, industrial production research, STI policy analysis, evaluation and foresight. In the three vertical fields of renewable energy, energy demand and modelling, and water research as well, science and technology were seen as engines to address societal challenges and contribute to solutions. However, there is a significant difference between these three vertical research fields and the horizontal ones with regard to the second zeitgeist wave. Even during the neo-liberal era, traditional neo-classical economic theory supported a strong role of the state in these three sectoral research fields, although not to generate new innovation dynamics. Environmental economics clearly stated that environmental problems constitute a separate class of market failure, which justifies state action and calls for directionality towards environmentally-friendly technologies. Regulatory economics added sunk costs as an additional requirement to a natural monopoly for sectors to be classified as forming a monopolistic bottleneck. Even though this school of thought limited the number of sectors which should be subject to economic regulation, electricity, natural gas and water systems still belonged to this shrinking list of sectors. Taken together, the resulting environmental and regulatory policies in these sectors triggered a demand for technology innovations and therefore simultaneously acted as a demand-side innovation policy. Consequently, during the second zeitgeist wave, the three sectoral research systems shifted in another direction with a greater number of de facto innovation policies than the horizontal research fields.

The tide turned again in the late 2010s, catalysed by the severe financial crisis from 2007 to 2009 and the intensifying awareness of the severity of societal challenges, above all the climate crisis. Once again, the calls grew louder for a more pro-active role of the state to support economic and technological development. From the perspective of the research fields in the energy and water sector, it became increasingly clear that technology development alone does not suffice. In the horizontal research fields, the perception of the severity of societal challenges led to policy approaches that sought to give direction to technological development and deployment and to support transformations in functional systems such as the energy or mobility sector much more systematically. Thus, the zeitgeist in this phase brought all nine research fields closer together again. This led to a number of major changes. The development of the innovation system was now strongly linked to our transformational ambitions. As a consequence, new STI policy concepts were developed and strongly influenced STI policy making. STI evaluation broadened its approaches to understand the transformational impact of policy and policy mixes. Foresight, STI policy evaluation and some areas of technology assessment strengthened their formative role, supporting reflexivity and broad, inclusive discourse moderation. For research in the sectoral fields of energy and water, it became clear that more systemic transformations beyond the existing sectoral logic were and are required. Analysis in these areas increasingly took the user perspective into consideration and in doing so integrated questions of the acceptance of innovations and behavioural change. In particular in the research field of renewable energies, this led to an uptake of broader innovation system approaches as well.

Determinants of Field Developments

Within the broad, long-term phases outlined above, the fields developed in complex and often idiosyncratic ways over the five decades of our analyses. However, it is worth noting that there are a number of patterns in the mechanisms of change across the field. These patterns we can now express through the various dimensions of our SIR development model outlined above. While the various drivers for the development of the fields can be classified in one of those dimensions, we often see a combination of drivers from two or even three dimensions in complex interactions and feedback loops.

At first sight, the dominant drivers of the fields have been *disruptive developments in the societal, political, technological or economic context*. Obviously, if the context changes significantly, new challenges arise, new questions are posed. There are numerous examples in the history of our fields. For instance, the fact that science and large parts of society have realised how our economic model threatens to violate the planetary boundaries, endangers our water resources and changes the climate in dramatic ways has kick-started the fields of renewable energy and energy demand analysis and the sustainable use of water resources. Economic development and especially the structural changes brought about by globalisation have shaped the economic challenges perceived by national governments, and consequently the questions they ask. Worldwide developments towards a platform economy have slowly but steadily led to the need to develop new theoretical models and empirical methods to understand the development and role of innovations in an innovation system. The political momentum triggered by the formulation of the Sustainable Development Goals has called for new analyses in terms of broader dimensions of sustainability across our fields, led to new, broader forms of foresight activities and inspired mission-driven policies that have resulted in demands for conceptual and evaluative support. The geo-political disruptions seen since the mid-2010s, culminating in systems competition and military conflict, have driven politicians, policy makers and businesses to call for technology sovereignty as well as energy sovereignty, which poses new questions in the field of innovation monitoring and analysis as well as the fields of renewables and energy demand. The Covid pandemic has, among other things, provoked the question of how science systems can be organised to better support crisis management across a wide range of dimensions. This has meant new challenges in the field of STI policy conceptualisation, as STI policy was called upon to help tackle an imminent crisis.

However, the mechanisms through which those—exemplary—contextual changes are translated into developments in our fields are diverse. They depend on how and by whom contextual developments are interpreted and which interpretations become dominant in societal and political discourses. Developments in socio-technical systems can create societal pressure, they can be seen as political opportunities or threats, or they can trigger the research community itself to ask new questions or highlight new aspects of the system it analyses.

The most important mechanism through which the perception and interpretation of contextual changes are translated into demands for all nine fields is the formulation of concrete *policy concerns,* which is the second dimension in our model. This underlines the application orientation of systems and innovation research. One of the most prominent examples here is Technology Assessment, which grew out of the need of US parliamentarians for an independent source of evidence and reflection to support decision-making vis-à-vis the executive in relation to new, disruptive technological developments. Similar dynamics can be observed in STI foresight and STI evaluation. The latter only exists because policy makers expressed a need for accountability and learning; the former is an expression of policy planning ambition within the framework of increasing contextual uncertainty and complexity. The water and energy research fields display a similar pattern of interaction between policy concerns and research fields in their reaction to the changing environmental situation. Policy and research both influenced the environmental footprint of the water system, which sparked numerous feedback loops. Significant improvements in energy efficiency technologies led to new policy ambitions in terms of standard setting, which in turn triggered new research to achieve these ambitions on a wider scale. The field of production research also illustrates that the timing and direction of this mutual influence can change over time, with research and policy concerns preceding the diffusion of new production paradigms in earlier phases, and then responding to it in later phases. The same research field also illustrates that the interaction between context and policy concerns can be both mutually reinforcing and contradictory. All of these examples clearly demonstrate that the rather trivial insight that contextual factors, i.e. the combination of technological, economic, societal and political developments influence research needs closer examination to identify feedback loops and translation mechanisms through policy concerns.

New concepts and theories constitute the third dimension driving change in our fields and their role vis-à-vis stakeholders and especially policy makers. One early example is the linear model of the impact of science on the economy and societal welfare, a concept that led to the development of a number of indicators in the area of science and innovation monitoring and dominated analytical perspectives. Another prominent example is the development of the innovation systems approach. This approach has evolved into a powerful heuristic to understand and compare the functioning of national, and subsequently regional, sectoral and technological systems. While some fields in this volume specifically supported the construction and operationalisation of this approach, all the fields were affected in terms of the questions asked and the empirical research conducted. Equally, policy makers increasingly understood the need to contextualise their activities and in order to do so needed to ask broader, system-specific questions. In the vertical research fields, the effect of changes in theoretical concepts within the epistemic community has been much more indirect. This perhaps reflects the more heterodox approach of these research fields, where insights from different epistemic communities tend to be recombined, rather than following new insights from one epistemic community.

Methodological advances and *new data availabilities* are the final driver of change in our fields. We saw a very concrete example of this in water research, where a series of methodological improvements enabled the identification of previously undetected forms of pollution, triggering policy concerns and spurring research on technologies to reduce pollution even further. The most pervasive developments, however, were those in information processing. Again and again, this has allowed novel perspectives on systems performance and dynamics and also led to new questions that were subsequently taken up by policy makers. This is true for mining new sources of empirical evidence on the one hand, but also with regard to continuous improvements in empirical modelling, on the other hand. Particularly in the energy research fields, this has enabled huge advances in the ability to model complex energy system developments. This has led to research evolving from looking at individual technologies to a more systemic analysis of energy system developments.

Finally, the relationship between data and methods, on the one hand, and theories underpinning the field, on the other hand, is a complex one in our fields. Not in all cases did data analysis and method development build on an existing theory in a given field. Often, the methodological advances were underpinned by theoretical concepts in established academic fields. These included transaction and experimental economics and motivation theories, which delivered the conceptual background for analysing data on energy behaviour and the acceptance of new energy technologies. Another example is complexity theory, which was the impetus for new advanced approaches in modelling energy demand. In other cases, new data analysis and method development have long taken place without any distinct theoretical underpinning. For example, the availability of vast amounts of unstructured data since the 2010s that can be mobilised through new techniques using AI-supported data analytics has resulted in new maps of innovation performance. This has generated a whole list of new questions, and subsequently new analytical possibilities. However, much of the data, which can now be used for the construction of indicators have long lacked, and in parts still lack, a clear conceptual, let alone a theoretical base.

Reflections on the Epistemic Developments of and Relationship Between Fields

Having interpreted the developments of the fields along the four major edges of our model, we can now turn to reflections on the *epistemic developments*, i.e. the evolving disciplinary nature of our fields and the changing relationship between fields. We can make three major observations. The first relates to the relationship of fields and epistemic communities over time. All of our fields are, in the traditional categorisation of academic fields, multidisciplinary, fed by different disciplines and have developed interdisciplinary dynamics over time. At the same time, the composition of the disciplines in our fields and their interplay has been evolving along with the dynamics of the conceptual SIR research change model just outlined. Those dynamics led to (sometimes rather drastic) changes in research requirements, which called for contributions from those disciplines that promised to address the changing nature of problems or methodological advances. For example, the research fields on energy and water were much more engineering based during their earlier phases than the other, more social science-oriented research fields. The more systemic nature of the research questions and the need to increasingly look at mechanisms of market diffusions led to the integration of concepts from disciplines such as economics, political science and psychology. Equally, the more recent ambitions of STI policy to design and implement transformative missions has intensified the need for researchers in STI policy or innovation monitoring to understand the nature of the underlying technologies, markets and behaviours in the fields such as energy, mobility or health. The latter observation highlights a major finding of this volume: Overall, the broadening ambition of STI policy, on the one hand, and the need to transform sectoral systems, on the other, raise more holistic questions in all fields. The increasingly systemic nature of policy concerns heightens complexity, which in turn increases the need for in-depth interdisciplinary analysis. In fact, the vertical fields, such as energy, water, production, and the horizontal fields of policy analysis, evaluation, TA and foresight show more and more overlapping agendas. Thus, the need for an increasing level of in-depth interdisciplinary cooperation is also accompanied by a need to integrate vertical, i.e. sectoral, and horizontal competences.

This leads to a second observation, which is the changing pattern of development within horizontal and vertical fields and consequently also the changing relationship patterns between these fields. It appears that the sectoral fields and the horizontal ones have been rather isolated from each other[4] for quite some time. This is true for both policy practice and research. For example, a wide range of demand-side innovation policy instruments have been used in sectoral policies. Financial incentives, such as demand subsidies, environmental tax schemes or even new market designs, as well as command and control policies or information programmes have long been used as a major instrument for the production and diffusion of energy or manufacturing technologies. Here, the policies were framed and labelled as sectoral ones, not as innovation demand policy. This also meant that sectoral policies, by their very nature, have always been directional, seeking to steer their sector in certain directions. In contrast, in STI policy, demand-side instruments lay dormant for many years, only to make a comeback in the 2010s, and STI policies and STI policy analysis only discovered directionality and what it means operationally in late 2010. Both analysts and STI policy makers only slowly turned to the sectoral experience for inspiration.

At the same time, the importance of functional innovation systems for explaining improvements and diffusion in technologies has long been somewhat neglected in the vertical research fields. Water research is perhaps the field in which a broad innovation systems approach was embraced most recently and is still only partially so. As pointed out above, we see the neo-liberal zeitgeist as one driver that has contributed to a drifting apart of development in sectoral policies and STI policies in general. A second possible explanation for the different developments in vertical and horizontal fields are changes in the mode of research over time. All nine research fields were in "expert mode" at the beginning of the 1970s. In the following phase, however, the horizontal research fields in particular (e.g. foresight, technology assessment and evaluation) moved towards a much more participatory approach, whereas the vertical fields remained in expert mode for longer and only started to integrate participatory approaches considerably later. Thus, we can first observe some form of drifting apart again, certainly between the vertical and horizontal research fields, followed by an acceleration of interaction and mutual influence over the last decade, during which the vertical fields have opened up more to participatory approaches. This, we believe, is a pattern that will continue to be important.

A third observation as to the development of our fields is the *interplay with established academic fields*. Our fields—perhaps with the exception of water research—began outside the established arena of academic scientific disciplines, and all of them were problem-oriented, largely multidisciplinary and showed interdisciplinary dynamics. However, to varying degrees, all of them have been grown out of or are linked to established academic fields such as evolutionary economics,

[4] As noted in the introduction to this volume, we do not discuss the growing field of transition studies, where vertical and horizontal perspectives have often been connected. However, this is being done to understand and conceptualise the dynamics of change and resistance to change rather than analysing the nature of sectoral fields as such.

sociology of science, political science, operation management and operations research in business administration and engineering. This epistemic origin had major implications for the development of our fields. Initially, at least in the first decades, scientific credibility in our fields had to be built on the quality criteria of established scientific areas. To give just one example, the standards set by evolutionary economics were important for the establishment of innovation monitoring and some areas of STI policy. Furthermore, scientific reputation had to be based on established journals in those academic fields. Only in a few instances, such as in water research, were the research fields associated with established academic institutions, which made it easier to access corresponding existing journals or establish new ones. For most of our nine fields, specialised journals have only recently emerged and gained scientific reputation—prominent examples are Research Policy (STI policy, Innovation Dynamics) and Energy Policy, or Renewable and Sustainable Energy Review (renewable energy, energy efficiency).

However, as our fields have always been problem- and application-oriented, they have generally been faster than established scientific fields to absorb the new demands of society and policy. In doing so, they have also influenced the quality criteria and directions of established fields, in particular by emphasising the importance of relevance in conjunction with traditional excellence. The prominent frame of the third mission or the inclusion of impact cases in the UK's Research Excellence Framework are two obvious institutionalised expressions of this shift. In some instances, this has also led to a growing acceptance of the application-oriented fields in established academic arenas. We can only speculate here, but we are currently observing a stronger and more rapid convergence of some of our application-oriented fields with established academic fields. This may well have to do with the fact that in other fields, such as evaluation of STI policy, Technology Assessment and Foresight, their difference to the established fields is the very mode of research, the idea of engaging and co-creating with stakeholders. This participatory approach is characterised by a very different rationale of the mode and purpose of research and may therefore reinforce rather than dismantle the walls to established academic disciplines; scientific reputation is harder to gain and established academics are less willing to open up. In contrast, the use of methodologies such as modelling and statistical analysis, which are highly accepted in various established academic communities, might have contributed to the closer links being forged between the vertical research fields and established academic fields.

4.2 Outlook

The last 50 years have seen major developments in our nine selected fields of systems and innovation research (SIR). Above, we have tried to capture and explain those developments in a highly condensed and simplified manner using our SIR developmental model. We have seen the changing role in all four dimensions of that model for the development of the fields over time, how our fields interacted with established academic fields and, in parts, developed traditional academic qualities themselves. The complex relationships between the fields are also apparent. These can be generally characterised by the limited convergence, in particular between the horizontal fields and the vertical ones. Finally, we have seen that the role of our problem-oriented fields vis-à-vis policy has also changed over time. Sometimes, academics were ahead of policy with conceptual and normative initiatives, then policy concerns again drove the fields to find evidence and develop models to support new challenges and demands.

We can only speculate as to the future developments of the fields considered in this volume. Applying our simplified model, we have the impression that, at the time of writing in early 2024, major contextual developments—translated mainly through growing policy concerns—will dominate the demands on our application-oriented fields and shape their dynamics. The strong push exerted by contextual developments relates, for example, to disruptive technological developments such as artificial intelligence or quantum technologies with rather unpredictable consequences in terms of economic and societal developments. It also relates to the growing pressure to mitigate and adapt to climate change and to move our systems towards sustainable development goals more broadly. It finally also relates to major geo-political power shifts, conflicts and tensions as well as societal and political fragmentation within all of the major democracies.

Whether these contextual dynamics will be embedded in yet another zeitgeist shift, we do not know. However, we can already see how developments in STI and in our vertical fields are being re-interpreted. A framework for sectoral as well STI policy is emerging whereby geo-political developments interfere with and dominate societal concerns in terms of SDGs, leading to competition between different systems and the re-nationalisation of policy initiatives. Dynamics in science, technology and innovation as well as in sectoral systems more generally are being interpreted and supported, not only in relation to economic growth or transformations towards SDGs, but with an overarching concern for national security. There are strong indications that these developments may very well intensify the current interventionist, transformative role of the state, albeit in a much more inward-looking and in parts defensive or even aggressive manner. While this does not render the SDG goals less urgent, it does place new constraints on their pursuit.

Against this background, our fields are being challenged in various ways. First, how will the profile and relationships of the fields develop? To start with, problem-oriented research, in which the very purpose of research is to contribute to solutions, may become more—not less—important in the future. The demand for supporting evidence and conceptual perspective may grow as a result. While this may fundamentally challenge more traditional scientific fields that have a different understanding of the purpose and criteria of excellence to reflect on the way they define and exert responsibility, it may further reinforce the SIR fields in their problem-oriented role to provide evidence and support the governance of future developments.

However, there are also a number of open questions ahead for SIR fields. The increasing urgency in terms of societal challenges in combination with growing technological and contextual complexity will demand more advanced, in-depth understanding of technological and sectoral dynamics. This will require even more flexibility in analysis and interaction and the combination of multiple perspectives in terms of understanding problem fluidity. Thus, it is essential that scholars in horizontal fields link more intelligently to scholars in vertical policy fields, as is done to various degrees in some centres across Europe (e.g. Utrecht/Kopernikus; SPRU, Fraunhofer ISI), but is still too limited. This compartmentalisation, as well as the compartmentalisation between policy areas, needs to be broken down if the relationship between the scholarly community in its broadest sense and policy making towards SDGs is to be conducive to supporting the socio-technical transformations needed. This also holds for the fragmentation of democratic societies and the social effects of innovations, which affect each of our nine research fields. At the same time, all the fields will have to further improve the way in which national—and EU level—policies are supported by reflecting upon and including geo-political and geo-economic developments conceptually and analytically, and they need to link this understanding to a broader variety of policy fields. In addition, the current trend towards a re-nationalisation of policy concerns may put further pressure on the freedom of science in terms of its openness to global knowledge flows and cooperation. To find the right balance between openness and international cooperation, on the one hand, and the necessary limitations with regard to the exposure to geo-political threats, on the other hand, will be a task not only, but especially for our application-oriented SIR fields.

However, it is far from given that the different SIR fields will continue to strengthen their links and converge to a greater degree. For example, will STI policy continue to be seen as a major means to steer societies and economies towards SDGs, or will it, in contrast, be once again defined as a major battleground for the competition of nations and systems, and thus revert to a focus on economic competitiveness? Against the background of those tensions, will the field of innovation monitoring support the development of indicators that help to map and model the system dynamics for transformation, or will it revert to economic dimensions and further differentiate the analytical portfolio to understand the economic dynamics of the platform economy? Will the growing importance of indicators and modelling approaches stemming from the vertical fields, with their system-wide, long-term perspective continue, and will vertical fields increasingly deliver those indicators in isolation from or in interaction with the innovation monitoring field? Will technology assessment be able to turn to transformation dynamics more broadly, or will the urgency of technologically-defined systems competition focus on the economic threats and opportunities of distinct technologies? The answers to these and many related questions are far from clear, but how they play out will define the purpose and identity of the fields in the years to come.

A second challenge is related to the need to increase true interdisciplinary cooperation and to integrate vertical and horizontal expertise as already outlined above. Such integration is also related to the increasingly heterodox approaches taken, while these approaches are simultaneously challenged with regard to their internal consistency. We can only speculate here if this will inhibit the ability of our nine SIR fields to establish links to traditional academic fields. But we see a need for SIR research fields to clarify how their field relates to the underlying concepts and theories, and how the different research modes and methodologies of the fields relate to each other as a prerequisite to entering a more integrated mode of research. Perhaps the development of explicit quality criteria for such heterodox research might help to confirm the rigorous quality standards to be met by such research. Finally, we have to phrase the question in the wider context of how academic excellence will be defined in the future. We see the first signs of questioning the established definition of research excellence, measured mostly by journal impact factors and citations. Increasingly, science is called upon to be more reflective on its societal responsibility, and we can only speculate that this might encourage traditional academic fields to be more open to heterodox approaches.

A third challenge is the way in which the fields define and develop their normativity. It has been a major legitimacy claim of the fields that they deliver evidence-based advice that is not biased or filtered by the normative claims of researchers. However, this claim has always been under pressure. In fact, the very origin of systems and innovation research was driven by normative concerns, not only of society and policy, but also of academics. For example, the concern about the competitiveness of European countries vis-à-vis the USA (late 1960s) or Japan (1980s) was a driver for the foundation and further development of fields such as STI policy and innovation monitoring. Academic concerns about exceeding planetary boundaries through energy production and consumption, among other things, supported the growth of energy efficiency and renewable energy research. As the need for and urgency of transformations have grown, so has the awareness of academics, especially in the vertical fields, about the importance of and necessity for research to support transformation. Equally, should science, technology and innovation continue to be framed within a competition between systems that includes a competition between value systems, any analytical work in our horizontal fields may be defined in much more normative terms, supporting shared values against external threats. It will become increasingly important for researchers to be very clear about their own role and how they deal with their own normativity.

In this respect, the development towards a participatory and formative turn we have seen in a number of our fields in different manifestations may have a number of effects. On the one hand, the more researchers accept the importance of input from and interaction with stakeholders and thus a growing plurality of normative claims in the systems, the more their own normative claims may reduce, or they may at least become more aware of their own biases in the process. On the other hand, broad participation may water down the specific role and responsibility of researchers. As they are participating in participatory normative processes themselves, they may, unconsciously or consciously, develop even stronger normative claims themselves. Thus, there will be a growing need for researchers, especially in the application-oriented SIR fields, to reflect on their normativity and their specific mode of responsibility. This will be a major task in order to retain legitimacy. This task will involve finding ways in which the fields can communicate with the general public that reflects this specific responsibility and role and supports, rather than endangers their legitimacy—a key prerequisite given the very close interaction between the application-oriented SIR research fields, politics and society.

Jakob Edler is Managing Director of the Fraunhofer Institute for Systems and Innovation Research ISI since October 2018. The Anglo-German is also Professor of Innovation Policy and Strategy at the Manchester Institute of Innovation Research (MIoIR), which he led from 2011 to September 2018. His research focuses on analysing and conceptualising governance, policy and management of innovation. J. Edler is Speaker of the Fraunhofer Group Innovation Research, member of the German Science Council, the German Academy of Science and Engineering (Acatech) and the Royal Society for the Encouragement of Arts, Manufactures and Commerce (UK).

Rainer Walz is Deputy Director of the Institute and Head of the Competence Centre Sustainability and Infrastructure Systems at the Fraunhofer Institute for Systems and Innovation Research ISI in Karlsruhe. He obtained his PhD and "habilitation" in economics. In 1991, Prof. Dr. Walz joined the Fraunhofer ISI. Prior employment includes the University of Wisconsin and the Enquête Commission "Protecting the Earth's Atmosphere" of the German Bundestag. He is also an adjunct professor of economics at the Karlsruhe Institute of Technology (KIT) and member of the EU expert group "Economic and Social Impact of Research".

SPRINGER NATURE

GPSR Compliance

The European Union's (EU) General Product Safety Regulation (GPSR) is a set of rules that requires consumer products to be safe and our obligations to ensure this.

If you have any concerns about our products, you can contact us on ProductSafety@springernature.com

In case Publisher is established outside the EU, the EU authorized representative is:

Springer Nature Customer Service Center GmbH
Europaplatz 3
69115 Heidelberg, Germany

The manufacturer's authorised representative in the EU is Springer Nature Customer Service Centre GmbH, Europaplatz 3, 69115 Heidelberg, Germany. If you have any concerns regarding our products, please contact ProductSafety@springernature.com

Printed and bound by CPI Group (UK) Ltd, Croydon, CR0 4YY
01/10/2026
02225897-0011